Lectures on Applied Mathematics

Springer-Verlag Berlin Heidelberg GmbH

Hans-Joachim Bungartz
Ronald H. W. Hoppe
Christoph Zenger

Editors

Lectures on Applied Mathematics

Proceedings of the Symposium
Organized by the Sonderforschungsbereich 438
on the Occasion
of Karl-Heinz Hoffmann's 60th Birthday,
Munich, June 30 – July 1, 1999

With 98 Figures

 Springer

Editors

Hans-Joachim Bungartz
Institut für Informatik
Technische Universität München
80290 München, Germany
e-mail: bungartz@in.tum.de

Ronald H. W. Hoppe
Institut für Mathematik
Universität Augsburg
86135 Augsburg, Germany
e-mail: hoppe@math.uni-augsburg.de

Christoph Zenger
Institut für Informatik
Technische Universität München
80290 München, Germany
e-mail: zenger@in.tum.de

Front cover photo designed by Ingo Eichenseher, Zentrum Mathematik,
Technische Universität München

Cataloging-in-Publication Data applied for
Die Deutsche Bibliothek - CIP-Aufnahme
Lectures on applied mathematics : proceedings of the Symposium on the
Occasion of Karl-Heinz Hoffmann's 60th Birthday, Munich, June 30 -
July 1, 1999 / organized by the Sonderforschungsbereich 438.
Hans-Joachim Bungartz ... ed.. - Berlin ; Heidelberg ; New York ;
Barcelona ; Hong Kong ; London ; Milan ; Paris ; Singapore ; Tokyo :
Springer, 2000

Mathematics Subject Classification (1991): 65-06, 65Cxx, 65Kxx, 65Lxx, 65Mxx, 65Nxx

ISBN 978-3-642-64094-0   ISBN 978-3-642-59709-1 (eBook)
DOI 10.1007/ 978-3-642-59709-1

Springer-Verlag is a company in the BertelsmannSpringer publishing group.
© Springer-Verlag Berlin Heidelberg 2000
Softcover reprint of the hardcover 1st edition 2000

Cover Design: design & production GmbH, Heidelberg
Typeset by the author using a Springer TeX macro package
            SPIN 10735110   46/3143/LK - 5 4 3 2 1 0

Dedicated to

Prof. Dr. Dr. h. c. Karl-Heinz Hoffmann

on the occasion of his 60th birthday

Preface

When the DFG (Deutsche Forschungsgemeinschaft) launched its collaborative research centre or SFB (Sonderforschungsbereich) 438 "Mathematical Modelling, Simulation, and Verification in Material-Oriented Processes and Intelligent Systems" in July 1997 at the Technische Universität München and at the Universität Augsburg, southern Bavaria got its second nucleus of the still young discipline scientific computing. Whereas the first and older one, FORTWIHR, the Bavarian Consortium for High Performance Scientific Computing, had put its main emphasis on the supercomputing aspect, this new initiative was now expected to focus on the mathematical part. Consequently, throughout all of the five main research topics (A) adaptive materials and thin layers, (B) adaptive materials in medicine, (C) robotics, aeronautics, and automobile technology, (D) microstructured devices and systems, and (E) transport processes in flows, mathematical aspects play a predominant role.

The formation of the SFB 438 and its scientific program are inextricably linked with the name of Karl-Heinz Hoffmann. As full professor for applied mathematics in Augsburg (1981 – 1991) and in München (since 1992) and as dean of the faculty of mathematics at the TU München, he was the driving force of this fascinating, but not always easy-to-realize idea of bringing together scientists from mathematics, physics, engineering, informatics, and medicine for joint efforts in modern applied mathematics. However, scarcely work had begun when the successful captain was called to take command on a bigger boat. The new foundation caesar (center of advanced european studies and research), part of the governmental program to console Bonn and its surrounding countryside about their loss of importance due to Berlin having become the new capital of the reunited Germany, needed and got a renowned founding director. Yet the SFB 438 lost its chairman after only a year. Fortunately, Karl-Heinz Hoffmann is still an active member of the research centre and involved in its work, and the new axis München-Bonn has already turned out to be a more than fruitful connection.

The two-day symposium on applied mathematics, held end of June 1999 in Munich on the occasion of Karl-Heinz Hoffmann's 60th birthday, was the second international meeting organized by the SFB 438. The conference brought together researchers from the SFB's groups and leading experts from several countries. Most of the invited speakers of this event are represented with an article in the first part of this volume. In the book's second part, members of the SFB provide an overview of its research program and, thus, complete this collection of current topics of applied mathematics and scientific computing and of fields Karl-Heinz Hoffmann has made and makes important contributions to.

At that point, we feel the need to express our thanks to several persons and institutions having contributed in many different ways to the symposium and to this volume.

First, we like to thank our guests from nearly all over Europe and from the USA who accepted our invitation to take part in the symposium, to prepare a chapter for this book, and to do this on time. Although, often, deadlines seem to be fixed just in order to be ignored afterwards, nobody caused us to throw out the schedule.

Next, let us thank Springer-Verlag and, especially, Dr. Martin Peters for their immediate enthusiasm for our book project. From the very beginning and throughout the preparation and the realization of this volume, the co-operation was very fruitful and pleasant.

Of course, we must not forget the helping hands of Marlis and Christian Clason. Marlis did a great job in organizing the symposium and kept in touch with all authors. Her polite and friendly, but nevertheless firm personality was, certainly, crucial for everyone's impressive punctuality. It is thanks to Christian that the LaTeX layout of all contributions converged in such a short time. Having in mind the first printed versions we got, many others would have taken to their heels.

Johannes Zimmer organized a good part of the reviewing and proof-reading process. We are very grateful to him and to all members of the SFB's groups involved in that for their commitment.

Finally, and needless to say, neither the symposium and this volume nor an important portion of the results reported in the second part of it would have been possible without the financial support of the DFG via the SFB 438.

München, November 1999   Hans-Joachim Bungartz
                                           Ronald H. W. Hoppe
                                           Christoph Zenger

Table of Contents

Part I

Invited Contributions

Pure, Applied, and Industrial Mathematics: Strength Through Connections

A. Friedman

School of Mathematics, University of Minnesota, 537 Vincent Hall, 206 Church Street SE, Minneapolis, MN 55455, USA

Dedicated to Professor Karl-Heinz Hoffmann
on the occasion of his 60th birthday

It gives me great pleasure to participate in the 60th birthday celebration of Karl-Heinz Hoffmann. Professor Hoffmann is known worldwide through his research in applied mathematics. He is also recognized for his leadership role in making connections between mathematics and the sciences, and between mathematical research in academia and industrial problems which lend themselves to mathematical modeling. I recall travelling with Professor Hoffmann to companies such as BASF and SIEMENS to discuss and identify problems of interest to these companies. I also recall attending his lectures in international conferences. In these lectures, he developed models in areas such as smart materials and phase transition, and presented innovative mathematical analysis as well as effective numerical simulations.

I want to make a confession: the title of my contribution today is not original. I "stole" it from the conference that was held last year in Minneapolis which was a celebration for my 65th birthday, and my wife's. The speakers at that conference came from both universities and industry, and the topics included stochastic fluid dynamics, new algorithms for codes, Markov chains in photographic film processing, inverse problems in PDE, mathematics education, mathematics in the textile industry, and free boundary problems. I was not "allowed" to give a talk in that conference. So now it is my chance to deliver a speech under the same title of that conference.

Until recently, the distinction between pure mathematics and applied mathematics was fairly clear. In pure mathematics one develops new concepts, new methods and proves new theorems in one or several of the mathematical disciplines such as algebra, topology, number theory, or analysis. By contrast, applied mathematics is motivated by questions that arise in science and engineering, and it tries to resolve such questions by using known mathematics, or by developing new mathematics as needed. With the computer revolution of the last 20-30 years such a distinction became less clear. For example, abstract mathematical results in areas such as algebraic topology, algebraic geometry, and number theory can be used, with the aid of computational facilities, to analyse and resolve questions that arise in protein folding, coding, and cryptography. Similarly, computational capabilities make it possible to apply theoretical results in discrete mathematics and graph theory to practical problems in communication and networks.

At this point, I would like to introduce the concept of "industrial mathematics". This is a topic close to my heart. During the 10-year period 1987-1997, as the Director of the Institute for Mathematics and its Applications (IMA), I worked hard to make connections between academic mathematicians and industry. This included continuous visits to companies to explore areas where mathematics can be useful, organizing workshops where industrial speakers presented problems to academic experts, and conducting a regular seminar at the IMA where the weekly speaker was a researcher from industry. We tried to provide a follow-up to such speakers in terms of analytic solutions, general understanding of qualitative nature, and numerical experiments.

As you may know, the complexity of processes and products in the manufacturing and service industries continue to increase, while at the same time, in order to remain competitive, industry must shorten the time from concept to product. Experiments are both costly and time consuming, and thus the demand for mathematical/statistical modeling and simulation increases.

The creation of mathematical/statistical models and the development of algorithms for computer simulation to obtain solutions of problems in industry is what we call **industrial mathematics** (see [3] for more details). How does industrial mathematics differ from "applied mathematics"? First, you must travel to industry and talk to its scientists and engineers in order to identify their mathematical problems; you cannot pick up problems just from the literature. Secondly, you cannot tell industry people: "Sorry, I cannot help you with your mathematical problems because I specialize in another field of mathematics"; when in industry, your speciality is mathematics, all of it! Thirdly, you have a limited time for solving problems; sometimes just a few weeks. However, you need not provide a full mathematical solution; partial timely solutions are usually all that is required. Finally, you need to be aware of the fact that the goals in university and in industry are different; whereas university's goals are teaching, research, and publications, industry's goal is to make profits.

What is the difference between industrial mathematics and engineering? (1) Engineers build tools and products. They use any scientific knowledge, theoretical or experimental, available to them. They may use physics, chemistry, mathematics, biology, etc., but they are not experts in any one of these disciplines. (2) Not every industrial/engineering problem can be formulated as a mathematical problem. (3) The most fundamental component in industrial mathematics is the development of a mathematical model of a problem from industry, after which mathematical analysis and numerical algorithms are used, or developed, to solve the problem. (4) Industrial mathematicians are strong in mathematics. Mathematicians in industry are viewed as having highly developed skills in abstraction, analysis of underlying structures, and logical thinking; as having the best tools for formulating and solving problems. They are often viewed as consultants; for more details see [6].

Building on the IMA contacts and experience with industry, the University of Minnesota established, four years ago, a new Minnesota Center for Industrial Mathematics (MCIM), including three new faculty positions. The mathematics department also introduced M.S. and Ph.D. mathematics degrees in the Program of Industrial and Applied Mathematics. For the Master's program, students spend one summer internship in a company and upon returning to the university they write a Master's thesis based on their work in the company; this usually includes modeling, analysis and numerical results.

The Ph.D. program includes courses such as numerical analysis, applied mathematics, mathematical modeling, and scientific computing, but also core mathematics courses. The Ph.D. thesis is based on a project from industry, and is expected to provide results useful to the company, but also to have original mathematics that can be published. Students in both programs are supported by the companies.

A number of mathematics departments in the U.S. are currently developing programs in industrial mathematics. The goals of these programs are to help industry by solving problems, to develop new mathematics as needed, and to broaden career opportunities for students and postdoctorates. Reference [3] describes how to start such a program.

I was asked by DMV-Mitteilungen (the German analogue of the AMS Notices) to write an article for the last International Congress of Mathematicians on the topic: "Reflections on the Future of Mathematics." After examining the research achievements and likely future outlook of some famous mathematicians such as Archimedes and Newton, I concluded that the history of mathematics shows how futile it is to predict long term future discoveries from the present. Indeed, new fields of mathematics, unimaginable today, may arise quite unexpectedly. Thus instead of predicting the future of mathematics for the next century, I offered several examples of key fields in science and technology where mathematics is emerging as a vital component [2]. These include materials science, and the life sciences.

Materials science is concerned with the properties and the use of materials. The objectives are the synthesis and manufacture of new materials, the modification of materials, the understanding and prediction of material properties, and the evolution and control of these properties over a time period.

Let me say a few words about just one subdiscipline in materials science, namely, the fast growing area of composites, which offers great opportunities for mathematical research. This is, incidentally, an area where Professor Hoffmann and his group have been very active.

If we insert into one material grains of another material, we get a composite material which may exhibit properties that are radically different from those of its constituents. Automobile companies, for example, are working with composites of aluminum and silicon-carbon grains which provide light-weight alternative to steel. Fluid with magnetic particles or electrically charged particles will enhance the effects of brake fluid and shock absorbers in the automobile.

Over the last decade, mathematicians have made important discoveries in the study of composites. They have developed new tools in functional analysis, PDE, and numerical analysis, by which they have been able to estimate or compute the effective properties of composites. But the list of new composites is ever increasing and new materials are constantly being developed.

Another emerging field of research for mathematics is the life sciences. The highly publicized genome project, i.e., mapping the DNA, and functional genomics, i.e., identifying the functionality of genes, require statistics, pattern recognition, and large-scale optimization methods. Less publicized but longer-term challenges are emerging in other areas of biology such as physiology. Consider, for instance, the kidney, whose function is to regulate the composition of the blood by maintaining a desired level of concentration of critical substances like salt. This function is performed by approximately one million tiny tubes around this kidney, called nephrons, whose task is to absorb salt from the blood into the kidney. It is an extremely complicated system of sensors and actuators. A complete model may include PDE, stochastic equations, fluid dynamics, elasticity theory, filtering theory, and control theory, and perhaps other tools that we do not yet have.

Heart dynamics, calcium dynamics, the auditory process, cell adhesion and motility (vital for physiological processes such as inflammation and wound healing) and biofluids are other topics in physiology where recent mathematical studies have already made some progress; much more is still to come. Other areas where mathematics is poised to make important progress include the growth process in general and embryology in particular, cell signaling, and immunology.

I would like to conclude by giving just one very recent example where interesting new mathematical ideas arise in trying to analyze a model in biology. The model is that of tumor growth. Let

$$\sigma = \text{nutrient concentration,}$$
$$q = \text{particle velocity,}$$
$$p = \text{internal pressure,}$$
$$S(\sigma) = \mu(\sigma - \tilde{\sigma}) = \text{cell proliferation rate.}$$

If the tumor occupies a region Ω then, after non-dimensionalization,

$$\begin{aligned}
\Delta\sigma - \lambda\sigma &= 0 & &\text{in } \Omega, \\
\Delta p &= -\mu(\sigma - \tilde{\sigma}) & &\text{in } \Omega, \\
\sigma &= \bar{\sigma} & &\text{on } \partial\Omega, \\
\frac{\partial p}{\partial n} &= 0 & &\text{on } \partial\Omega, \\
p &= \gamma k & &\text{on } \partial\Omega,
\end{aligned} \qquad (0.1)$$

where k is the mean curvature and $\tilde{\sigma}, \bar{\sigma}, \mu, \lambda$ and γ are positive constants [1,6].

Note that $\partial\Omega$ is a free boundary, unknown in advance. The above system is a simple model of dormant tumor and one can easily check that if $\tilde{\sigma}/\bar{\sigma} < \frac{1}{3}$ (or if $\tilde{\sigma}/\bar{\sigma} < \frac{1}{2}$ in two dimensions) then there exists a unique radial solution, up to translation.

The interesting question is whether there are also non-radial solutions and, in particular, whether there are symmetry breaking solutions near the spherical solution.

Friedman and Reitich [4] recently proved, for the two-dimensional case, that indeed there exist bifurcation branches of solutions with free boundary

$$r = R_0 + \epsilon \cos \ell\theta + \sum_{n=2}^{\infty} \epsilon^n f_n(\theta) \equiv f(\theta, \epsilon)$$

about any $R_0 > 0$ and integer $\ell \geq 2$ with

$$\gamma = \gamma_0 + \sum_{n=1}^{\infty} \epsilon^n \gamma_n$$

where γ_0 satisfies a bifurcation equation which depends on l; the functions σ, p, and f are all analytic in ϵ and the spatial variable, and uniqueness holds under the orthogonality conditions

$$\int_0^{2\pi} f(\theta) \cos \theta d\theta = 0, \quad \int_0^{2\pi} \sigma \cos \theta d\theta = \int_0^{2\pi} p \cos \theta d\theta = 0 \qquad (0.2)$$

and

$$\int_0^{2\pi} f_n(\theta) \cos(\ell\theta) d\theta = 0, \ n \geq 2. \qquad (0.3)$$

Condition (0.3) is needed in order to eliminate trivial new solutions obtained by translation of the origin, and the condition (0.2) is the usual orthogonality condition in bifurcation theory, associated with the eigenspace (here generated by $\cos \ell\theta$).

There is no general theory of free boundary problems which covers the system (0.1). It would be very interesting to study this stationary system beyond the scope of bifurcation branches of solutions, in general, and its time dependent analog.

References

1. Byrne, H. M., Chaplain, M. A. J.: Free Boundary Value Problems Associated with Growth and Development of Multicellular Spheroids. European J. Appl. Math. 8 (1997), 639–658

2. Friedman, A.: Reflections on the Future of Mathematics, in Mitteilungen der Deutschen Mathematiker-Vereinigung, Heft 2/98, 30–32
3. Friedman, A., Lavery, J.: How to Start an Industrial Mathematics Program in the University, Society for Industrial and Applied Mathematics, Philadelphia, Pennsylvania, 1993
4. Friedman, A., Reitich, F.: Symmetry-Breaking Bifurcation of Analytic Solutions to Free Boundary Problems: An Application to a Model of Tumor Growth, to appear
5. Greenspan, H. P.: On the Growth and Stability of Cell Cultures and Solid Tumors, Theoretical Biology, **56** (1976), 229–242
6. SIAM Report on Mathematics in Industry, Society for Industrial and Applied Mathematics, Philadelphia, Pennsylvania, 1997

On $\mathcal{H}^2$-Matrices

W. Hackbusch[1], B. Khoromskij[1], and S. A. Sauter[2]

[1] Max-Planck-Institut Mathematik in den Naturwissenschaften, Inselstr. 22-26, D-04103 Leipzig, Germany
[2] Universität Zürich, Institut für Mathematik, Winterthurer Str. 190, CH-8057 Zürich, Switzerland

Dedicated to Professor Karl-Heinz Hoffmann
on the occasion of his 60th birthday

Abstract. A class of matrices ($\mathcal{H}$-matrices) has recently been introduced by one of the authors. These matrices have the following properties: (i) They are sparse in the sense that only few data are needed for their representation. (ii) The matrix-vector multiplication is of almost linear complexity. (iii) In general, sums and products of these matrices are no longer in the same set, but their truncations to the $\mathcal{H}$-matrix format are again of almost linear complexity. (iv) The same statement holds for the inverse of an $\mathcal{H}$-matrix.

The term "almost linear complexity" used above means that estimates are given by $O(n \log^\alpha n)$. The logarithmic factor can be avoided by a further improvement, which is described in the present paper. We prove that the storage requirements and the cost of the matrix-vector multiplication is strictly linear in the dimension n, while still (full) system matrices of the boundary element method can be approximated up to the discretization error.

AMS Subject Classifications: 65F05, 65F30, 65F50, 65N38, 68P05, 45B05, 35C20
Key words: Hierarchical matrices, hierarchical bases, full matrices, fast matrix-vector multiplication, BEM, FEM.

1 Introduction

For linear systems with sparse $n \times n$-matrices several optimal iteration methods are known, where optimality is characterized by an estimation of the arithmetic operations by $O(n)$. A different situation is given in the case of full matrices. Then standard techniques require a storage amount of $O(n^2)$ and $O(n^2)$ operations for the matrix-vector multiplication. Other arithmetic operations like matrix-matrix multiplications or the inversion even lead to $O(n^3)$ operations.

Full matrices are directly obtained by the discretization of integral equations as they are common in the boundary element method (BEM; cf. [2]). Another source of a full matrix is the inverse of a sparse FEM matrix which, e.g., appears in the Schur complement of any saddle point problem (cf. [1, Sect. 11.7]). In both cases, the matrices are affected with a discretization error. Therefore, one may replace the full matrix M by a more convenient matrix M', provided that the error $M - M'$ is of the size of the discretization error.

The hierarchical matrices (abbreviated as $\mathcal{H}$-matrices) define a set of matrices which provides the approximations M' discussed above. As described in detail in [3] and [4], $\mathcal{H}$-matrices have the following properties:

(i) They are data-sparse in the sense that the size data to be stored is almost linear in the dimension n.
(ii) The matrix-vector multiplication is of almost linear complexity.
(iii) In general, sums and products of these matrices are no longer in the same set, but their truncations to the $\mathcal{H}$-matrix format are again of almost linear complexity.
(iv) The same statement holds for the inverse of an $\mathcal{H}$-matrix.

The basic (hierarchical) structure of $\mathcal{H}$-matrices is the *cluster tree* which is already introduced in the *panel clustering method* (see [6] and [7] or [2, Section 9.7]).

The term "almost linear complexity" used above means that estimates are given by $O(n \log^\alpha n)$. The logarithmic factor can be avoided by a further improvement which leads to the $\mathcal{H}^2$-matrices (hierarchical bases $\mathcal{H}$-matrices) introduced and analysed in this paper. These matrices are already mentioned in [3] under the name "uniform $\mathcal{H}$-matrices". The essential analysis is given in Sect. 4. The approximation by a Taylor polynomial of fixed degree is replaced by a variable degree. Although we use lower approximation degrees for most of the matrix blocks, the overall error estimate does not deteriorate.

2 Hierarchical Bases $\mathcal{H}$-Matrices

After presenting the introductory example (Sect. 2.1), we define the cluster tree (Sect. 2.2), which is the basis of the standard $\mathcal{H}$-matrices (Sect. 2.3). Finally, in Sect. 2.4, we introduce the hierarchical bases $\mathcal{H}$-matrices, which we call $\mathcal{H}^2$-matrices.

2.1 Introductory Example

The matrices we have in mind may stem from integral or differential equations. In the latter case, it is of interest to represent the inverse matrix as an $\mathcal{H}$-matrix[1]. Here, we consider the example of the integral operator

$$(Ku)(x) = \int_0^1 \log(|x - y|)u(y)dy. \tag{2.1}$$

---

[1] Formally, the inverse can be considered as a discretization of an integral operator with the Green function as kernel.

Its finite element discretization with piecewise constant basis functions corresponding to the interval partitioning

$$J_i = [(i-1)h, ih], \quad h := 1/n, \quad 1 \le i \le n \tag{2.2}$$

leads to the matrix

$$M = (m_{ij})_{i,j \in I}, \quad m_{ij} := \int_{J_i} \int_{J_j} \log(|x-y|) dx dy, \tag{2.3}$$

where

$$I = \{1, \dots, n\} \tag{2.4}$$

is the underlying *index set*. As further simplification, we assume that n is a power of 2:

$$n = 2^p. \tag{2.5}$$

In boundary element applications (BEM), one has to replace the unit interval by a surface, the equidistant partitioning by a general triangulation and the kernel function $\log(|x-y|)$ by some appropriate singularity function (cf. [2]). However, in order not to distract the attention of the reader from the main ideas, we consider the matrix M from (2.3). The kernel $\log(|x-y|)$ shares typical properties with the kernels arising in general BEM applications. The results of this paper can be extended to general BEM problems as well.

The matrix M from (2.3) is a full matrix, i.e., the usual storage amount is $O(n^2)$ instead of $O(n)$ for standard sparse matrices. Furthermore, a simple matrix-vector multiplication $M \cdot x$ requires $O(n^2)$ operations. The aim of the $\mathcal{H}^2$-matrix concept is to replace M by an approximation M' such that the error $M - M'$ is of the size of the discretization error (therefore negligible), while *storage*(M) and *cost*$(M \cdot x)$ amounts to $O(n)$ instead of $O(n^2)$.

The discussion of the error $M - M'$ is performed in Sect. 4. The details about the storage and matrix-vector multiplication cost are given in Sect. 3.

2.2 The Cluster Tree

We start with the full index set $I_1^0 := I = \{1, \dots, n\}$ from (2.4) and split it into the parts $I_1^1 := \{1, \dots, \frac{n}{2} = 2^{p-1}\}$, $I_2^1 := \{2^{p-1} + 1, \dots, n\}$. Similarly, the new sets are divided so that, in general,

$$I_i^\ell := \{(i-1)2^{p-\ell} + 1, \dots, i\, 2^{p-\ell}\} \quad \text{for } 0 \le \ell \le p, \ 1 \le i \le 2^\ell. \tag{2.6}$$

The superscript ℓ indicates the *level*. At level p, we reach the one-element sets $I_1^p = \{1\}, \dots, I_n^p = \{n\}$.

Obviously, these sets form a tree T (the so-called *cluster tree*).

Remark 2.1. (a) I is the root of T. (b) The sets I_i^ℓ are the vertices ("clusters") of T at level ℓ. (c) T is a binary tree: I_i^ℓ has two sons $I_{2i-1}^{\ell+1}$ and $I_{2i}^{\ell+1}$ if $\ell < p$. (d) The sets I_i^p are the leaves of T. (e) The cardinality of I_i^ℓ is $\#I_i^\ell = 2^{p-\ell}$.

In the following, we use the variables τ and σ for the vertices of the tree T and call $\tau \in T$ a *cluster*. Usually, the sons of $\tau \in T$ are denoted by τ', τ''.

An isomorphic description occurs when we replace the index i by the interval J_i from (2.2) which is the support of the ith basis function. Then a cluster $\tau \in T$ corresponds to the interval

$$J(\tau) := \bigcup\{J_\alpha : \alpha \in \tau\}. \tag{2.7}$$

The partitioning of the set I into I_1^0, I_1^1 corresponds to the definition of a block structure of a vector (over the index set I). The tree structure of T allows to continue the block decomposition in a hierarchical way. The *hierarchical matrices* based on this tree structure are abbreviated as $\mathcal{H}$-*matrices*.

2.3 $\mathcal{H}$-Matrices

The Model Partitioning. Since we are dealing with matrices, we have to consider the index set $I \times I$. In the following, we describe a particular partitioning P_2 of $I \times I$ such that

$$I \times I = \bigcup\{b : b \in P_2\}, \tag{2.8}$$

where each block $b \in P_2$ is of the form $b = I_i^\ell \times I_j^\ell$ for some $0 \le \ell \le p$, $1 \le i,j \le 2^\ell$. The subscript 2 in P_2 should indicate that P_2 partitions the *two*fold product index set $I \times I$. In the interesting case, the blocks $b \in P_2$ corresponding to the block matrix $M^b := (m_{\alpha\beta})_{(\alpha,\beta)\in b}$ do not all belong to only one level ℓ. The level number ℓ of a block b is written as *level*(b).

The easiest way to introduce the partitioning P_2 is by a recursive description of the matrix block structure. For this purpose we consider four different matrix formats: $\mathcal{R}$-, $\mathcal{N}$-, $\mathcal{N}^*$-, and, finally, the $\mathcal{H}$-matrices.

$\mathcal{R}$-*matrices* are matrices of rank $\le k$. The value of k and its possible dependence on b will be discussed later. These $\mathcal{R}$-matrices can be represented in the form

$$\sum_{i=1}^k [a_i, c_i], \quad \text{where } [a_i, c_i] := a_i \cdot c_i^H \tag{2.9}$$

with column vectors a_i and row vectors c_i^H. The set of $\mathcal{R}$-matrices is denoted by $\mathcal{M}_\mathcal{R}$.

The $\mathcal{N}$-*matrices* correspond to off-diagonal blocks $b = I_i^\ell \times I_{i+1}^\ell$ ($\mathcal{N}$ abbreviates "neighbourhood"). For $\ell = p$, $\mathcal{N}$-matrices are simple 1×1-matrices.

For $\ell = p - 1, \ldots, 1$, the following recursive definition holds: Abbreviate $m = 2^{p-\ell}$. An $m \times m$ matrix M has the $\mathcal{N}$-format if

$$M = \begin{bmatrix} M_{11} & M_{12} \\ M_{21} & M_{22} \end{bmatrix} \text{ with } \tfrac{m}{2} \times \tfrac{m}{2} \text{ } \mathcal{R}\text{-matrices } M_{11}, M_{12}, M_{22}$$

$$\text{and } \mathcal{N}\text{-matrix } M_{21}. \tag{2.10}$$

Similarly, we define the transposed type: M is an $\mathcal{N}^*$-*matrix* if M^T is of $\mathcal{N}$-type, i.e., in (2.10) M_{11}, M_{21}, M_{22} are $\mathcal{R}$-matrices and M_{12} is an $\mathcal{N}^*$-matrix. The sets of $\mathcal{N}$- and $\mathcal{N}^*$-matrices are denoted by $\mathcal{M}_{\mathcal{N}}$ and $\mathcal{M}_{\mathcal{N}^*}$.

Finally, the $\mathcal{H}$-*matrices* ("hierarchical matrices") are defined in

Definition 2.2. Let M be an $n \times n$-matrix with $n = 2^p$. Then M is an $\mathcal{H}$-matrix (notation: $M \in \mathcal{M}_{\mathcal{H}}$) if either $n = 1$ ($p = 0$) or if the partitioning into 2×2 blocks of size $\tfrac{n}{2} \times \tfrac{n}{2}$ leads to

$$M = \begin{bmatrix} M_{11} & M_{12} \\ M_{21} & M_{22} \end{bmatrix} \text{ with } M_{11}, M_{22} \in \mathcal{M}_{\mathcal{H}}, \text{ } M_{12} \in \mathcal{M}_{\mathcal{N}},$$

$$M_{21} \in \mathcal{M}_{\mathcal{N}^*}. \tag{2.11}$$

In the case of $p = 3$, the resulting block structure of an 8×8-matrix is

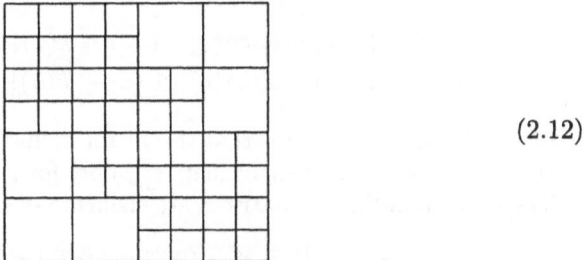

$$\tag{2.12}$$

Let $P_2 \subset \mathbb{P}(I \times I)$ be the set of the finally resulting blocks in (2.11). In the case of (2.12), P_2 consists of 40 1×1-blocks and 6 2×2-blocks.

Here, we remark that we need a partitioning with two properties: On the one hand side, the partitioning should contain as few blocks as possible to reduce the costs for storage and operations, while on the other hand the blocks should be small enough so that the resulting matrix is a sufficiently good approximation of the true matrix. We shall see that P_2 from (2.11) is a good compromise.

The rank k involved in $\mathcal{M}_{\mathcal{R}}$ is not necessarily a constant. In the following, $k : P_2 \to \mathbb{N}$ is a function of the block $b \in P_2$. Then, a submatrix M^b over the index block $b \in P_2$ belongs to $\mathcal{M}_{\mathcal{R}}$ if the block M^b satisfies $\text{rank}(M^b) \leq k(b)$. The following definition is equivalent to Definition 2.2, if we choose the partitioning P_2 from above.

Definition 2.3. Let P_2 be a block partitioning of $I \times I$ and $k : P_2 \to \mathbb{N}$. The underlying field of the matrices is $\mathbb{K}$. The set of $\mathcal{H}$-matrices induced by P_2 and k is

$$\mathcal{M}_{\mathcal{H},k}(I \times I, P_2) := \{M \in \mathbb{K}^{I \times I} : \text{ each block } M^b, \, b \in P_2,$$
$$\text{satisfies } \text{rank}(M^b) \leq k(b)\}. \tag{2.13}$$

2.4 Definition of $\mathcal{H}^2$-Matrices

Up to now, we made use of the cluster tree T, which yields a hierarchy among the clusters and leads to the optimal partitioning P_2. Next, we introduce another hierarchical structure connected with the vectors a_i, c_i from (2.9). This second hierarchy gives rise to the exponent 2 in the name $\mathcal{H}^2$-matrices (or[2] *hierarchical basis $\mathcal{H}$-matrices*).

Hierarchical Bases for Row and Column Vectors of $\mathcal{H}^2$-Matrices.
So far, an $\mathcal{R}$-matrix $\sum_{i=1}^{k(b)} [a_i, c_i]$ from (2.9) could be formed with arbitrary vectors a_i, c_i. Another situation occurs if we fix two bases $\{a_i\}, \{c_i\}$ depending on the block $b \in P_2$. Any block b has the form $b = \tau \times \sigma$ with clusters $\tau, \sigma \in T$. We require that $\{a_i\}$ depends only on the row-index cluster τ, while $\{c_i\}$ depends only on the column-index cluster σ :

$$V_a(\tau) = \text{span}\{a_i^\tau : 1 \leq i \leq k(\tau)\} \subseteq \mathbb{K}^\tau,$$
$$V_c(\sigma) = \text{span}\{c_j^\sigma : 1 \leq j \leq k(\sigma)\} \subseteq \mathbb{K}^\sigma. \tag{2.14}$$

The notation $a_i^\tau \in \mathbb{K}^\tau$ means that the vector a_i^τ has components $a_{i,\nu}^\tau$ only for $\nu \in \tau$, while $c_j^\sigma \in \mathbb{K}^\sigma$ has coefficients $c_{j,\nu}^\sigma$ only for $\nu \in \sigma$.

The corresponding $\mathcal{R}$-matrices are elements of the tensor vector space

$$V(b) = \text{span}\{[a_i^\tau, c_j^\sigma] : 1 \leq i \leq k(\tau), 1 \leq j \leq k(\sigma)\} =$$
$$= V_a(\tau) \times V_c(\sigma) \quad \text{for } b = \tau \times \sigma. \tag{2.15}$$

In our model case, the clusters τ, σ of $b = \tau \times \sigma$ belong to the same level. If we make the natural assumption that the rank k is a function k_ℓ of the level ℓ only, $k(\tau) = k(\sigma) = k_{level(b)} =: k(b)$ follows.

The storage requirements are less than for $\mathcal{H}$-matrices. Since the vectors a_i^τ, c_j^σ are pre-defined, only the coefficients with respect to the basis $\{[a_i^\tau, c_j^\sigma]\}$ for $V(b)$ from (2.15) are to be stored.

Remark 2.4. Let (2.14) be given. It needs $k(\tau) \cdot k(\sigma)$ coefficients ζ_{ij} to code an $\mathcal{R}$-matrix $\sum_{i,j} \zeta_{ij}[a_i^\tau, c_j^\sigma]$.

---

[2] The "hierarchical bases" which appear in our context have another hierarchical structure than the hierarchical bases known from the finite element method. Also the hierarchical structure of wavelet bases is different.

Restrictions. Consider a cluster $\tau \in T$ being not a leaf. Its sons are denoted by τ', τ'' (the tree of the model problem is binary). The decomposition $\tau = \tau' \cup \tau''$ describes a block partitioning of the vector a_i^τ into the block vectors

$$(a_{i,\nu}^\tau)_{\nu \in \tau'} = R_a^{\tau',\tau} a_i^\tau \quad \text{and} \quad (a_{i,\nu}^\tau)_{\nu \in \tau''} = R_a^{\tau'',\tau} a_i^\tau. \tag{2.16}$$

The *restriction operator* $R_a^{\tau',\tau}$ denotes the mapping from the full vector into a block vector. Conversely, we can represent the vector a_i^τ as the composition

$$a_i^\tau = \begin{pmatrix} R_a^{\tau',\tau} a_i^\tau \\ R_a^{\tau'',\tau} a_i^\tau \end{pmatrix}, \tag{2.17}$$

if we first enumerate the indices of τ' and then those of τ''.

Similarly, the restrictions $R_c^{\sigma',\sigma}$ are defined for the row vectors c_j^σ and yield

$$c_j^\sigma = \begin{pmatrix} R_c^{\sigma',\sigma} c_j^\sigma \\ R_c^{\sigma'',\sigma} c_j^\sigma \end{pmatrix}. \tag{2.18}$$

Consistency Conditions. Let $\tau, \tau', \tau'' \in T$ be as before. The *consistency relation* between the spaces $V_a(\tau)$ and $V_a(\tau'), V_a(\tau'')$ is

$$V_a(\tau') = R_a^{\tau',\tau} V_a(\tau), \qquad V_a(\tau'') = R_a^{\tau'',\tau} V_a(\tau). \tag{2.19}$$

Similarly, we require the analogous relations for the spaces $V_c(\sigma), V_c(\sigma')$, $V_c(\sigma'')$:

$$V_c(\sigma') = R_c^{\sigma',\sigma} V_c(\sigma), \qquad V_c(\sigma'') = R_c^{\sigma'',\sigma} V_c(\sigma). \tag{2.20}$$

One important conclusion from $R_a^{\tau',\tau} V_a(\tau) \subseteq V_a(\tau')$ is

Remark 2.5. It is not necessary to store the vectors a_i^τ explicitly. Instead, one can store the coefficients $\alpha_{ij}^{\tau,\tau'}$ of the representation

$$R_a^{\tau',\tau} a_i^\tau = \sum_{j=1}^{k(\tau')} \alpha_{ij}^{\tau,\tau'} a_j^{\tau'} \qquad \text{for } 1 \leq i \leq k(\tau) \tag{2.21}$$

and the analogously defined coefficients $\alpha_{ij}^{\tau,\tau''}$.

The other direction $V_a(\tau') \subseteq R_a^{\tau',\tau} V_a(\tau)$ implies

Remark 2.6. The dimension k must be a monotone function of the vertices, i.e., $k(\tau') \leq k(\tau)$ if τ' is a son of τ. If $k = k_\ell$ depends on the level ℓ as discussed above, $k_{\ell+1} \leq k_\ell$ holds.

Normal Form. Among all vectors $\{a_i^\tau : 1 \le i \le k(\tau), \tau \in T\}$ satisfying (2.19) and (2.21), we can choose a basis in such a way that the sum in (2.21) runs over $1 \le j \le \min(i, k(\tau'))$, i.e.,

$$R_a^{\tau',\tau} a_i^\tau = \sum_{j=1}^{\min(i,k(\tau'))} \alpha_{ij}^{\tau,\tau'} a_j^{\tau'} \qquad \text{for } 1 \le i \le k(\tau). \qquad (2.22)$$

Similarly,

$$R_c^{\sigma',\sigma} c_j^\sigma = \sum_{i=1}^{\min(j,k(\sigma'))} \gamma_{ji}^{\sigma,\sigma'} c_i^{\sigma'} \qquad \text{for } 1 \le j \le k(\sigma). \qquad (2.23)$$

Furthermore, the vectors could be chosen to be orthonormal, i.e., $\langle a_i^\tau, a_j^\tau \rangle = \delta_{ij}$ for $1 \le i, j \le k(\tau)$ with Kronecker's symbol δ_{ij}. However, it is even more convenient if the respective bases $\{a_i^\tau\}$ and $\{c_j^\tau\}$ of $V_a(\tau)$ and $V_c(\tau)$ (which may be different!) are *bi-orthonormal*, i.e.,

$$\langle a_i^\tau, c_j^\tau \rangle = \delta_{ij} \qquad \text{for } 1 \le i, j \le k(\tau). \qquad (2.24)$$

Finally, we remark that for level p, where all clusters τ contain only one index, $k(\tau) = 1$ holds and all basis vectors are the unit vector $a_1^\tau = c_1^\tau = (1)$.

Case of Constant $k(b)$. The simplest case is a constant rank $k(\tau) = k_{const}$. Since we called $\{a_i^\tau : 1 \le i \le k(\tau)\}$ a basis, $\dim V_a = k(\tau)$ holds and thus $\#\tau \ge k(\tau)$ is required. Therefore $k(\tau) = k_{const}$ cannot hold for small blocks of level ℓ, where $\#\tau = 2^{p-\ell} < k_{const}$. Hence, the exact requirement is

$$k(\tau) = \min\{k_{const}, 2^{p-level(\tau)}\} \qquad \text{for all } \tau \in T. \qquad (2.25)$$

If the $\mathcal{H}$-matrix M' has to approximate a BEM matrix M up to the error $O(h^\gamma)$ with γ being the consistency order, the choice of k_{const} should be of the order $k_{const} = \log n = p = p - level(I)$.

Case of Variable $k(b)$. Formula (2.25) gives a first advice to choose a smaller rank $k(\tau)$ for small blocks. As we shall see later, it is reasonable to choose the rank $k(\tau)$ due to a rule like $k(\tau) := p - level(\tau) + 1$ or, more general, $k(\tau) := \alpha(p - level(\tau)) + \beta$ for some $\alpha, \beta \ge 1$ (see (4.8) below).

It is a result of the approximation considerations in Sect. 4 that this choice does not deteriorate the approximation quality. On the other hand, it is obvious that for the larger number of smaller blocks we have to deal with less coefficients ζ_{ij} (see Remark 2.4) and $\alpha_{ij}^{\tau,\tau'}$, $\gamma_{ji}^{\sigma,\sigma'}$ (see (2.22), (2.23)). Therefore, a smaller rank yields lower costs for the storage and for the various arithmetic operations. In this context, the key inequality is (2.26) expressing the fact that the sum over all vertices weighted by $k(\tau)^\gamma$ for any (fixed) γ remains bounded linearly in n :

$$\sum_{\tau \in T} k(\tau)^\gamma = \sum_{\ell=0}^p 2^\ell (\alpha(p-\ell) + \beta)^\gamma \sim n \quad \text{for all } \gamma \in \mathbb{N}. \qquad (2.26)$$

In the variable case, conditions (2.19) and (2.20) are nontrivial, since the restriction of $k(\tau)$ must lead to a vector space $V_a(\tau')$ of a *lower* dimension $k(\tau')$. Here, it is interesting to consider equations (2.17) and (2.21) as the fundamental construction of the basis vectors of $V_a(\tau)$.

3   Storage and Complexity Bounds

Next we prove that the storage size is $O(n)$ without any logarithmic factor (Sect. 3.1). Then we describe the matrix-vector multiplication algorithm in Sect. 3.2 and show its $O(n)$ complexity.

3.1   Storage Requirements

According to Remark 2.5, we have to store the matrices

$$
\begin{aligned}
A^{\tau,\tau'} &:= (\alpha_{ij}^{\tau,\tau'})_{1\leq i\leq k(\tau),1\leq j\leq k(\tau')} && \text{for } \tau,\tau' \in T,\ \tau' \text{ son of } \tau, \\
C^{\sigma,\sigma'} &:= (\gamma_{ji}^{\sigma,\sigma'})_{1\leq j\leq k(\sigma),1\leq i\leq k(\sigma')} && \text{for } \sigma,\sigma' \in T,\ \sigma' \text{ son of } \sigma,
\end{aligned}
\tag{3.1}
$$

of the size $k_\ell \times k_{\ell+1}$, where $\ell = level(\tau) = level(\sigma) \in \{0,1,\ldots,p-1\}$ (cf. (2.22), (2.23)). There are $2^{\ell+1}$ different pairs τ,τ' with $\ell = level(\tau)$ and τ' son of τ. Assuming $k_\ell \leq \alpha\,(p-\ell) + \beta$ as proposed in Sect. 2.4 and used in Definition 4.7 below, the required storage amounts to

$$
\sum\nolimits_{\ell=0}^{p-1} 2^{\ell+1} k_\ell k_{\ell+1} \leq \sum\nolimits_{\ell=0}^{p-1} 2^{\ell+1}(\alpha\,(p-\ell) + \beta)(\alpha\,(p-\ell-1) + \beta).
$$

Thanks to (2.26), we obtain

Remark 3.1. The storage needed for all transfer matrices $A^{\tau,\tau'}$, $C^{\sigma,\sigma'}$ is proportional to n.

Next we consider the storage of the coefficients

$$
Z^b = (\zeta_{ij}^b)_{1\leq i,j\leq k(b)} \qquad \text{for } b = \tau \times \sigma \in P_2
$$

of the block matrix $M^b = \sum_{i,j} \zeta_{ij}^b [a_i^\tau, c_j^\sigma]$ (cf. Remark 2.4). The total storage is $\sum_{b\in P_2} k(b)^2$. Let $P_2(\ell) := \{b \in P_2 : level(b) = \ell\}$. The recursions discussed in [3] show $\#P_2(\ell) \sim 2^\ell$. Hence, $\sum_{b\in P_2} k(b)^2 \leq C \sum_{\ell=0}^{p}(p-\ell)^2 2^\ell \sim n$ proves

Remark 3.2. The storage needed for all block coefficients matrices Z^b, $b \in P_2$, is bounded by $O(n)$.

3.2   Description of the Fast Matrix-Vector Multiplication Algorithm

The fast matrix-vector multiplication algorithm is performed in three steps: (i) forward transformation (see Sect. 3.2), (ii) block multiplication phase (see Sect. 3.2), and (iii) backward transformation (see Sect. 3.2). All steps are shown to require only $O(n)$ operations, hence the matrix-vector multiplication algorithm has linear complexity.

Block Matrix Times Vector. First we consider the multiplication of a block M^b, $b = \tau \times \sigma \in P_2$, with a vector $\hat{x}_\sigma \in \mathcal{V}_a(\sigma)$. We denote the coefficient vector with respect to the basis $\{a_i^\sigma : 1 \leq i \leq k(\sigma)\}$ by $\hat{\mathbf{x}}_\sigma$, i.e.,

$$\hat{x}_\sigma = \sum\nolimits_{i=1}^{k(\sigma)} \hat{\mathbf{x}}_{\sigma,i}\, a_i^\sigma. \tag{3.2}$$

Remark 3.3. (a) Let $b = \tau \times \sigma \in P_2$, $M^b = \sum_{i,j} \zeta_{ij}^b [a_i^\tau, c_j^\sigma]$ with $Z^b = (\zeta_{ij}^b)_{1 \leq i,j \leq k(b)}$ and $\hat{x}_\sigma \in \mathcal{V}_a(\sigma)$. Then $y_\tau = M^b \hat{x}_\sigma$ has the coefficient vector $\mathbf{y}_\tau = Z^b \hat{\mathbf{x}}_\sigma$ with respect to the basis $\{a_i^\tau : 1 \leq i \leq k(\tau)\}$.

(b) Let $x_\sigma \in \mathbb{K}^\sigma$ have the decomposition $x_\sigma = \hat{x}_\sigma + x_\sigma^\perp$ with $\hat{x}_\sigma \in \mathcal{V}_a(\sigma)$ and $x_\sigma^\perp \perp \mathcal{V}_c(\sigma)$. Then $M^b x_\sigma = M^b \hat{x}_\sigma$ holds and part (a) applies to $M^b \hat{x}_\sigma$.

Proof. Part (b) is trivial. For (a) note that $M_\sigma^b \hat{x}_\sigma = \left(\sum_{i,j} \zeta_{ij}^b [a_i^\tau, c_j^\sigma] \right) \cdot$ $\cdot (\sum_h \hat{\mathbf{x}}_{\sigma,h}\, a_h^\sigma) = \sum_{i,j} \zeta_{ij}^b \hat{\mathbf{x}}_{\sigma,j} a_i^\tau$ because of $\langle c_j^\sigma, \sum_h \hat{\mathbf{x}}_{\sigma,h}\, a_h^\sigma \rangle = \hat{\mathbf{x}}_{\sigma,j}$ (cf. (2.24)). $\qquad\square$

Forward Transformations. Let a vector $x \in \mathbb{K}^I$ be given. Mx is to be computed, where M is an $\mathcal{H}^2$-matrix. Due to Remark 3.3, we have to represent the block vector $x_\sigma := (x_i)_{i \in \sigma}$ as the sum $x_\sigma = \hat{x}_\sigma + x_\sigma^\perp$, where the coefficient vector $\hat{\mathbf{x}}_\sigma$ of $\hat{x}_\sigma$ must be available. Since M contains blocks of all levels, we need the coefficient vectors $\hat{\mathbf{x}}_\sigma$ for all $\sigma \in T$.

We introduce the notation $T(\ell) := \{\tau \in T : \text{level}\,(\tau) = \ell\}$ for all $0 \leq \ell \leq p$. The following computations start at level p and proceed to level 0 :

- *Start at level $\ell = p$.* Let $\sigma = \{s\} \in T(p)$. The one-dimensional block vector $x_\sigma = (x_i)_{i \in \sigma} = (x_s)$ is identical to the coefficient vector $\mathbf{x}_\sigma$, since the basis is the unit vector $a_1^\sigma = (1)$. Hence, $\hat{\mathbf{x}}_\sigma$ is known without any computation and $x_\sigma^\perp = 0$ holds.

- *Recursion $\ell+1 \to \ell$ for $p > \ell \geq 0$.* Assume that the coefficient vectors $\hat{\mathbf{x}}_\tau$ of the first summand in $x_\tau = \hat{x}_\tau + x_\tau^\perp$ are already computed for all $\tau \in T(\ell+1)$. For all $\sigma \in T(\ell)$ the new coefficient vectors $\hat{\mathbf{x}}_\sigma$ are constructed as follows. Let $\sigma', \sigma'' \in T(\ell+1)$ be the sons of σ. The decomposition from level $\ell+1$ yields

$$x_\sigma = x_\sigma^I + x_\sigma^{II} \quad \text{with } x_\sigma^I := \begin{bmatrix} \hat{x}_{\sigma'} \\ \hat{x}_{\sigma''} \end{bmatrix} \text{ and } x_\sigma^{II} := \begin{bmatrix} x_{\sigma'}^\perp \\ x_{\sigma''}^\perp \end{bmatrix}.$$

The latter term x_σ^{II} is orthogonal to $\mathcal{V}_c(\sigma)$, since $\langle c_j^\sigma, x_\sigma^{II} \rangle = \langle R_c^{\sigma',\sigma} c_j^\sigma, x_{\sigma'}^\perp \rangle$ $+ \langle R_c^{\sigma'',\sigma} c_j^\sigma, x_{\sigma''}^\perp \rangle = 0 + 0 = 0$. The first term is to be split into $x_\sigma^I = \hat{x}_\sigma + x_\sigma^{III}$ determined by $\hat{x}_\sigma \in \mathcal{V}_a(\sigma)$, $x_\sigma^{III} \perp \mathcal{V}_c(\sigma)$; then $x_\sigma = \hat{x}_\sigma + x_\sigma^\perp$ with $x_\sigma^\perp := x_\sigma^{II} + x_\sigma^{III}$ is the desired decomposition. The entries of the coefficient vectors $\hat{\mathbf{x}}_\sigma$ are determined by $\hat{\mathbf{x}}_{\sigma,j} = \langle c_j^\sigma, \hat{x}_\sigma \rangle = \langle c_j^\sigma, x_\sigma^I \rangle$. Using the construction (2.18) of c_j^σ (cf. (2.24)), we obtain

$$\hat{x}_{\sigma,j} = \langle c_j^\sigma, x_\sigma^I \rangle = \left\langle \begin{bmatrix} R_c^{\sigma',\sigma} c_j^\sigma \\ R_c^{\sigma'',\sigma} c_j^\sigma \end{bmatrix}, \begin{bmatrix} \hat{x}_{\sigma'} \\ \hat{x}_{\sigma''} \end{bmatrix} \right\rangle$$

$$= \left\langle R_c^{\sigma',\sigma} c_j^\sigma, \hat{x}_{\sigma'} \right\rangle + \left\langle R_c^{\sigma'',\sigma} c_j^\sigma, \hat{x}_{\sigma''} \right\rangle.$$

Inserting the representation (2.23) of $R_c^{\sigma',\sigma} c_j^\sigma$ and $\hat{x}_{\sigma'} = \sum_h \hat{x}_{\sigma',h} a_h^{\sigma'}$, we result in[3]

$$\left\langle R_c^{\sigma',\sigma} c_j^\sigma, \hat{x}_{\sigma'} \right\rangle = \left\langle \sum_i \gamma_{ji}^{\sigma,\sigma'} c_i^{\sigma'}, \sum_h \hat{x}_{\sigma',h} a_h^{\sigma'} \right\rangle = \sum_i \gamma_{ji}^{\sigma,\sigma'} \hat{x}_{\sigma',i}$$

$$= (C^{\sigma,\sigma'} \hat{x}_{\sigma'})_j$$

with $C^{\sigma,\sigma'}$ defined in (3.1). Since the second term is similar, the final representation is

$$\hat{x}_\sigma = C^{\sigma,\sigma'} \hat{x}_{\sigma'} + C^{\sigma,\sigma''} \hat{x}_{\sigma''}. \tag{3.3}$$

By assumption, the coefficient vectors $\hat{x}_{\sigma'}, \hat{x}_{\sigma''}$ are known. Therefore, only two matrix-vector multiplications by the $k_\ell \times k_{\ell+1}$-matrices $C^{\sigma,\sigma'}, C^{\sigma,\sigma''}$ are needed to compute the desired coefficient vectors $\hat{x}_\sigma$ for $\sigma \in T(\ell)$.
Since the number of operations needed in (3.3) is proportional to the number of entries in $C^{\sigma,\sigma'}, C^{\sigma,\sigma''}$, Remark 3.1 implies

Remark 3.4. The performance of (3.3) for all $\sigma \in T(\ell)$, $\ell = p - 1, \ldots, 0$, requires $O(n)$ operations and yields the coefficient vectors $\hat{x}_\sigma$ for all $\sigma \in T$.

Multiplication Phase. For all blocks M^b, $b = \tau \times \sigma \in P_2$, the intermediate products $y_\tau^b := M^b x_\sigma$ are to be computed, i.e., according to Remark 3.3 the coefficient vectors $\mathbf{y}_\tau^b = Z^b \hat{x}_\sigma$ of y_τ^b are to be computed. The upper index b is used in $\mathbf{y}_\tau^b$, since for the same $\tau \in T$ several $\mathbf{y}_\tau^b$ for different b may occur (namely $b = \tau \times \sigma$ and $b^* = \tau \times \sigma^*$ with $\sigma \neq \sigma^*$).
    The number of operations for all products $Z^b \hat{x}_\sigma$, $b = \tau \times \sigma \in P_2$, is again proportional to the entries in all matrices Z^b. Therefore, Remark 3.2 implies

Remark 3.5. The matrix-vector multiplications $\mathbf{y}_\tau^b := Z^b \hat{x}_\sigma$ for all $b = \tau \times \sigma \in P_2$ requires $O(n)$ operations.

Backward Transformations. In the final step we have to gather all partial results $\mathbf{y}_\tau^b$ obtained in the previous phase. Here we use a backward transformation starting at level $\ell = 0$ and proceeding to $\ell = p$. On each level ℓ, we

---

[3] Without the biorthogonality (2.24), equation (3.3) is obtained with another matrix $\hat{C}^{\sigma,\sigma'}$.

compute $\mathbf{y}_\tau$ for all $\tau \in T(\ell)$, where $\mathbf{y}_\tau$ is the coefficient vector for the sum y_τ defined by

$$y_{\tau,i} := \sum_{b'=\tau'\times\sigma'\in P_2 \text{ with } \tau'\supseteq\tau} (y^{b'}_{\tau'})_i \qquad \text{for } i \in \tau.$$

Note that all $\tau' \supseteq \tau$ belong to some $T(\ell')$ with $\ell' \leq \ell$. As before, we set $P_2(\ell) := \{b \in P_2 \text{ with } level(b) = \ell\}$.

– *Start at level $\ell = 0$.* Since the partitioning P_2 contains no block of level 0 (the only level-0-block is $I \times I$ and not admissible), the start is given by

$$\mathbf{y}_I := 0,$$

where $I \in T(0)$ is the only cluster of level 0.
– *Recursion $\ell \to \ell+1$ for $p > \ell \geq 0$.* Assume that the coefficient vectors $\mathbf{y}_\tau$ for all $\tau \in T(\ell)$ are already computed. Let $\tau', \tau'' \in T(\ell+1)$ be the sons of some $\tau \in T(\ell)$. The vector $y_\tau = \sum_i y_{\tau,i} a^\tau_i$ corresponding to $\mathbf{y}_\tau$ equals $\begin{bmatrix} R^{\tau',\tau}_a \, y_\tau \\ R^{\tau'',\tau}_a \, y_\tau \end{bmatrix}$ by the definition of $R^{\tau',\tau}_a$ and $R^{\tau'',\tau}_a$. The coefficient vectors $\hat{\mathbf{y}}_{\tau'}$ and $\hat{\mathbf{y}}_{\tau''}$ of $R^{\tau',\tau}_a \, y_\tau$ and $R^{\tau'',\tau}_a \, y_\tau$ are given by

$$\hat{\mathbf{y}}_{\tau'} = (A^{\tau,\tau'})^T \mathbf{y}_\tau, \qquad \hat{\mathbf{y}}_{\tau''} = (A^{\tau,\tau''})^T \mathbf{y}_\tau, \tag{3.4}$$

as one concludes from $R^{\tau',\tau}_a \, y_\tau = R^{\tau',\tau}_a \sum_i y_{\tau,i} a^\tau_i = \sum_i y_{\tau,i} R^{\tau',\tau}_a a^\tau_i = \sum_{i,j} y_{\tau,i} \alpha^{\tau,\tau'}_{ij} a^{\tau'}_j$.
Next, we have to add all contributions from blocks of level $\ell + 1$:

$$\mathbf{y}_{\tau'} = \hat{\mathbf{y}}_{\tau'} + \sum_{\sigma' \text{ with } b'=\tau'\times\sigma'\in P_2(\ell+1)} \mathbf{y}^{b'}_{\tau'}. \tag{3.5}$$

Remark 3.6. The number of operations involved in the backward transformations (3.4) and (3.5) is $2\sum_{\ell=0}^{p-1} k_\ell k_{\ell+1} \#T(\ell+1) + \sum_{\ell=0}^{p-1} \#P_2(\ell+1) = 2\sum_{\ell=0}^{p-1} k_\ell k_{\ell+1} 2^{\ell+1} + \#P_2 \sim n$.

– *Result at level $\ell = p$.* The resulting coefficient vectors $\mathbf{y}_\tau$ for all one-element clusters $\tau = \{i\} \in T(p)$ coincide with the component y_i of $y = Mx$. Therefore, the matrix-vector multiplication is completed.

3.3   Other Matrix Operations

Different from general $\mathcal{H}$-matrices, the sum of two $\mathcal{H}^2$-matrices (with the same partitioning and the same hierarchical bases) can be performed *exactly*. Since only the matrices Z^b, $b \in P_2$, are to be added, the cost is clearly $O(n)$.

We do not discuss the matrix-matrix multiplications in detail, but it may be mentioned that the product of two blocks is rather cheap since we have to perform scalar products of the form $\langle c^\sigma_j, a^\sigma_i \rangle$, which are trivial because of (2.24).

3.4   Constant $k(\tau)$

The proof of the following statement is left to the reader.

Proposition 3.7. *Let $k_{const} \in \{1,\dots,n\}$. Choose the rank $k(\tau)$ according to (2.25). Then the storage size of $A^{\tau,\tau'}$, $C^{\sigma,\sigma'}$, Z^b (see Remarks 3.1-2) as well as the matrix-vector multiplication cost amounts to $O(n \cdot k_{const})$.*

4   Approximation by Variable Order

In this section, we will explain how the approximation of the integral operator (2.1) via $\mathcal{H}^2$-matrices with variable order k can be realized.

4.1   Galerkin Matrix

Let $b = \tau \times \sigma \in P_2$ and $(i,j) \in b$. The matrix element m_{ij} is defined in (2.3) by $\int_{J_i} \int_{J_j} s(x,y) dx dy$, where $s(x,y) = \log(|x-y|)$. If we find an expansion $s(x,y) \approx \tilde{s}(x,y) := \sum_{\alpha,\beta=1}^{k(b)} \gamma_{\alpha,\beta}\varphi_\alpha(x)\psi_\beta(y)$ which is sufficiently accurate on the rectangle $J(\tau) \times J(\sigma)$ (cf. (2.7)), the Galerkin matrix based on $\tilde{s}$ instead of s has the entries $\tilde{m}_{ij} := \sum_{\alpha,\beta=1}^{k(b)} \gamma_{\alpha,\beta} a_{\alpha,i} c_{\beta,j}$ for $(i,j) \in b$ with $a_{\alpha,i} = \int_{J_i} \varphi_\alpha(x) dx$ and $c_{\beta,j} = \int_{J_j} \psi_\beta(y) dy$. Obviously, the block matrix $(\tilde{m}_{ij})_{(i,j)\in b} = \sum_{\alpha,\beta=1}^{k(b)} \gamma_{\alpha,\beta} a_\alpha c_\beta^T$ is of the desired form, if the spaces spanned by a_α or c_β satisfy the respective consistency conditions (2.19) or (2.20).

Concerning the consistency conditions, we state the following criterion.

Remark 4.1. For $b = \tau \times \sigma$, let $\varphi_\alpha^\tau(x)$ and $\psi_\beta^\sigma(y)$, $1 \le \alpha, \beta \le k(b)$, be the functions involved in the approximation of $s(x,y)$ by $\tilde{s}(x,y) := \sum_{\alpha,\beta=1}^{k(b)} \gamma_{\alpha,\beta}\varphi_\alpha^\tau(x) \cdot \psi_\beta^\sigma(y)$ for $(x,y) \in J(\tau) \times J(\sigma)$. Set $\mathcal{W}_a(\tau) := \mathrm{span}\{\varphi_\alpha^\tau : 1 \le \alpha \le k(\tau)\}$ and $\mathcal{W}_c(\sigma) := \mathrm{span}\{\psi_\beta^\sigma : 1 \le \alpha \le k(\sigma)\}$. Then the condition

$$\mathcal{W}_a(\tau') = \{\varphi|_{J(\tau')} : \varphi \in \mathcal{W}_a(\tau)\} \qquad \text{for all sons } \tau' \text{ of } \tau \qquad (4.1)$$

implies the consistency condition (2.19). Similar for (2.20).

Note that $\dim \mathcal{V}_a(\tau) \le \dim \mathcal{W}_a(\tau)$ holds, where the strict inequality may occur. Since nothing is to be discussed about the 1×1-blocks b in P_2 (belonging to level p), we restrict our considerations to the subset of the "far field" blocks:

$$P_2^{far} := \{b \in P_2 : \; level(b) < p\}. \qquad (4.2)$$

In the following, the approximation $\tilde{s}(x,y)$ is based on Taylor expansions[4] of the kernel functions. However, in order to satisfy (4.1), the arising polynomials of degree $k(\tau) - 1$ must be replaced by very particular functions.

---

[4] Other expansions based on projections or interpolations are possible.

4.2   Taylor Expansions

First, we have to introduce some notations. $J := [0,1]$ denotes the integration domain.

In order to simplify the notation, we replace the rank $k(\tau)$ by $k(\tau) + 1$, since then $k(\tau)$ also coincides with the polynomial degree. In other words, the summation $\sum_{i=1}^{k(\tau)}$ is replaced by $\sum_{i=0}^{k(\tau)}$.

Definition 4.2. Let $\omega \subseteq J$ and $\check{c}_\omega$ be the smallest interval containing ω. The Čebyšev centre z_ω of ω is the midpoint of $\check{c}_\omega$ and the Čebyšev radius r_ω equals the halved interval length of $\check{c}_\omega$.

Definition 4.3. For $\omega, w \subseteq J$, the difference domain $d_{\omega,w}$ is given by

$$d_{\omega,w} := \omega - w := \{z \in \mathbb{R} : \exists (x,y) \in \omega \times w \text{ with } z = x - y\}.$$

Definition 4.4. Let $\omega \subseteq J$ and $f : \omega \to \mathbb{R}$ be sufficiently smooth. The Taylor operator $T_\omega^{(m)}$ of order $m \in \mathbb{N}$ is given by

$$T_\omega^{(m)}[f](x) = \sum_{\nu=0}^{m-1} \frac{1}{\nu!} f^{(\nu)}(z_\omega)(x - z_\omega)^\nu.$$

Lemma 4.5. *Let $\omega, w \subseteq J$ satisfy $\eta \operatorname{dist}(\omega, w) \geq r_\omega + r_w$ with some $\eta \in (0,1]$. On the difference domain $d = d_{\omega,w}$, the Taylor expansion of the function $s : d \to \mathbb{R}$,*

$$s(z) = \log|z| \tag{4.3}$$

about z_d of order m satisfies

$$|s(z) - T_d^{(m)}[s](z)| \leq \begin{cases} |\log \operatorname{dist}(\omega,w)| & m = 0 \\ \eta^m/m & m > 0 \end{cases} \qquad \text{for all } z \in d. \tag{4.4}$$

Proof. The remainder $R_d^{(m)}(z)$ of the Taylor expansion can be estimated (by using $r_d = r_\omega + r_w$)

$$|R_d^{(m)}(z)| \leq r_d^m \sup_{\xi \in d} \frac{|s^{(m)}(\xi)|}{m!} = r_d^m \sup_{\xi \in d} \begin{cases} |\log|\xi|| & \text{if } m = 0 \\ \frac{1}{m}|\xi|^{-m} & \text{if } m > 0 \end{cases}$$

$$\leq (r_\omega + r_w)^m \begin{cases} |\log \operatorname{dist}(\omega,w)| & (m = 0) \\ \frac{1}{m} \operatorname{dist}^{-m}(\omega,w) & (m > 0) \end{cases}$$

$$\leq \begin{cases} |\log \operatorname{dist}(\omega,w)| & (m = 0) \\ \eta^m/m & (m > 0) \end{cases}. \qquad \square$$

Corollary 4.6. *Let ω, w be as in Lemma 4.5. The Taylor approximation of the kernel function $\log(|x - y|)$ is denoted by $T_d^{(m)} [s] (x - y)$. The explicit representation*

$$T_d^{(m)} [s] (x - y) = \sum_{\nu + \mu \leq m-1} \kappa_{\omega,w}^{(\nu,\mu)} \Phi_\omega^{(\nu)} (x) \Phi_w^{(\mu)} (y) \qquad (4.5)$$

holds with $\Phi_\omega^{(\nu)} (x) = (x - z_\omega)^\nu$ and $\kappa_{\omega,w}^{(\nu,\mu)} = (-1)^{\nu+\mu} s^{(\nu+\mu)} (z_d) / (\mu! \nu!)$.

Proof. Reorganizing sums and products results in

$$T_d^{(m)} [s] (x - y) = \sum_{\nu=0}^{m-1} \frac{1}{\nu!} s^{(\nu)} (z_d) (z - z_d)^\nu$$

$$= \sum_{\nu=0}^{m-1} \sum_{\mu=0}^{\nu} \frac{(-1)^{\nu-\mu}}{\nu!} s^{(\nu)} (z_d) \binom{\nu}{\mu} (x - z_\omega)^\mu (y - z_w)^{\nu-\mu}$$

$$= \sum_{\mu=0}^{m-1} \sum_{\nu=0}^{m-1-\mu} \frac{(-1)^{\nu+\mu} s^{(\nu+\mu)} (z_d)}{\mu! \, \nu!} (x - z_\omega)^\mu (y - z_w)^\nu. \qquad \square$$

4.3   The Hierarchical Bases Construction

The essential step in the definition of the variable order approximation method is the definition of the restriction operators. In order to illustrate the underlying idea, we consider a cluster τ of the cluster tree with sons τ', τ''. Assume that the spaces $\mathcal{W}_a (\tau')$ and $\mathcal{W}_a (\tau'')$ with bases $\left\{ \tilde{\Phi}_{\tau'}^{(i)} \right\}_{i=0}^{k(\tau')}$ and $\left\{ \tilde{\Phi}_{\tau''}^{(i)} \right\}_{i=0}^{k(\tau'')}$ are already defined. Then, every function in $\mathcal{W}_a (\tau)$ has the representation

$$u (x) = \begin{cases} \sum_{\mu=0}^{k(\tau')} a_{\tau'}^{(\mu)} \tilde{\Phi}_{\tau'}^{(\mu)} (x) & \text{for } x \in J(\tau'), \\ \sum_{\mu=0}^{k(\tau'')} a_{\tau''}^{(\mu)} \tilde{\Phi}_{\tau''}^{(\mu)} (x) & \text{for } x \in J(\tau''), \end{cases} \qquad (4.6)$$

and Remark 4.1 guarantees the consistency condition (2.19). Because of $\mathcal{W}_c (\tau) = \mathcal{W}_a (\tau)$, also (2.20) holds.

The difficulty is that (4.6) is not able to represent monomials of degree $k (\tau)$ if $k (\tau) > k(\tau') = k(\tau'')$. Instead, we are looking for approximations of $\Phi_\tau^{(\nu)} := (x - z_\tau)^\nu$ (see Corollary 4.6) by functions $\tilde{\Phi}_\tau^{(\nu)}$ belonging to $\mathcal{W}_a (\tau)$, i.e., having a representation (4.6).

The basis functions $\tilde{\Phi}_\tau^{(\nu)}$ will be chosen as the composition of the *local* Taylor polynomials of degree $k (\tau)$ in $J(\tau')$ and $J(\tau'')$ obtained by expansions of the true monomial $\Phi_\tau^{(\nu)}$ around the respective Čebyšev centres $z_{\tau'}$ and $z_{\tau''}$. The coefficients $a_t^{(\nu,\mu)} := a_t^{(\mu)} [\Phi_\tau^{(\nu)}]$, $t \in \{\tau', \tau''\}$, in (4.6) are

$$a_t^{(\nu,\mu)} := a_t^{(\mu)} [\Phi_\tau^{(\nu)}] = \frac{1}{\mu!} \partial_x^{(\mu)} [(x - z_\tau)^\nu] |_{x=z_t}. \qquad (4.7)$$

We repeat that this construction yields the true monomial $\tilde{\Phi}_\tau^{(\nu)} = \Phi_\tau^{(\nu)}$ if the functions $\tilde{\Phi}_t^{(\mu)}$, $t \in \{\tau', \tau''\}$, in the right-hand side of (4.6) are the true monomials $\Phi_t^{(\mu)}$ and if $\nu \leq k(t)$, $t \in \{\tau', \tau''\}$. However, our variable order assumption leads to the case $\nu = k(\tau) > k(t)$.

Definition 4.7. The polynomial degree distribution depends on constants $\alpha, \beta \geq 0$ and is given by

$$k(\tau) := \alpha \, (p - level\,(\tau)) + \beta. \tag{4.8}$$

For the leaves $\tau \in T$ (i.e., $level\,(\tau) = p$), we put

$$\tilde{\Phi}_\tau^{(\nu)}\,(x) = (x - z_\tau)^\nu \qquad \text{for all } \nu \in \{0, 1, \ldots, \beta\}.$$

Assume that the basis functions $\tilde{\Phi}_t^{(\nu)}$ are defined on all clusters $t \in T$ with $level\,(t) \geq \ell$. Then, for $\tau \in T$ with $level\,(\tau) = \ell - 1$, the basis functions $\tilde{\Phi}_\tau^{(\nu)}$ are given by

$$\tilde{\Phi}_\tau^{(\nu)}\,(x) := \begin{cases} \sum_{\mu=0}^{k(\tau')} a_{\tau'}^{(\nu,\mu)} \tilde{\Phi}_{\tau'}^{(\mu)}\,(x) & \text{for } x \in J(\tau'), \\ \sum_{\mu=0}^{k(\tau'')} a_{\tau''}^{(\nu,\mu)} \tilde{\Phi}_{\tau''}^{(\mu)}\,(x) & \text{for } x \in J(\tau''), \end{cases} \tag{4.9}$$

with $a_t^{(\nu,\mu)}$ from (4.7), where τ', τ'' are the sons of τ.

The approximation of the kernel function is given by replacing the Taylor polynomials $\Phi_\tau^{(\nu)}$ in (4.5) by the functions $\tilde{\Phi}_\tau^{(\nu)}$.

Definition 4.8. Let $\eta \in (0, 1]$. A block $b = \tau \times \sigma \in P_2$ is η-admissible if the following condition holds:

$$\eta \, dist\,(\tau, \sigma) \geq \max\{\operatorname{diam} \tau, \operatorname{diam} \sigma\}. \tag{4.10}$$

We remark that for the partitioning in (2.11) and (2.12) all $b \in P_2^{far}$ (cf. (4.2)) satisfy (4.10) with $\eta = 1$.

Definition 4.9. Let $b = \tau \times \sigma$ denote an η-admissible block. The approximation to the function s as in (4.3) is given by

$$\tilde{s}_b\,(x, y) := \sum_{\nu + \mu \leq k(b)} \kappa_{\tau,\sigma}^{(\nu,\mu)} \tilde{\Phi}_\tau^{(\nu)}\,(x)\,\tilde{\Phi}_\sigma^{(\mu)}\,(y) \tag{4.11}$$

where $\kappa_{\tau,\sigma}^{(\nu,\mu)}$ are the Taylor coefficients from (4.5).

We remark that the approximation of s by $\tilde{s}_b$ in $J(\tau) \times J(\sigma)$ is not the optimal one. First, we could find better coefficients $a_t^{(\nu,\mu)}$ in (4.9) when we look for the best expansion with respect to the basis $\tilde{\Phi}_t^{(\mu)}$ (instead of $\Phi_t^{(\mu)}$). Second, in (4.11) we could allow all indices $0 \leq \nu, \mu \leq k\,(b)$ instead of $\nu + \mu \leq k\,(b)$.

4.4 Error Analysis

The error analysis of the variable order approximation algorithm consists of a local estimate of the error $s - \tilde{s}_b$ on admissible blocks and a global estimate of the consistency error. We begin with the local estimates. On an η-admissible block $b = \tau \times \sigma \in P_2$, the local approximation error is defined by

$$e_b(x,y) = s(x,y) - \tilde{s}_b(x,y) \quad \text{for } (x,y) \in J(\tau) \times J(\sigma) \tag{4.12}$$

and its maximum norm by

$$\varepsilon_b := \|e_b\|_{\infty,b} = \sup_{(x,y) \in J(\tau) \times J(\sigma)} |e_b(x,y)|. \tag{4.13}$$

The expansion with the true Taylor polynomials defines the function

$$s_b(x,y) = \sum_{\nu + \mu \le k(b)} \kappa_{\tau,\sigma}^{(\nu,\mu)} \Phi_\tau^{(\nu)}(x) \Phi_\sigma^{(\mu)}(y).$$

The error is split into

$$e_b = (s - s_b) + (s_b - \tilde{s}_b) =: e_b^I + e_b^{II}, \tag{4.14}$$

where e_b^I is already estimated by (4.4). The error e_b^{II} has the representation:

$$
\begin{aligned}
e_b^{II}(x,y) = & \sum_{\nu+\mu \le k(b)} \kappa_{\tau,\sigma}^{(\nu,\mu)} \left(\Phi_\tau^{(\nu)}(x) - \tilde{\Phi}_\tau^{(\nu)}(x) \right) \Phi_\sigma^{(\mu)}(y) \\
& + \sum_{\nu+\mu \le k(b)} \kappa_{\tau,\sigma}^{(\nu,\mu)} \tilde{\Phi}_\tau^{(\nu)}(x) \left(\Phi_\sigma^{(\mu)}(y) - \tilde{\Phi}_\sigma^{(\mu)}(y) \right) \\
=: & \ e_b^{III}(x,y) + e_b^{IV}(x,y).
\end{aligned}
\tag{4.15}
$$

The estimate of the difference $\Phi_\tau^{(\nu)} - \tilde{\Phi}_\tau^{(\nu)}$ plays the key role in the following error estimation.

Lemma 4.10. *Choose $\alpha \ge 2$ and $\beta \ge \max\left\{\frac{6}{5}\alpha + 1, 2\alpha - 1\right\}$ in (4.8) and put*

$$\omega \ge \max\{2.5, \alpha - 1\}. \tag{4.16}$$

Then, for $\tau \in T$ with $\ell = \text{level}(\tau)$, the difference $\Phi_\tau^{(\nu)} - \tilde{\Phi}_\tau^{(\nu)}$ can be estimated by

$$\|\Phi_\tau^{(\nu)} - \tilde{\Phi}_\tau^{(\nu)}\|_{\infty,\tau} \le \lambda^{-k_\ell} \left(\omega 2^{-\ell}\right)^\nu, \qquad \text{where } \lambda = 3/2.$$

Proof. We define an intermediate approximation $\check{\Phi}_\tau^{(\nu)}$ to $\Phi_\tau^{(\nu)}$ by using the true Taylor polynomials in the right-hand side of

$$\check{\Phi}_\tau^{(\nu)}(x) := \begin{cases} \sum_{\mu=0}^{k(\tau')} a_{\tau'}^{(\nu,\mu)} \Phi_{\tau'}^{(\mu)}(x) & \text{for } x \in J(\tau'), \\ \sum_{\mu=0}^{k(\tau'')} a_{\tau''}^{(\nu,\mu)} \Phi_{\tau''}^{(\mu)}(x) & \text{for } x \in J(\tau''). \end{cases} \tag{4.17}$$

The estimate of $\Phi_\tau^{(\nu)} - \tilde{\Phi}_\tau^{(\nu)}$ will be performed on all leaves contained in τ separately. Let t be a leaf satisfying $t \subseteq \tau$ and define the sequence $(\tau_\ell, \tau_{\ell+1}, \ldots, \tau_p)$ by $t = \tau_p \subset \tau_{p-1} \subset \ldots \subset \tau_{\ell+1} \subset \tau_\ell = \tau$ and $level\,(\tau_i) = i$. For $x \in J(t)$, the representation

$$\Phi_{\tau_i}^{(\nu)} - \tilde{\Phi}_{\tau_i}^{(\nu)} = \Phi_{\tau_i}^{(\nu)} - \check{\Phi}_{\tau_i}^{(\nu)} + \sum_{\mu=0}^{k_{i+1}} a_{\tau_{i+1}}^{(\nu,\mu)} \left(\Phi_{\tau_{i+1}}^{(\mu)} - \tilde{\Phi}_{\tau_{i+1}}^{(\mu)} \right)$$

holds, where $k_{i+1} = k\,(\tau_{i+1})$. Furthermore, we replace, for ease of notation, indices τ_i by i, e.g., z_i instead of z_{τ_i}. Define the error quantities

$$\delta_i^{(\nu)} := \|\Phi_i^{(\nu)} - \tilde{\Phi}_i^{(\nu)}\|_{\infty,t}, \qquad \varepsilon_i^{(\nu)} := \|\Phi_i^{(\nu)} - \check{\Phi}_i^{(\nu)}\|_{\infty,t}.$$

Then,

$$\delta_i^{(\nu)} \le \varepsilon_i^{(\nu)} + \sum_{\mu=0}^{k_{i+1}} \left| a_{i+1}^{(\nu,\mu)} \right| \delta_{i+1}^{(\mu)}. \tag{4.18}$$

$\check{\Phi}_i^{(\nu)}$ is the Taylor expansion of $\Phi_i^{(\nu)}$ on τ_{i+1} of order k_{i+1}. Estimating the remainder of the Taylor expansion results in $\varepsilon_i^{(\nu)} = 0$ for all $\nu \le k_{i+1}$ and, for $\nu > k_{i+1}$,

$$\varepsilon_i^{(\nu)} \le \frac{r_{i+1}^{k_{i+1}+1}}{(k_{i+1}+1)!} \frac{\nu!}{(\nu - k_{i+1} - 1)!} \left\| (\cdot - z_i)^{\nu - k_{i+1} - 1} \right\|_{\infty,\tau_{i+1}} =$$

$$= \binom{\nu}{k_{i+1}+1} r_i^{\nu - k_{i+1} - 1} r_{i+1}^{k_{i+1}+1} = \binom{\nu}{k_{i+1}+1} 2^{-i\nu - k_{i+1} - 1}.$$

The coefficients $a_{i+1}^{(\nu,\mu)}$ in (4.17) vanish for $\mu > \nu$ and, for $\mu \le \nu$, we have

$$\left| a_{i+1}^{(\nu,\mu)} \right| = \binom{\nu}{\mu} |z_{i+1} - z_i|^{\nu-\mu} = \binom{\nu}{\mu} r_{i+1}^{\nu-\mu} = \binom{\nu}{\mu} 2^{-(i+1)(\nu-\mu)}.$$

The coefficients $\delta_i^{(\nu)}$ in (4.18) are bounded from above by $\tilde{\delta}_i^{(\nu)}$ defined by

$$\tilde{\delta}_p^{(\nu)} = 0 \qquad \text{for all } \nu \le k_p \tag{4.19}$$

and, for $i = p-1, p-2, \ldots, \ell$, by

$$\tilde{\delta}_i^{(\nu)} = \sum_{\mu=0}^{\nu} \binom{\nu}{\mu} 2^{-(i+1)(\nu-\mu)} \tilde{\delta}_{i+1}^{(\mu)} \qquad \text{for } \nu \le k_{i+1},$$

$$\tilde{\delta}_i^{(\nu)} = \binom{\nu}{k_{i+1}+1} 2^{-(i+1)\nu - k_{i+1} - 1} +$$

$$+ \sum_{\mu=0}^{k_{i+1}} \binom{\nu}{\mu} 2^{-(i+1)(\nu-\mu)} \tilde{\delta}_{i+1}^{(\mu)} \qquad \text{for } k_{i+1} < \nu \le k_i. \tag{4.20}$$

The estimate $\tilde{\delta}_i^{(\nu)} \le \lambda^{i-p} \left(\omega 2^{-i} \right)^\nu$ stated in Lemma 4.11 yields the proof. $\square$

Lemma 4.11. *Let α, β, and ω be as in Lemma 4.10 and set $\lambda = 3/2$. Then, the coefficients $\tilde{\delta}_i^{(\nu)}$ can be estimated by $\tilde{\delta}_i^{(\nu)} \leq \lambda^{-k_i} \left(\omega 2^{-i} \right)^{\nu}$.*

The rather technical proof of this lemma can be found in the Appendix of [5].

We have all ingredients to estimate the approximation error on a local block.

Theorem 4.12. *Let α, β, and ω be as in Lemma 4.10. Let $\eta \leq (8\omega)^{-1}$. Then, there exists a constant $C < \infty$ so that, for every η-admissible block $b = \tau \times \sigma$ with $\mathrm{level}(\tau) = \mathrm{level}(\sigma) = \ell$, the error ε_b as in (4.13) can be estimated by*

$$\varepsilon_b \leq C \left(\tfrac{2}{3} \right)^{k(b)} .$$

Proof. In view of the splittings (4.14) and (4.15) it remains to estimate $e_b^{III,IV}$. The coefficients $\kappa_{\tau,\sigma}^{(\nu,\mu)}$ from (4.15) can be estimated by

$$\left| \kappa_{\tau,\sigma}^{(\nu,\mu)} \right| \leq \binom{\nu+\mu}{\nu} \begin{cases} |\log \mathrm{dist}(\tau,\sigma)| & \text{if } \nu+\mu = 0 \\ \mathrm{dist}^{-\nu-\mu}(\tau,\sigma) / (\nu+\mu) & \text{if } \nu+\mu > 0 \end{cases} .$$

By using $\left| \Phi_\sigma^{(\mu)}(y) \right| = |y - z_\sigma|^\mu \leq r_\sigma^\mu = 2^{-(\ell+1)\mu}$ for $y \in J(\sigma)$, we obtain (with $\lambda = 3/2$)

$$\left| e_b^{III}(x,y) \right| \leq \sum_{\nu+\mu \leq k(b)} \left| \kappa_{\tau,\sigma}^{(\nu,\mu)} \right| \left| \Phi_\tau^{(\nu)}(x) - \tilde{\Phi}_\tau^{(\nu)}(x) \right| \left| \Phi_\sigma^{(\mu)}(y) \right|$$

$$\leq \sum_{1 \leq \nu+\mu \leq k(b)} \frac{1}{\nu+\mu} \binom{\nu+\mu}{\nu} \frac{\lambda^{-k(b)} \left(\omega 2^{-\ell} \right)^\nu}{\mathrm{dist}^{\nu+\mu}(\tau,\sigma)} 2^{-(\ell+1)\mu}$$

$$\leq \lambda^{-k\ell} \sum_{1 \leq \nu+\mu \leq k(b)} \frac{1}{\nu+\mu} \binom{\nu+\mu}{\nu} (2\omega)^\nu \, \eta^{\nu+\mu} \, 2^{-\nu-\mu}.$$

Choosing $\eta \leq (4\omega)^{-1}$, we obtain

$$\left| e_b^{III}(x,y) \right| \leq \lambda^{-k(b)} \sum_{\mu=0}^{k} \sum_{\nu=0}^{\mu} \frac{\mu!}{\nu! \, (\mu-\nu)!} 2^{-\nu-\mu} = \lambda^{-k(b)} \left(4 - 3 \left(\tfrac{3}{4} \right)^{k(b)} \right)$$

$$\leq 4 \, \lambda^{-k(b)}.$$

The estimate of e_b^{IV} can be obtained in the same fashion by using

$$\left| \tilde{\Phi}_\tau^{(\nu)}(x) \right| \leq \left| \Phi_\tau^{(\nu)}(x) \right| + \left| \tilde{\Phi}_\tau^{(\nu)}(x) - \Phi_\tau^{(\nu)}(x) \right| \leq 2^{-(\ell+1)\nu} + \lambda^{-k(b)} \left(\omega 2^{-\ell} \right)^\nu$$

$$\leq \left(\tilde{\omega} 2^{-\ell-1} \right)^\nu$$

with $\tilde{\omega} = 2\omega$ and proceeding in an analogous way. $\qquad \square$

We come now to the global estimate of error caused by replacing the kernel function by the variable order approximation. Let S denote the space of piecewise constant functions on the mesh $\{J_i\}_{i=1}^{n}$. The L^2-norm on an interval $t = (a, b)$ is denoted by $\|\cdot\|_{0,t}$ while we skip the index t for $t = (0, 1)$. The global error of the approximation is given by

$$E(u, v) := \sum_{b \in P_2^{far}} \int_b u(x)\, e_b(x, y)\, v(y)\, dy\, dx \qquad \text{for } u, v \in S.$$

Theorem 4.13. *Let α, β, ω, and η be as in Theorem 4.12. There exists a constant C so that, for all $u, v \in S$,*

$$E(u, v) \leq Ch\, \|u\|_0\, \|v\|_0.$$

Proof. Let $P_2(\ell) := \{b \in P_2 : level(b) = \ell\}$ and $T(\ell) := \{\tau \in T : level(\tau) = \ell\}$ for all $0 \leq \ell \leq p - 1$. For $b = \tau \times \sigma \in P_2(\ell)$, we have $|\tau| = |\sigma| = 2^{-\ell}$. Introduce

$$N_1 = \sup_{0 \leq \ell < p} \sup_{\tau \in T(\ell)} \sum_{\sigma \text{ with } \tau \times \sigma \in P_2(\ell)} 1, \qquad N_2 = \sup_{0 \leq \ell < p} \sup_{\sigma \in T(\ell)} \sum_{\tau \text{ with } \tau \times \sigma \in P_2(\ell)} 1.$$

Using Theorem 4.12, we obtain with $\lambda = 3/2$

$$E = \sum_{\ell=0}^{p-1} \sum_{b \in P_2(\ell)} \int_b u(x)\, e_b(x, y)\, v(y)\, dy\, dx$$

$$\leq \sum_{\ell=0}^{p-1} \sum_{b=\tau \times \sigma \in P_2(\ell)} C\lambda^{-k_\ell} \sqrt{|\tau|\,|\sigma|}\, \|u\|_{0,\tau}\, \|v\|_{0,\sigma}$$

$$\leq Ch \sum_{\ell=0}^{p-1} \lambda^{-k_\ell} 2^{p-\ell} \sqrt{\sum_{b=\tau \times \sigma \in P_2(\ell)} \|u\|_{0,\tau}^2} \sqrt{\sum_{b=\tau \times \sigma \in P_2(\ell)} \|v\|_{0,\sigma}^2}$$

$$\leq Ch \sum_{\ell=0}^{p-1} \lambda^{-k_\ell} 2^{p-\ell} \sqrt{\sum_{\tau \in T(\ell)} \|u\|_{0,\tau}^2 \sum_{\sigma: \tau \times \sigma \in P_2(\ell)} 1} \sqrt{\sum_{\sigma \in T(\ell)} \|v\|_{0,\sigma}^2 \sum_{\tau: \tau \times \sigma \in P_2(\ell)} 1}$$

$$= Ch\, \|u\|_0\, \|v\|_0 \sqrt{N_1 N_2} \sum_{\ell=0}^{p} \lambda^{-k_\ell} 2^{p-\ell}.$$

For $\alpha \geq 2$, we have $\lambda^{-k_\ell} 2^{p-\ell} \leq \left(\frac{9}{8}\right)^{\ell-p}$ and $\sum_{\ell=0}^{p} \lambda^{-k_\ell} 2^{p-\ell} \leq \sum_{\ell=0}^{p} \left(\frac{9}{8}\right)^{\ell-p} \leq 9$. One can see from the construction in Definition 2.2 that the numbers N_1, N_2 can be estimated from above by a constant as well. $\qquad\square$

References

1. Hackbusch, W.: Iterative Solution of Large Sparse Systems. Springer Verlag, New York, 1994

2. Hackbusch, W.: Integral Equations, Theory and Numerical Treatment. ISNM 128. Birkhäuser, Basel, 1995
3. Hackbusch, W.: A Sparse Matrix Arithmetic Based on $\mathcal{H}$-Matrices. Part I: Introduction to $\mathcal{H}$-Matrices. Computing **62** (1999), 89–108
4. Hackbusch, W., Khoromskij, B. N.: A Sparse $\mathcal{H}$-Matrix Arithmetic. Part II: Application to Multi-Dimensional Problems. Preprint Nr 22/1999, Max-Planck-Institut für Mathematik, Leipzig, 1999, to appear in Computing
5. Hackbusch, W., Khoromskij, B. N., Sauter, S. A.: On $\mathcal{H}^2$-Matrices. Preprint, Max-Planck-Institut für Mathematik, Leipzig, 1999
6. Hackbusch, W., Nowak, Z. P.: On the Fast Matrix Multiplication in the Boundary Element Method by Panel Clustering. Numer. Math. **54** (1989), 463–491
7. Sauter, S. A.: Über die effiziente Verwendung des Galerkin-Verfahrens zur Lösung Fredholmscher Integralgleichungen. Dissertation, Universität Kiel, 1992

On the Stability of Microstructure for General Martensitic Transformations*

M. Luskin

School of Mathematics, University of Minnesota, 206 Church Street SE,
Minneapolis, MN 55455, USA

*Dedicated to Professor Karl-Heinz Hoffmann
on the occasion of his 60th birthday*

Abstract. We describe a general theory for the stability of the laminated microstructure for martensitic crystals. Our theory has been applied to the orthorhombic to monoclinic transformation, the cubic to tetragonal transformation, the tetragonal to monoclinic transformation, and the cubic to orthorhombic transformation.

1 Introduction

We describe recent results for the stability of laminated microstructure for crystals that undergo a symmetry reducing solid-solid phase transformation. In the geometrically nonlinear theory of martensite [2,3,11,25], the energy density is minimized on multiple energy wells $SO(3)U_1 \cup \ldots \cup SO(3)U_N$ where $U_1, \ldots, U_N \in \mathbb{R}^{3 \times 3}$ for $N > 1$ are symmetry-related transformation strains (variants) and $SO(3)$ is the set of all 3×3 real orthogonal matrices with determinant equal to one.

An early version of the stability theory for crystal microstructure was first developed for some one-dimensional models in [13,14]. Results allowing the theory to be extended to the multi-dimensional geometrically nonlinear theory of crystals were first given for a rotationally invariant double well energy density ($N = 2$) in [24]. These results apply directly to the orthorhombic to monoclinic transformation. The theory has since been applied to the cubic to tetragonal transformation ($N = 3$) [20], the tetragonal to monoclinic transformation ($N = 4$) [5], and the cubic to orthorhombic transformation ($N = 6$) [4]. In general, the analysis of stability becomes more difficult for larger N since the additional wells give the crystal more freedom to deform without the cost of additional energy. In fact, for the tetragonal to monoclinic transformation ($N = 4$) and the cubic to orthorhombic transformation ($N = 6$) we have shown that there are special lattice constants for which the laminated microstructure is not stable.

---

* This work was supported in part by NSF DMS 95-05077, by AFOSR F49620-98-1-0433, by ARO DAAG55-98-1-0335, by the Institute for Mathematics and Its Applications, and by the Minnesota Supercomputer Institute.

The stability theory can also be used to analyze laminates with varying volume fraction [21] and conforming and nonconforming finite element approximations [22,24]. We also note that the stability theory was used to analyze the microstructure in ferromagnetic crystals [26]. Related results on the numerical analysis of nonconvex variational problems can be found, for example, in [6–10,12,16–19,23,27–29].

In Sect. 2, we describe the geometrically nonlinear theory of martensite. We refer the reader to [2,3] and to the introductory article [25] for a more detailed discussion of the geometrically nonlinear theory of martensite. In Sect. 3, we prove a condition that allows a reduction to an approximate mixture of two strains. In Sect. 4, we show how this condition can be verified for the cubic to tetragonal transformation. In Sect. 5, we give results for the stability and uniqueness of the microstructure that follows from the estimate for the reduction to an approximate mixture to two strains. Finally, in Sect. 6, we give convergence results for finite element methods that following directly from an approximation result and the results in Sect. 5.

2   The Continuum Model

We denote deformations by functions $y : \Omega \to \mathbb{R}^3$ and corresponding deformation gradients by $\nabla y : \Omega \to \mathbb{R}^{3\times3}$ where Ω is a bounded domain with a Lipschitz continuous boundary $\partial\Omega$. We consider the minimization of the total energy

$$\mathcal{E}(y) = \int_{\Omega} \phi(\nabla y(x))\, dx$$

over an admissible class $\mathcal{A}$ of deformations where $\phi(F) : \mathbb{R}^{3\times3} \to \mathbb{R}$ is the energy density at a fixed temperature. Following the geometrically nonlinear theory of elasticity, we assume that the free energy density is frame-indifferent

$$\phi(RF) = \phi(F) \qquad \text{for all } F \in \mathbb{R}^{3\times3} \text{ and } R \in \mathrm{SO}(3) \ . \qquad (2.1)$$

We specialize the general geometrically nonlinear theory to martensitic crystals by taking the reference configuration Ω to be the high-symmetry phase (austenite) of the crystal at the transformation temperature. Following the geometrically nonlinear theory of martensite [2,3,25], we then have that

$$\phi(R_i^T F R_i) = \phi(F) \qquad \text{for all } F \in \mathbb{R}^{3\times3} \text{ and } R_i \in \mathcal{G} \qquad (2.2)$$

where $\mathcal{G}$ is the symmetry group of the high-symmetry phase.

We consider the deformation of the martensitic crystal at a temperature below the transformation temperature. The free energy density is then minimized at a transformation (Bain) strain U_1, so it follows by the frame-indifference (2.1) and the symmetry (2.2) of the energy density that the energy density is minimized on the union $\mathcal{U} = \mathcal{U}_1 \bigcup \cdots \bigcup \mathcal{U}_N$ of the N energy wells

$$\mathcal{U}_i = \mathrm{SO}(3)U_i = \{RU_i : R \in \mathrm{SO}(3)\} \qquad \text{for } i = 1, \dots, N$$

where the symmetry-related transformation strains (variants) $U_2, \ldots U_N$ satisfy

$$\{R_i^T U_1 R_i : R_i \in \mathcal{G}\} = \{U_1, \ldots, U_N\} . \tag{2.3}$$

By adding a constant, we may assume that the minimum value of ϕ is 0. Finally, we shall assume that ϕ is continuous and satisfies the growth condition

$$\phi(F) \geq \kappa \|F - \pi(F)\|^2 \qquad \text{for all } F \in \mathbb{R}^{3 \times 3} , \tag{2.4}$$

where $\kappa > 0$ is a constant and $\pi : \mathbb{R}^{3 \times 3} \to \mathcal{U}$ is a projection defined by

$$\|F - \pi(F)\| = \min_{G \in \mathcal{U}} \|F - G\| \qquad \text{for all } F \in \mathbb{R}^{3 \times 3} . \tag{2.5}$$

This projection exists for any $F \in \mathbb{R}^{3 \times 3}$ since the set $\mathcal{U}$ is compact.

There exists a continuous deformation $y(x) \in C(\mathbb{R}^3; \mathbb{R}^3)$ such that [2,25]

$$\nabla y(x) = \begin{cases} F_1 \text{ for all } x \text{ such that } x \cdot n < s, \\ F_0 \text{ for all } x \text{ such that } x \cdot n > s, \end{cases}$$

where $n \in \mathbb{R}^3$, $n \neq 0$, and $s \in \mathbb{R}$, if and only if there exists $a \in \mathbb{R}^3$ such that

$$F_1 = F_0 + a \otimes n . \tag{2.6}$$

Thus, if (2.6) holds for $a \neq 0$, then $x \cdot n = s$ is an interface plane with normal n.

In the following, we will be interested in a simple laminate. We suppose that for fixed $i, j \in \{1, \ldots, N\}$ with $i \neq j$, and for Q, a, and n with $a, n \neq 0$ the interface equation

$$QU_i = U_j + a \otimes n \tag{2.7}$$

is satisfied. For any fixed $\lambda \neq 0, 1$, we denote

$$F_\lambda = \lambda Q U_i + (1 - \lambda) U_j = U_j + \lambda a \otimes n . \tag{2.8}$$

We shall assume that the energy density $\phi(F)$ satisfies the growth condition

$$\phi(F) \geq C_1 \|F\|^p - C_0 \qquad \text{for all } F \in \mathbb{R}^{3 \times 3} ,$$

where C_0 and C_1 are positive constants independent of $F \in \mathbb{R}^{3 \times 3}$ and where we assume $p > 3$ to ensure that deformations with finite energy are uniformly continuous [1]. We can then denote the set of deformations of finite energy by

$$W^\phi = \{y \in C(\bar{\Omega}; \mathbb{R}^3) : \int_\Omega \phi(\nabla y(x)) \, dx < \infty\} ,$$

and we can define the set $\mathcal{A}$ of admissible deformations as

$$\mathcal{A} = \{y \in W^\phi : y(x) = y_0(x) \text{ for all } x \in \partial\Omega\} \tag{2.9}$$

where

$$y_0(x) = F_\lambda x \quad \text{for all } x \in \Omega \ .$$

We can prove the following lemma by constructing laminates with length scale converging to zero whose deformation gradients oscillate with volume fraction λ at QU_i and $1 - \lambda$ at U_j [10,25].

Lemma 2.1. *Let $\mathcal{A}$ be defined as in (2.9). Then the total energy $\mathcal{E}(y)$ satisfies*

$$\inf_{y \in \mathcal{A}} \mathcal{E}(y) = 0 \ .$$

3 Reduction to the Approximate Mixture of Two Strains

Recall the definitions (2.5) and (2.9) of π and $\mathcal{A}$, respectively. For each $k \in \{1, \dots, N\}$ and each $y \in \mathcal{A}$, we define

$$\Omega_k(y) = \{x \in \Omega : \pi(\nabla y(x)) \in \mathcal{U}_k\}$$

and the volume fraction with respect to the k-th energy well $\mathcal{U}_k$ to be

$$\tau_k(y) = \frac{\text{meas } \Omega_k(y)}{\text{meas } \Omega} \ .$$

Since every $x \in \Omega$ is in $\Omega_k(y)$ for some $k \in \{1, \dots, N\}$, we have that

$$\sum_{k=1}^{N} \tau_k(y) = 1 \quad \text{for all } y \in \mathcal{A} \ . \tag{3.1}$$

By the rank-one connection (2.7) and the definition of F_λ (2.8) we have

$$F_\lambda = QU_i \left(I - (1 - \lambda)(QU_i)^{-1} a \otimes n) \right) = U_j (I + \lambda U_j^{-1} a \otimes n) \ ,$$

so

$$|F_\lambda w| = |U_i w| = |U_j w| \quad \text{for all } w \in \mathbb{R}^3 \text{ such that } w \cdot n = 0. \tag{3.2}$$

Since $\det(QU_i) = \det U_i = \det U_j > 0$ by (2.3), we have that $U_j^{-1} a \cdot n = 0$. Hence, we have that

$$\text{Cof } F_\lambda = (\text{Cof } U_j) (I - \lambda n \otimes U_j^{-1} a) \tag{3.3}$$

where the cofactor of a nonsingular matrix $A \in \mathbb{R}^{3 \times 3}$ is defined by $\text{Cof } A = (\det A) A^{-T}$. We then obtain from (3.3) that

$$|(\text{Cof } F_\lambda) w| = |(\text{Cof } U_i) w| = |(\text{Cof } U_j) w|$$
$$\text{for all } w \in \mathbb{R}^3, \ w \cdot U_j^{-1} a = 0 \ . \tag{3.4}$$

We next recall that since the subdeterminant of the gradient is a null-Lagrangian [15], we have for $y \in \mathcal{A}$ that

$$\int_\Omega \nabla y(x)\, dx = \int_\Omega F_\lambda\, dx,$$
$$\int_\Omega \operatorname{Cof} \nabla y(x)\, dx = \int_\Omega \operatorname{Cof} F_\lambda\, dx \; . \tag{3.5}$$

We note that it follows from (2.4) that

$$\int_\Omega \|\nabla y(x) - \pi\,(\nabla y(x))\|^2\, dx \le \kappa^{-1}\mathcal{E}(y) \quad \text{for all } y \in \mathcal{A} \; . \tag{3.6}$$

Next, for $y \in \mathcal{A}$, we set $F(x) = \nabla y(x)$ for $x \in \Omega$, so it follows from (3.5) that $F(x) = (F_{kl}(x)) \in L^2(\Omega; \mathbb{R}^{3\times3})$. Now $\pi(\nabla y(x)) \in \mathcal{U}$ for all $x \in \Omega$, so if we set $P(x) = \pi(\nabla y(x))$ for $x \in \Omega$ we have that $P(x) = (P_{kl}(x))$ is uniformly bounded in $L^\infty(\Omega; \mathbb{R}^{3\times3})$ for all $y \in \mathcal{A}$. We have

$$F_{kl}F_{pq} - P_{kl}P_{pq} = (F_{kl} - P_{kl})P_{pq} + P_{kl}(F_{pq} - P_{pq}) + (F_{kl} - P_{kl})(F_{pq} - P_{pq})$$

for any $k,\, l,\, p,\, q \in \{1,2,3\}$. Hence, we have by the Cauchy-Schwarz inequality and (3.6) that

$$\int_\Omega \Big| \left[\operatorname{Cof} \nabla y(x) - \operatorname{Cof} \pi\,(\nabla y(x))\right] w \Big|\, dx$$
$$\le C\left[\left(\int_\Omega \|\nabla y(x) - \pi\,(\nabla y(x))\|^2\, dx \right)^{1/2} \right.$$
$$\left. + \int_\Omega \|\nabla y(x) - \pi\,(\nabla y(x))\|^2\, dx \right] \tag{3.7}$$
$$\le C\left[\mathcal{E}(y)^{1/2} + \mathcal{E}(y)\right] \; .$$

The following result was proven in [4] for the cubic to orthorhombic transformation and in [5] for the tetragonal to monoclinic transformation. In the estimates below, C will denote a generic positive constant that is independent of $y \in \mathcal{A}$ and is allowed to change from equation to equation.

Lemma 3.1. *Given $i, j \in \{1, \ldots, N\}$, $Q \in SO(3)$, and $a, n \in \mathbb{R}$, $a, n \ne 0$ satisfying the interface equation (2.7), there exists a constant $C > 0$ such that*

$$\rho_1(y; w) \equiv \sum_{k \ne i, j} \tau_k(y)\left(|U_i w|^2 - |U_k w|^2\right) \tag{3.8}$$
$$\le C\mathcal{E}(y)^{1/2} \quad \text{for all } w \in \mathbb{R}^3,\ |w| = 1,\ w \cdot n = 0,$$

$$\rho_2(y; w) \equiv \sum_{k \ne i, j} \tau_k(y)\left[|(\operatorname{Cof} U_i)w|^2 - |(\operatorname{Cof} U_k)w|^2\right] \tag{3.9}$$
$$\le C\left[\mathcal{E}(y)^{1/2} + \mathcal{E}(y)\right] \quad \text{for all } w \in \mathbb{R}^3,\ |w| = 1,\ w \cdot U_j^{-1}a = 0$$

for any $y \in \mathcal{A}$.

Proof. We have by (3.1), (3.2), (3.4), and (3.5) that for any $w \in \mathbb{R}^3$ with $|w| = 1$

$$\rho_1(y; w) = \sum_{k=1}^{N} \tau_k(y) \left(|U_i w|^2 - |U_k w|^2 \right)$$

$$= \sum_{k=1}^{N} \tau_k(y) \left(|F_\lambda w|^2 - |U_k w|^2 \right)$$

$$= \frac{1}{\text{meas } \Omega} \int_\Omega \left[|F_\lambda w|^2 - |\pi(\nabla y(x)) w|^2 \right] dx$$

$$= -\frac{1}{\text{meas } \Omega} \int_\Omega \left| \left[F_\lambda - \pi(\nabla y(x)) \right] w \right|^2 dx \qquad (3.10)$$

$$+ \frac{2}{\text{meas } \Omega} \int_\Omega \left[\nabla y(x) - \pi(\nabla y(x)) \right] w \cdot F_\lambda w \, dx$$

$$\leq \frac{2}{\text{meas } \Omega} \int_\Omega \left[\nabla y(x) - \pi(\nabla y(x)) \right] w \cdot F_\lambda w \, dx \ .$$

We obtain from the Cauchy-Schwarz inequality and the above inequality (3.6) that

$$\left| \int_\Omega \left[\nabla y(x) - \pi(\nabla y(x)) \right] w \cdot F_\lambda w \, dx \right| \leq C \mathcal{E}(y)^{1/2} \ .$$

So, it follows from (3.10) that for all $w \in \mathbb{R}^3$ with $|w| = 1$

$$\rho_1(y; w) = \sum_{k=1}^{N} \tau_k(y) \left(|U_i w|^2 - |U_k w|^2 \right) \leq C \mathcal{E}(y)^{1/2} \ . \qquad (3.11)$$

The result (3.9) then follows from the above inequality (3.11) and (3.2).

Next, we obtain similar estimates for the cofactor. We have from (3.1) and (3.5) that for any $w \in \mathbb{R}^3$, $|w| = 1$,

$$\rho_2(y; w) = \sum_{k=1}^{N} \tau_k(y) \left[|(\text{Cof } U_i) w|^2 - |(\text{Cof } U_k) w|^2 \right]$$

$$= \sum_{k=1}^{N} \tau_k(y) \left[|(\text{Cof } F_\lambda) w|^2 - |(\text{Cof } U_k) w|^2 \right]$$

$$= \frac{1}{\text{meas } \Omega} \int_\Omega \left[|(\text{Cof } F_\lambda) w|^2 - |(\text{Cof } \pi(\nabla y(x))) w|^2 \right] dx$$

$$= -\frac{1}{\text{meas } \Omega} \int_\Omega \left| \left[\text{Cof } F_\lambda - \text{Cof } \pi(\nabla y(x)) \right] w \right|^2 dx$$

$$+ \frac{2}{\text{meas } \Omega} \int_\Omega \left[\text{Cof } \nabla y(x) - \text{Cof } \pi(\nabla y(x)) \right] w \cdot (\text{Cof } F_\lambda) w \, dx$$

$$\leq \frac{2}{\text{meas } \Omega} \int_\Omega \left[\text{Cof } \nabla y(x) - \text{Cof } \pi(\nabla y(x)) \right] w \cdot (\text{Cof } F_\lambda) w \, dx \ .$$

The result (3.10) then follows from the above inequality, (3.4), and (3.7). □

We can use Lemma 3.1 to reduce the analysis of the stability of the laminated microstructure to an analysis of the stability of a mixture of two variants [4,5] by evaluating $\rho_1(y; w)$ and $\rho_2(y; w)$ for appropriate $w \in \mathbb{R}^3$ to establish the inequality:

$$\tau_k(y) \le C \left[\mathcal{E}(y)^{1/2} + \mathcal{E}(y)\right] \quad \text{for all } k \in \{1, \dots, N\} \backslash \{i, j\}, \ y \in \mathcal{A}. \ (3.12)$$

In the following section, we will show how this can be done for the cubic to tetragonal transformation [20].

We have also used (3.9) for $\rho_1(y; w)$ and (3.10) for $\rho_2(y; w)$ to prove the estimate (3.12) for the tetragonal to monoclinic transformation ($N = 4$) [5] and the cubic to orthorhombic transformation ($N = 6$) [4], except for special cases when the material parameters in the transformation strain satisfy certain identities, in which case it was shown that the inequality (3.12) does not hold.

4   An Example: The Cubic to Tetragonal Transformation

We consider the cubic to tetragonal transformation [20] which has three energy wells ($N = 3$) given by

$$U_1 = \nu_1 I + (\nu_2 - \nu_1)e_1 \otimes e_1, \qquad U_2 = \nu_1 I + (\nu_2 - \nu_1)e_2 \otimes e_2,$$
$$U_3 = \nu_1 I + (\nu_2 - \nu_1)e_3 \otimes e_3$$

for material parameters $0 < \nu_1$, $0 < \nu_2$, $\nu_1 \ne \nu_2$. We assume that $\{e_1, e_2, e_3\}$ is an orthonormal basis for $\mathbb{R}^3$.

The following two lemmas [2,3,25] state that for the cubic to tetragonal transformation each $F_0 \in \mathcal{U}_i$ is not rank-one connected to any $F_1 \in \mathcal{U}_i$ with $F_0 \ne F_1$, but that every $F_0 \in \mathcal{U}_i$ is rank-one connected to two distinct $F_1 \in \mathcal{U}_j$ for all $j \ne i$, $j \in \{1, 2, 3\}$.

Lemma 4.1. *If $F_0 \in \mathcal{U}_i$ for some $i \in \{1, 2, 3\}$, then there does not exist $F_1 \in \mathcal{U}_i$ with $F_0 \ne F_1$, such that F_0 and F_1 are rank-one connected.*

Lemma 4.2. *If $F_0 \in \mathcal{U}_i$ for some $i \in \{1, 2, 3\}$, then for any $j \ne i$, $j \in \{1, 2, 3\}$, there exist two distinct $F_1 \in \mathcal{U}_j$ such that F_0 and F_1 are rank-one connected. If $QU_i \in \mathcal{U}_i$ and $U_j \in \mathcal{U}_j$ are rank-one connected so that*

$$QU_i = U_j + a \otimes n \tag{4.1}$$

for $Q \in SO(3)$, $a \in \mathbb{R}^3$, and $n \in \mathbb{R}^3$; then (up to a scalar multiple)

$$n \in \{e_i + e_j \, e_i - e_j\} \ . \tag{4.2}$$

Further, if $n = e_i \pm e_j$, then

$$U_j^{-1}a \in \mathrm{Span}(e_i \mp e_j) \ . \tag{4.3}$$

By (4.2), we have that $e_k \cdot n = 0$ for $k \in \{1, 2, 3\}$ such that $i \neq j \neq k$, so we can take $w = e_k$ in (3.9) to obtain

$$\tau_k(y)(\nu_1^2 - \nu_2^2) \leq C\mathcal{E}(y)^{1/2} \ ,$$

and we can conclude the inequality (3.12) if $\nu_1 > \nu_2$. Similarly, by (4.3), we have that $e_k \cdot U_j^{-1}a = 0$ such that $k \in \{1, 2, 3\}$ and $i \neq j \neq k$, so we can take $w = e_k$ in (3.10) to obtain

$$\tau_k(y)(\nu_1^2 \nu_2^2 - \nu_1^4) \leq C \left[\mathcal{E}(y)^{1/2} + \mathcal{E}(y) \right] \ .$$

Hence, we can also conclude from the above inequality that the inequality (3.12) holds if $\nu_1 < \nu_2$.

We note that the algebra for the proof of the inequality (3.12) for the tetragonal to monoclinic transformation ($N = 4$) [5] and the cubic to orthorhombic transformation ($N = 6$) [4] is more difficult since multiple choices of $w \in \mathbb{R}^3$ must used to obtain (3.12) from (3.9) and (3.10).

5   The Stability of the Microstructure

We assume in what follows that for the laminated microstructure under consideration, the inequalities (3.9) and (3.10) imply the estimate

$$\tau_k(y) \leq C \left[\mathcal{E}(y)^{1/2} + \mathcal{E}(y) \right] \quad \text{for all } k \in \{1, \dots, N\} \backslash \{i, j\}, \ y \in \mathcal{A}. \tag{5.1}$$

We recall that

$$\mathcal{A} = \{y \in W^\phi : y(x) = y_0(x) \text{ for } x \in \partial\Omega\}$$

where

$$y_0(x) = [\lambda Q U_i + (1 - \lambda)U_j]x \qquad \text{for all } x \in \Omega \ .$$

The results in this section for the general martensitic transformation can be deduced from the inequality (5.1) by the identical arguments used to deduce the results from (5.1) for the cubic to orthorhombic case [4] by making the obvious modifications in the argument to change $N = 6$ to general N. For this reason, we state the results given in this section without proof.

We also recall that the energy density ϕ is minimized on the union $\mathcal{U}$ of the N energy wells

$$\mathcal{U}_i = \mathrm{SO}(3)U_i = \{RU_i : R \in \mathrm{SO}(3)\} \qquad \text{for } i = 1, \dots, N,$$

and that ϕ is continuous and satisfies the growth condition

$$\phi(F) \geq \kappa \|F - \pi(F)\|^2 \qquad \text{for all } F \in \mathbb{R}^{3\times 3} .$$

The following theorem gives estimates for the derivative of the limiting macroscopic deformation y in any direction tangential to the parallel layers of the laminate, for the L^2 approximation of the limiting macroscopic deformation, and for the weak convergence of the limiting macroscopic deformation.

Theorem 5.1. *We assume that the inequality (5.1) holds. Then the following results hold:*

(1) For any $w \in \mathbb{R}^3$ such that $w \cdot n = 0$ and $|w| = 1$, we have

$$\int_\Omega |[\nabla y(x) - \nabla y_0(x)]\, w|^2 \; dx \leq C \left[\mathcal{E}(y)^{1/2} + \mathcal{E}(y) \right] \qquad \text{for all } y \in \mathcal{A} .$$

(2) We have

$$\int_\Omega |y(x) - y_0(x)|^2 \; dx \leq C \left[\mathcal{E}(y)^{1/2} + \mathcal{E}(y) \right] \qquad \text{for all } y \in \mathcal{A} .$$

(3) For any Lipschitz domain $\omega \subset \Omega$, there exists a constant $C = C(\omega) > 0$ such that

$$\left\| \int_\omega [\nabla y(x) - \nabla y_0(x)] \; dx \right\| \leq C \left[\mathcal{E}(y)^{1/8} + \mathcal{E}(y)^{1/2} \right] \qquad \text{for all } y \in \mathcal{A} .$$

The following corollary states that the deformation gradients of energy-minimizing sequences of deformations must oscillate with a length scale that converges to zero.

Corollary 5.2. *If the inequality (5.1) holds, then there does not exist any $y \in \mathcal{A}$ such that*

$$\mathcal{E}(y) = \min_{z \in \mathcal{A}} \mathcal{E}(z) .$$

For fixed $i, j \in \{1, \ldots, N\}$ with $i \neq j$, we can define a projection $\pi_{ij} : \mathbb{R}^{3\times 3} \to \mathcal{U}_i \cup \mathcal{U}_j$ by

$$\|F - \pi_{ij}(F)\| = \min_{G \in \mathcal{U}_i \cup \mathcal{U}_j} \|F - G\| \qquad \text{for all } F \in \mathbb{R}^{3\times 3} .$$

We also define the operators $\Theta : \mathbb{R}^{3\times 3} \to SO(3)$ and $\Pi : \mathbb{R}^{3\times 3} \to \{QU_i, U_j\}$ by the unique decomposition

$$\pi_{ij}(F) = \Theta(F)\Pi(F) \qquad \text{for all } F \in \mathbb{R}^{3\times 3} .$$

The next theorem states that the deformation gradients of energy-minimizing sequences of deformations must oscillate between QU_i and U_j.

Theorem 5.3. *For a transformation such that (5.1) holds, we have*

$$\int_\Omega \|\nabla y(x) - \Pi(\nabla y(x))\|^2 \, dx \leq C \left[\mathcal{E}(y)^{1/2} + \mathcal{E}(y)\right] \qquad \text{for all } y \in \mathcal{A} \ .$$

For any subset $\omega \subset \Omega$, $\rho > 0$, and $y \in \mathcal{A}$, we define the sets

$$\omega_\rho^i(y) = \{x \in \omega : \Pi(\nabla y(x)) = QU_i \text{ and } \|\nabla y(x) - QU_i\| < \rho\},$$
$$\omega_\rho^j(y) = \{x \in \omega : \Pi(\nabla y(x)) = U_j \text{ and } \|\nabla y(x) - U_j\| < \rho\} \ .$$

The next theorem demonstrates that the deformation gradients of energy-minimizing sequences of deformations must oscillate with local volume fraction λ at QU_i and local volume fraction $1 - \lambda$ at U_j. It also demonstrates that the Young measure for this problem is unique [3,25] and is given by

$$\nu = \lambda \delta_{QU_i} + (1 - \lambda)\delta_{U_j} \ .$$

Theorem 5.4. *We suppose that the reduction (5.1) is valid. Then for any Lipschitz domain $\omega \subset \Omega$ and any $\rho > 0$, there exists a constant $C = C(\omega, \rho) > 0$ such that for all $y \in \mathcal{A}$*

$$\left|\frac{\text{meas } \omega_\rho^i(y)}{\text{meas } \omega} - \lambda\right| + \left|\frac{\text{meas } \omega_\rho^j(y)}{\text{meas } \omega} - (1 - \lambda)\right| \leq C \left[\mathcal{E}(y)^{1/8} + \mathcal{E}(y)^{1/2}\right] \ .$$

We now denote by $\mathcal{V}$ the Sobolev space of all measurable functions $f : \Omega \times \mathbb{R}^{3 \times 3} \to \mathbb{R}$ such that

$$\|f\|_{\mathcal{V}}^2 = \int_\Omega \left\{\left[\underset{F \in \mathbb{R}^{3 \times 3}}{\text{ess sup}} \|\nabla_F f(x, F)\|\right]^2 + |\nabla z_f(x) n|^2 + z_f(x)^2\right\} dx < \infty \ ,$$

where $z_f : \Omega \to \mathbb{R}$ is defined by

$$z_f(x) = f(x, QU_i) - f(x, U_j) \qquad \text{for all } x \in \Omega \ .$$

The final theorem in this section gives an estimate for the weak convergence of nonlinear functions of the deformation gradient.

Theorem 5.5. *We assume that the inequality (5.1) holds. Then we have*

$$\left|\int_\Omega \{f(x, \nabla y(x)) - [\lambda f(x, QU_i) + (1 - \lambda)f(x, U_j)]\} \, dx\right|$$
$$\leq C\|f\|_{\mathcal{V}} \left[\mathcal{E}(y)^{1/4} + \mathcal{E}(y)^{1/2}\right] \qquad \text{for all } f \in \mathcal{V} \text{ and all } y \in \mathcal{A} \ .$$

6   The Finite Element Approximation of Microstructure

We consider the finite element approximation of the variational problem

$$\inf_{v \in \mathcal{A}} \mathcal{E}(v)$$

given by

$$\inf_{v_h \in \mathcal{A}_h} \mathcal{E}(v_h)$$

where $\mathcal{A}_h$ is a finite-dimensional subspace of $\mathcal{A}$ defined for $h \in (0, h_0]$ for some $h_0 > 0$. The following approximation theorem for the energy can be proven for the most widely used P_k or Q_k type conforming finite elements on quasi-regular meshes, in particular for the P_1 linear elements defined on tetrahedra and the Q_1 trilinear elements defined on rectangular parallelepipeds [4,10,20–22,24,25].

Theorem 6.1. *For each $h \in (0, h_0]$, there exists $y_h \in \mathcal{A}_h$ such that*

$$\mathcal{E}(y_h) = \min_{z_h \in \mathcal{A}_h} \mathcal{E}(z_h) \leq Ch^{1/2} \ . \tag{6.1}$$

For the remainder of this section, we again recall that the energy density ϕ is minimized on the union $\mathcal{U}$ of the N energy wells

$$\mathcal{U}_i = \mathrm{SO}(3)U_i = \{RU_i : R \in \mathrm{SO}(3)\} \qquad \text{for } i = 1, \dots, N,$$

and that ϕ is continuous and satisfies the growth condition

$$\phi(F) \geq \kappa \, \|F - \pi(F)\|^2 \qquad \text{for all } F \in \mathbb{R}^{3 \times 3} \ .$$

We also assume that the inequality (5.1) describing the reduction of low energy deformation to two energy wells holds. The following corollaries for the finite element approximation follow directly from the above estimate for the approximation of the energy (6.1). We assume below that $y_h \in \mathcal{A}_h$ is a finite element approximation satisfying the quasi-optimality condition

$$\mathcal{E}(y_h) \leq \sigma \inf_{z_h \in \mathcal{A}_h} \mathcal{E}(z_h) \tag{6.2}$$

for some constant $\sigma \geq 1$ independent of h.

Corollary 6.2. *If the inequality (5.1) holds, then we have the following estimates:*

(1) There exists of positive constant C such that for any $y_h \in \mathcal{A}_h$ satisfying (6.2) we have

$$\int_{\Omega} |y_h(x) - y_0(x)|^2 \, dx \leq Ch^{1/4}$$

and

$$\int_{\Omega} \|\nabla y_h(x) - \Pi(\nabla y_h(x))\|^2 \, dx \leq Ch^{1/4} \ .$$

(2) For any $w \in \mathbb{R}^3$ such that $w \cdot n = 0$ and $|w| = 1$, we have

$$\int_\Omega |[\nabla y_h(x) - \nabla y_0(x)] \, w|^2 \, dx \le Ch^{1/4}$$

for any $y_h \in \mathcal{A}_h$ satisfying (6.2).

(3) If $\omega \subset \Omega$ is a Lipschitz domain, then there exists a constant $C = C(\omega) > 0$ such that for any $y_h \in \mathcal{A}_h$ satisfying (6.2) we have

$$\left\| \int_\omega [\nabla y_h(x) - \nabla y_0(x)] \, dx \right\| \le Ch^{1/16} .$$

Corollary 6.3. *We assume that the inequality (5.1) holds. Then we have the following results:*

(1) If $\omega \subset \Omega$ is a Lipschitz domain and $\rho > 0$, then there exists a constant $C = C(\omega, \rho) > 0$ such that for any $y_h \in \mathcal{A}_h$ satisfying (6.2)

$$\left| \frac{\text{meas}\,\omega_\rho^i(y_h)}{\text{meas}\,\omega} - \lambda \right| + \left| \frac{\text{meas}\,\omega_\rho^j(y_h)}{\text{meas}\,\omega} - (1 - \lambda) \right| \le Ch^{1/16} .$$

(2) We have

$$\left| \int_\Omega \{ f(x, \nabla y_h(x)) - [\lambda f(x, QU_i) + (1 - \lambda) f(x, U_j)] \} \, dx \right| \le C \|f\|_{\mathcal{V}} \, h^{1/8}$$

for any $f \in \mathcal{V}$ and any $y_h \in \mathcal{A}_h$ satisfying (6.2).

References

1. Adams, R.: *Sobolev Spaces.* Academic Press, New York, (1975)
2. Ball, J., James, R.: Fine Phase Mixtures as Minimizers of Energy. Arch. Rat. Mech. Anal. **100** (1987), 13–52
3. Ball, J., James, R.: Proposed Experimental Tests of a Theory of Fine Microstructure and the Two-Well Problem. Phil. Trans. R. Soc. Lond. A **338** (1992), 389–450
4. Bhattacharya, K., Li, B., Luskin, M.: The Simply Laminated Microstructure in Martensitic Crystals that Undergo a Cubic to Orthorhombic Phase Transformation. Arch. Rat. Mech. Anal., (2000)
5. Bělík, P., Luskin, M.: Stability of Microstructure for Tetragonal to Monoclinic Martensitic Transformations. Manuscript, (1999)
6. Carstensen, C., Plecháč, P.: Numerical Solution of the Scalar Double-Well Problem Allowing Microstructure. Math. Comp. **66** (1997), 997–1026
7. Carstensen, C., Plecháč, P.: Adaptive Algorithms for Scalar Non-Convex Variational Problems. Appl. Numer. Math. **26** (1998), 203–216
8. Chipot, M.: Numerical Analysis of Oscillations in Nonconvex Problems. Numer. Math. **59** (1991), 747–767
9. Chipot, M., Collins, C.: Numerical Approximations in Variational Problems with Potential Wells. SIAM J. Numer. Anal. **29** (1992), 1002–1019

10. Chipot, M., Collins, C., Kinderlehrer, D.: Numerical Analysis of Oscillations in Multiple Well Problems. Numer. Math. **70** (1995), 259–282
11. Chipot, M., Kinderlehrer, D.: Equilibrium Configurations of Crystals. Arch. Rat. Mech. Anal. **103** (1988), 237–277
12. Chipot, M., Müller, S.: Sharp Energy Estimates for Finite Element Approximations of Nonconvex Problems. Preprint, (1997)
13. Collins, C., Kinderlehrer, D., Luskin, M.: Numerical Approximation of the Solution of a Variational Problem with a Double Well Potential. SIAM J. Numer. Anal. **28** (1991), 321–332
14. Collins, C., Luskin, M.: Optimal Order Estimates for the Finite Element Approximation of the Solution of a Nonconvex Variational Problem. Math. Comp. **57** (1991), 621–6371
15. Dacorogna, B.: Direct Methods in the Calculus of Variations. Springer-Verlag Berlin, 1989
16. Dolzmann, G.: Numerical Computation of Rank-One Convex Envelopes. SIAM J. Numer. Anal., (2000)
17. French, D.: On the Convergence of Finite Element Approximations of a Relaxed Variational Problem. SIAM J. Numer. Anal. **28** (1991), 419–436
18. Gremaud, P.-A.: Numerical Analysis of a Nonconvex Variational Problem Related to Solid-Solid Phase Transitions. SIAM J. Numer. Anal. **31** (1994), 111–127
19. Kružík, M.: Numerical Approach to Double Well Problems. SIAM J. Numer. Anal. **35(5)** (1998), 1833–1849
20. Li, B., Luskin, M.: Finite Element Analysis of Microstructure for the Cubic to Tetragonal Transformation. SIAM J. Numer. Anal. **35** (1998), 376–392
21. Li, B., Luskin, M.: Nonconforming Finite Element Approximation of Crystalline Microstructure. Math. Comp. **67** (223) (1998), 917–946
22. Li, B., Luskin, M.: Approximation of a Martensitic Laminate with Varying Volume Fractions. Math. Model. Numer. Anal., (1999)
23. Li, Z.: Simultaneous Numerical Approximation of Microstructures and Relaxed Minimizers. Numer. Math. **78** (1997), 21–38
24. Luskin, M.: Approximation of a Laminated Microstructure for a Rotationally Invariant, Double Well Energy Density. Numer. Math. **75** (1996), 205–221
25. Luskin, M.: On the Computation of Crystalline Microstructure. Acta Numer., (1996), 191–257
26. Luskin, M., Ma, L.: Analysis of the Finite Element Approximation of Microstructure in Micromagnetics. SIAM J. Numer. Anal. **29** (1992), 320–331
27. Nicolaides, R., Walkington, N.: Strong Convergence of Numerical Solutions to Degenerate Variational Problems. Math. Comp. **64** (1995), 117–127
28. Pedregal, P.: On the Numerical Analysis of Non-Convex Variational Problems. Numer. Math. **74** (1996), 325–336
29. Roubíček, T.: Numerical Approximation of Relaxed Variational Problems. J. Convex Anal. **3** (1996), 329–347

Dissipative Evolution of Microstructure in Shape Memory Alloys

T. Roubíček

[1] Mathematical Institute, Charles University, Sokolovská 83, CZ-186 75 Praha 8, Czech Republic
[2] Institute of Information Theory and Automation, Academy of Sciences, Pod vodárenskou věží 4, CZ-182 08 Praha 8, Czech Republic

Dedicated to Professor Karl-Heinz Hoffmann
on the occasion of his 60th birthday

Abstract. This contribution surveys models for evolution of twinning and inelastic response of crystallic materials due to martensitic phase transformations. Then a focus is devoted to a "mesoscopical" evolution model in a scalar situation based on a simple 2nd-order evolution variational inequality. A rate-independent dissipation mechanism is involved and shown by computer experiments to make possible a modelling of inelastic response of crystallic materials like shape-memory effects.

Keywords Nonconvex variational problems, relaxation, Young measures, martensitic transformation, thermomechanical evolution, dissipation, activation, numerical tests.

1 Introduction – Plasticity in Crystallic Metals

Microstructures appear often in various multi-scale physical systems, e.g. in crystallic metals, micromagnetism, magnetostriction, etc., and represent a long lasting challenge for theoretical, numerical, and experimental investigation.

Here we focus on plasticity in metals which can typically be created by slip or/and by *twinning*. The former case, created by a slip (activated by stress typically hundreds of MPa) of one atomic layer on the adjacent one, leads on macroscopical level classically to Hencky or Prandtl-Reuss models. The latter case, related with so-called *quasi-plasticity* and *pseudo-elasticity* [27,28] (sometimes called also *super-* or *ferro-elasticity*), is created in *martensitic* structures by a *phase transformation* (activated by stress typically tens of MPa or less) and will be just discussed in this contribution.

We will survey only continuum models; models on atomic level do also exists, however. We will try to sort the continuum models in accord with the level on which the martensite is treated. See also Müller [39, Sect. 7.2] for a brief survey.

A standard model for a steady-state relies on (a zero activation energy and then) a minimum-energy variational principle, involving the elastic potential energy of the type

$$V(u) - \langle f, u \rangle \to \min,$$

$$V(u) := \int_\Omega \varphi(\nabla u) + \varepsilon |\nabla^2 u|^2 \mathrm{d}x + \int_\Gamma \zeta(x, u) \mathrm{d}S, \qquad (1.1)$$

where $\Omega \subset \mathbb{R}^n$ is the reference shape of the body in question with a boundary Γ, $u : \Omega \to \mathbb{R}^m$ is a displacement assumed to live in $W^{1,p}(\Omega; \mathbb{R}^m)$, the Sobolev space of functions $u : \Omega \to \mathbb{R}^m$ such that the distributional gradient ∇u belongs to the Lebesgue space $L^p(\Omega; \mathbb{R}^{m \times n})$, and $\varphi : \mathbb{R}^{m \times n} \to \mathbb{R}$ is a potential-energy density which is typically not convex (or, more precisely, even not quasiconvex), and ζ is the potential of the surface support, and f (a linear functional on u's) denotes external body and surface loading forces. The higher-order term with ε enables to describe finer effects due to surface energy between regions with nearly constant ∇u, cf. Kohn and Müller [34] and [39, Sect. 6]. However, except for microscopical models in Sect. 1.1, $\varepsilon = 0$ will be assumed here. Often, the potential energy depends also on temperature. Sometimes, this dependence changes very essentially the shape of φ which creates the so-called *shape-memory effect*, cf. also Sect. 6 below. For a consistent thermodynamical model, a departing point is a definition of a specific *Helmholtz free energy* in the simplest form as:

$$\psi(s, \vartheta) := \varphi(s, \vartheta) + \varepsilon |\nabla s|^2 - c \vartheta \ln \vartheta, \qquad s = \nabla u, \qquad (1.2)$$

where we indicated explicitly the dependence of φ on the absolute temperature ϑ.

1.1 Microscopical Models

The straightforward approach to evolution is to use the potential energy (1.1) and, by chosen kinetic and dissipative energy and by usage of a Hamilton principle, to derive the usual dynamics. The dissipation is usually considered only due to viscosity mechanism which transforms kinetic energy of elastic vibrations into vibration of the atomic grid, i.e. heat. In such a way, we arrive to equation

$$\varrho \frac{\partial^2 u}{\partial t^2} - \mathrm{div}\left(\frac{\partial \varphi}{\partial s}(\nabla u, \vartheta) + \mu \nabla \frac{\partial u}{\partial t}\right) + 2\varepsilon \nabla^4 u = f, \qquad (1.3)$$

with suitable boundary and initial conditions, $\varrho > 0$ denotes mass density and $\mu > 0$ the viscosity coefficient. In anisothermal situations, the standard requirement is the preservation (in an isolated system) of the total energy

$$\int_\Omega e + \frac{\varrho}{2}\left|\frac{\partial u}{\partial t}\right|^2 \mathrm{d}x + \int_\Gamma \zeta(x, u)\mathrm{d}S$$

$$= V(u, \vartheta) + \int_\Omega \frac{\varrho}{2}\left|\frac{\partial u}{\partial t}\right|^2 + c\vartheta - \vartheta \frac{\partial \varphi}{\partial \vartheta}(\nabla u, \vartheta)\,\mathrm{d}x \qquad (1.4)$$

where we used $V(\cdot, \vartheta)$ defined by (1.1) but with φ replaced by $\varphi(\cdot, \vartheta)$, and the specific *internal energy* $e := \psi + \vartheta\eta$ with the specific *entropy* $\eta := -\partial\psi/\partial\vartheta$ and ψ from (1.2). Then one can see that the equation for displacement (1.3) must be completed by a heat equation

$$c\frac{\partial\vartheta}{\partial t} - \mathrm{div}(\kappa\nabla\vartheta) = \mu\left|\nabla\frac{\partial u}{\partial t}\right|^2 + \vartheta\frac{\partial}{\partial t}\left(\frac{\partial\psi}{\partial\vartheta}(\nabla u, \vartheta)\right) \tag{1.5}$$

with κ thermal conductivity; here we used the isotropic Fourier law: heat flux $= -\kappa\nabla\vartheta$. Such kind of models has been proposed by Falk [15,16] and widely investigated by K.-H. Hoffmann and a large group of his collaborators, in particular by Alt, Bubner, Chen, Niezgódka, Sprekels, Zheng Songmu, Zochowski, Jun Zou and others, see e.g. [2,8,9,20,22–25,41,53] and [7, Chapter 5] and references therein. Besides, isothermal case (1.3) has been thoroughly studied by Ball et al. [4], Friesecke and McLeod [19], Rybka [50] and Hoffmann [52], and Swart and Holmes [55].

The interpretation of the model (1.3) is that the potential barrier between particular phases is overcome by a sufficiently large kinetic energy. Though this sort of models can be fitted quite well with experiments (see Bubner [8]), the philosophy of this model does not seem too much related with real phase transformation because the potential barrier is usually very high in comparison with kinetic energy appearing in the system, which is perhaps related with impossibility of formation of microstructure proved rigorously in [4,19] for 1D cases and in [51] for n-D cases. In fact, microstructural changes are activated in much lower energies than the potential barrier between particular phases.

The key idea seems to build in a proper mechanism that would make possible "tunelling" the potential barrier between particular phases on domains where $\varepsilon\nabla^2 u$ is large, reflecting an activation of phase transformation by stress or temperature.

Further disadvantage of such sort of models is that computer calculations can be performed only for domains of the scale of particular few layers, i.e. about 10^{-6}m, cf. Killough [31], Kloucek and Luskin [32,33], Morin and Spies [38]. Such scales are, however, not of the real engineering interest. An interesting attempt to overcome this drawback was made by Theil [57] who considered (1.3) for $\varepsilon = 0$ but with an additional elastical bonding term $+u$, defined a Young-measure solution, and proved existence and even uniqueness; this is already a rather mesoscopical-type model as in Sects. 1.3–1.4.

1.2 Macroscopical Models

The macroscopical steady-state model neglects, of course, the higher-order term (i.e. $\varepsilon = 0$) and moreover fine oscillations of minimizing sequences to (1.1) are not involved explicitly. In the isothermal case, this gives rise to a minimization problem which uses, instead of V, an effective energy:

$$V^{\#}(u) := \int_{\Omega} [Q\varphi](\nabla u) \mathrm{d}x \; + \; \int_{\Gamma} \zeta(x, u) \mathrm{d}S, \tag{1.6}$$

where $[Q\varphi]$ denotes the quasiconvex envelope defined by

$$[Q\varphi](s) = \inf_{\substack{u \in C^1(\bar{\Omega}; \mathbb{R}^m) \\ u|_{\Gamma}=0}} \int_{\Omega} \varphi(s + \nabla u) \, \mathrm{d}x \; . \tag{1.7}$$

In fact, $V^{\#}$ is a weakly lower semicontinuous envelope of the potential V on $W^{1,p}(\Omega; \mathbb{R}^m)$. This sort of relaxation of (1.1) has been widely investigated and there are hundreds of papers about it, see e.g. [5,6,39,47] and references therein.

However, it does not seem any direct usage of (1.6) for evolution because the solution to (1.6) forgets all information about microstructure just essential for determination of its evolution.

1.3 Mesoscopical Models

On a mesoscopical level, we want to see the character of fast oscillations of the gradient of minimizers ∇u to (1.1) in "the limit" if ε tends to zero. This can be described by a probability measure ν_x on $\mathbb{R}^{m \times n}$ possibly depending (i.e. being parameterized) on $x \in \Omega$; cf. e.g. [5,6,47]. We then call $\nu = \{\nu_x\}_{x \in \Omega}$ a Young measure if, in addition, $x \mapsto \nu_x$ is weakly measurable.

Thus we come to a finer relaxation of the minimization problem by a continuous extension of (1.1). For this, one need to enlarge sufficiently the original space $W^{1,p}(\Omega; \mathbb{R}^m)$. Supposing $p > 1$ and a coercivity/growth $c|A|^p \le \varphi(A) \le C(1 + |A|^p)$ with some $c > 0$, it can be proved that minimizing sequences of (1) do not concentrate energy and a correct relaxation of (1) can use the so-called L^p-Young measures $\mathcal{Y}^p(\Omega; \mathbb{R}^{m \times n}) := \{\nu \text{ is a Young measure;} \int_{\Omega} \int_{\mathbb{R}^n} |s|^p \nu_x(\mathrm{d}s) \mathrm{d}x < +\infty\}$. Then, as in [47, Chapter 6], the relaxed problem looks as

$$
\left.
\begin{array}{c}
\bar{V}(u,\nu) - \langle f, u \rangle \to \min, \qquad (u, \nu) \in Q, \\[2mm]
\text{where } \bar{V}(u,\nu) := \displaystyle\int_{\Omega} \int_{\mathbb{R}^n} \varphi(s) \nu_x(\mathrm{d}s) \mathrm{d}x + \int_{\Gamma} \zeta(x, u) \mathrm{d}S , \\[2mm]
Q := \big\{ (u, \nu) \in W^{1,p}(\Omega; \mathbb{R}^m) \times \mathcal{G}^p(\Omega; \mathbb{R}^n) : \\[2mm]
\nabla u(x) = \int_{\mathbb{R}^n} s\nu_x(\mathrm{d}s) \quad \text{for a.a. } x \in \Omega \big\} ,
\end{array}
\right\} \tag{1.8}
$$

where $\mathcal{G}^p(\Omega; \mathbb{R}^{m \times n})$ is the set of so-called gradient L^p-Young measures $\{\nu \in \mathcal{Y}^p(\Omega; \mathbb{R}^{m \times n}); \exists \{u_k\}_{k \in \mathbb{N}} \subset W^{1,p}(\Omega; \mathbb{R}^m) \text{ bounded: } \forall v \in C_0(\mathbb{R}^m) : \int_{\mathbb{R}^m} v(s) \nu_x(\mathrm{d}s) = \lim_{k \to \infty} v(\nabla u_k) \text{ weakly* in } L^{\infty}(\Omega)\}$.

In scalar cases (i.e. $m = 1$ or $n = 1$), an evolution of (u, ν) based on this variational problem and Hamilton's principle was proposed in [44–46], see also Hoffmann in [21]. A modification, proposed basically by Srinivasa in [49], of these models suitable to a proper description of shape-memory effects will be just presented, generalized for anisothermal case, and studied in Sects. 2–5.

1.4 Other Mesoscopical Models – Concept of Mixtures

There are a lot of models which treat martensitic/austenitic structures as *mixtures* of particular phases of martensite or/and austenite.

Frémond's model [17,18] uses the elastic energy $\sum_i c_i(x)\varphi_i(x, \nabla u)$ as a convex combination of energy of single phases φ_i, the coefficients $c_i(x)$ of this convex combination at a given point x expressing the ratio of particular phases at (an "infinitesimal neighbourhood" of) x. This is similar to description by the probability measure ν_x from Sect. 1.3. This model has been further investigated by Colli et al. [10–14], Hoffmann et al. [20], Tiihonen [56] and Wörsching [59].

Various other mixture-based models are often used among physicists and engineers, see e.g. Abeyaratne, Chu and James [1,29], Auricchio [3], Huo, Müller and Seelecke [26–28,40], Levitas [35], Rajagopal, Srinivasa and Wineman [42,43,54,58]. These models mostly refer to polycrystals, where the geometrical compatibility of different phases plays the role only within particular single-crystal grains, and after averaging is less important and neglected. For this sort of models, we also refer to Mielke, Theil and Levitas [36,37] whose model is rate independent likewise the model presented below (at least if the kinetic energy $T = 0$ and the dissipation potential R is positively homogeneous) and it seems that our continuous isothermal model (2.5) as well as the discrete one (4.1) may coincide in special situations with the model [36,37] under the apriori knowledge that the involved Young measures will have only as many atoms as the number of phases.

Some other philosophy of the concept of mixtures can be also found; e.g. Kafka's model [30] considers shape-memory alloys as mixtures of their chemical constituents, e.g. Nickel and Titan.

2 Mesoscopical Model of Antiplane Twinning Evolution

From now on, we will confine ourselves only to a scalar case, i.e. $m = 1$, which simplifies the situation very essentially. Such situation appears in an antiplane shear deformation. In a one-dimensional case, i.e. $n = 1$, we have in mind a configuration outlined on Fig. 1.

Fast spatial oscillations of the strain $s = \nabla u$ outlined on Fig. 1 are called fine structure. They are caused by a nonconvexity of the energy φ, their width influenced by a surface energy (cf. the term $\varepsilon|\nabla^2 u|^2$ in (1.1)) being supposed much smaller than the scale of the specimen. The situation on

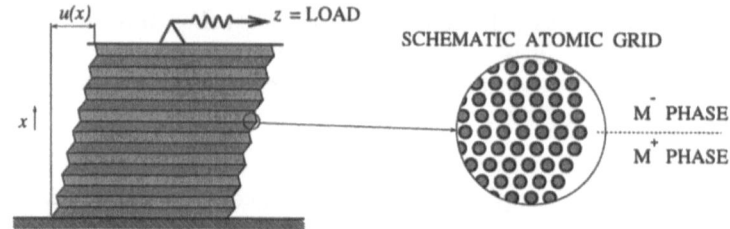

Fig. 1. Twinning and a one-dimensional antiplane shear deformation

Fig. 1 corresponds to the case when φ has two minima; then we speak about a double-well potential, outlined on Fig. 2a for the relative temperature $\theta = \vartheta - \vartheta_0$ negative.

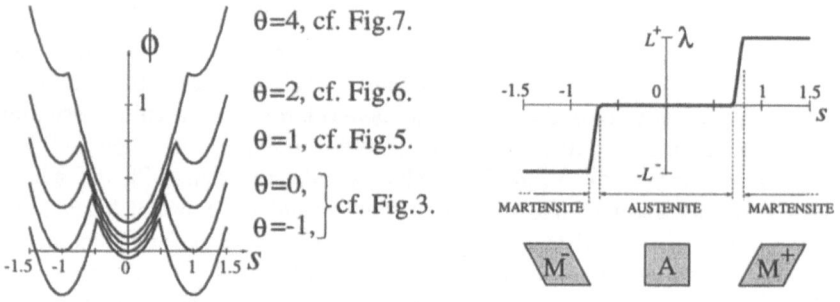

Fig. 2. a. Multi-well potential $\varphi(\cdot, \theta)$, cf. (5.1)   **Fig. 2. b.** Function λ; cf. (2.2)

The particular wells of φ express energetically advantageous crystal configurations, either cubic (called austenite, $s = 0$) or "monoclinic" or "tetragonal" (called martensite, $s = \pm 1$). High temperatures prefer high-symmetry configurations (i.e. cubic) and vice versa, which causes shape memory effects, cf. e.g. [3,18,28,40,15,28].

If $\zeta(x, \cdot)$ is convex, the relaxed functional $\bar{V}$ defined in (1.8) is convex no matter whether φ is convex or not. Moreover, in the scalar case $m = 1$, also Q defined in (1.8) is convex. Thus (1.8) can equivalently be written in the form of the inclusion (i.e. variational inequality):

$$\bar{V}'(q) + N_Q(q) \ni \bar{f}, \tag{2.1}$$

where $\bar{V}'$ is the Gâteaux derivative of $\bar{V}$, $N_Q(q)$ denotes the normal cone to Q at the point $q \equiv (u, \nu)$, and $\bar{f} := (f, 0)$.

When outer loading conditions changes, the system is suppose to start to evolve. However, this is a so far not much understood process at least in context of applied mathematics. In fact, microstructural changes are activated by (a sufficiently large) temperature or stress already in much lower energies than the macroscopical potential barrier between particular phases; one may imagine that the potential barrier can be overcome by particular atoms much more easily than by a whole macroscopical volume. Thermal activation of the phase transformation is due to chaotic oscillations of atoms, cf. [28], and it may interact with stress-activation mechanism, which might be the reason for making the stress-activation threshold temperature-dependent, cf. (6.1) below. The *activation* of phase transformation causes *hysteresis* effects in stress/strain diagram (cf. Figs. 3–7) which further determines a rate-independent energy *dissipation* or vice versa. Here we rather prescribe the energy dissipated by (or, in other words, needed for) a structure change within the phase transformation; for this we use a continuous function (=energy) λ which is constant on particular phases and has jumps when e moves between the regions M^+/A and M^-/A. We mean

$$\lambda(e) = \begin{cases} -L^- & \text{for } e < e_1 - \varepsilon, \\ 0 & \text{for } e \in [e_1 + \varepsilon, e_2 - \varepsilon], \\ L^+ & \text{for } e > e_2 + \varepsilon, \end{cases} \qquad (2.2)$$

cf. Fig. 2b, where L^+ (resp. L^-) is the energy dissipated within phase transformation $A \leftrightarrow M^+$ (resp. $A \leftrightarrow M^-$) per unit volume; $0 < \varepsilon << 1$ is just to allow $\lambda(\cdot)$ to be continuous.

If we want to define the evolution $t \mapsto q(t) \equiv (u(t), \nu(t))$, we must postulate the *impulse* $\dot{q} \equiv (\dot{u}, \dot{\nu})$ with the dot indicating the time derivative, for which we need some geometry. In [44,45] the convex geometry of Q was taken and then the Rayleigh *dissipation potential* was taken (with $\alpha = 2$) as follows:

$$R(\dot{q}) := \int_\Omega |\lambda \bullet \dot{\nu}|^\alpha \mathrm{d}x \qquad (2.3)$$

where $[\lambda \bullet \dot{\nu}](x) := \int_{\mathbb{R}^n} \lambda(s)\dot{\nu}_x(\mathrm{d}s)$ and $\lambda : \mathbb{R}^n \to \mathbb{R}^k$ defines a dissipation. E.g. for $\alpha = 2$, $k = n$ and $\lambda(s) = \mu s$, we get the "standard" dissipation $R(\dot{q}) = \mu \int_\Omega |\nabla \dot{u}|^2 \mathrm{d}x$ which leads to the usual term $\mu \nabla \dot{u}$ in the evolution equation, cf. (1.3). If $\alpha = 1$, the dissipation is *rate independent* (i.e. monotone transformation of time scale has no influence on energy dissipation).

Moreover, it is natural to define the *kinetic energy* as the quadratic form

$$T(\dot{q}) := \frac{1}{2} \int_\Omega \varrho(x) |\dot{u}|^2 \mathrm{d}x \qquad (2.4)$$

where $\varrho > 0$ is the mass density; note that changes of only macroscopical displacement u (but not the microstructure ν) contribute to kinetic energy.

By a Hamilton variational principle for dissipative systems, in [44,45] the following evolution model has been proposed:

$$T'\ddot{q} + R'(\dot{q}) + \bar{V}'(q) + N_Q(q) \ni \bar{f}(t), \quad q(0) = q_0, \quad \dot{q}(0) = p_0, \qquad (2.5)$$

where T', R', and $\bar{V}'$ are (in general nonlinear) operators being the (sub)differentials of the potentials T, R, and $\bar{V}$, respectively. Note that now we consider a time-varying external loading $f = f(t)$.

For R' linear (i.e. for $\alpha = 2$), the existence (and, under some additional conditions, also a uniqueness) of a solution to (2.5) has been shown in [44]. For $\alpha > 1$, the analysis of (2.5) seems to be possible by rather standard convex-analytical tools, while the case $\alpha = 1$ will probably need a suitable regularization; this analysis is out of the scope of this contribution, however.

3   An Anisothermal Model

Often, temperature varies significantly and then, beside modifying $\varphi = \varphi(s, \vartheta)$ and $\bar{V} = \bar{V}(q, \vartheta)$ and also (2.5) by replacing $\bar{V}'(q)$ with the partial (sub)differential $\bar{V}_q'(q, \vartheta)$, a consistent thermodynamics is to be taken into account. Again, a standard departing point is the specific "mesoscopic" *Helmholtz free energy*, which can be considered in view of (1.2) as

$$\psi(q, \vartheta) = \varphi(\cdot, \vartheta) \bullet \nu - c\vartheta \ln\vartheta, \qquad (3.1)$$

where ϑ denotes the absolute temperature. Then, specific *entropy* is defined by

$$\eta = -\frac{\partial \psi}{\partial \vartheta} = c(1 + \ln\vartheta) - \frac{\partial \varphi}{\partial \vartheta} \bullet \nu, \qquad (3.2)$$

while specific *internal energy* is

$$e = \psi + \vartheta\eta = c\vartheta + \left(\varphi(\cdot, \vartheta) - \vartheta\frac{\partial \varphi}{\partial \vartheta}\right) \bullet \nu. \qquad (3.3)$$

Then the total free energy is

$$\Psi(q, \vartheta) = \int_\Omega \psi(q, \vartheta)\mathrm{d}x = \bar{V}(q, \vartheta) - \int_\Omega c\vartheta \ln\vartheta \, \mathrm{d}x. \qquad (3.4)$$

From the requirement of conservation of the total energy

$$T(\dot{q}) + \int_\Omega e \, \mathrm{d}x = T(\dot{q}) + \bar{V}(q, \vartheta) + \int_\Omega c\vartheta \, \mathrm{d}x - \int_\Omega \vartheta\frac{\partial \varphi}{\partial \vartheta} \bullet \nu \, \mathrm{d}x \qquad (3.5)$$

for isolated systems and assuming the thermally linear isotropic medium so that the heat flux is of the form $-\kappa\nabla\vartheta$, one gets the equation for the temperature in the form

$$c\dot{\vartheta} - \mathrm{div}(\kappa\nabla\vartheta) = \alpha|\lambda \bullet \dot{\nu}|^\alpha + \vartheta\frac{\partial}{\partial t}\left(\frac{\partial \varphi}{\partial \vartheta} \bullet \nu\right), \quad \vartheta(\cdot, 0) = \vartheta_0. \qquad (3.6)$$

Then, we have also $\vartheta\dot{\eta} - \mathrm{div}(\kappa\nabla\vartheta) = \alpha|\lambda\bullet\dot{\nu}|^\alpha$, from which we can get by Green's formula the Clausius-Duhem *entropy inequality*

$$\frac{\mathrm{d}}{\mathrm{d}t}\int_\Omega \eta(\vartheta,q)\,\mathrm{d}x = \int_\Omega \alpha\frac{|\lambda\bullet\dot{\nu}|^\alpha}{\vartheta} + \kappa\frac{|\nabla\vartheta|^2}{\vartheta^2}\,\mathrm{d}x \geq 0 \qquad (3.7)$$

valid for isolated systems, i.e. zero heat flux through the boundary Γ of Ω, and for $\vartheta_0 \geq 0$ which ensures, at least for smooth solutions to (3.6), that $\vartheta \geq 0$.

If the absolute temperature ϑ does not vary too much far from the initial temperature ϑ_0 which is assumed constant, it is also usual to linearize partly the problem by introducing *relative temperature* θ and by putting

$$\varphi(s,\vartheta) = \varphi_0(s) + \theta\varphi_1(s)\,, \qquad \theta = \vartheta - \vartheta_0. \qquad (3.8)$$

In our case in Fig. 2a, one could take φ_0 a triple-well potential and φ_1 constant on martensite and on austenite areas of the strains s, which gives equal effects as the original φ (cf. (3.7) below) provided nonappearance of the support of the Young measure ν near transient regions. Formula (3.8) simplifies the internal energy (3.3) into the form $e = c\theta + \varphi_0\bullet\nu$ (up to the constant $c\vartheta_0$) as well as the adiabatic-heat term in (3.6), so that the heat equation in terms of relative temperature θ now turns into

$$c\dot{\theta} - \mathrm{div}(\kappa\nabla\theta) = \alpha|\lambda\bullet\dot{\nu}|^\alpha + (\theta + \vartheta_0)\,\varphi_1\bullet\dot{\nu}\,, \quad \theta(\cdot,0) = 0, \qquad (3.9)$$

and the evolution inclusion (2.5) modified for $\bar{V} = \bar{V}(q,\vartheta)$ turns into

$$T'\ddot{q} + R'(\dot{q}) + \bar{V}_0'(q) + \theta\bar{V}_1'(q) + N_Q(q) \ni \bar{f}(t),$$
$$q(0) = q_0, \qquad \dot{q}(0) = p_0, \quad (3.10)$$

where $\bar{V}_0(u,\nu) = \int_\Omega \varphi_0\bullet\nu\,\mathrm{d}x + \int_\Gamma \zeta(x,u)\,\mathrm{d}S$ and $\bar{V}_1(u,\nu) = \int_\Omega \varphi_1\bullet\nu\,\mathrm{d}x$.

4 Numerical Approximation

Considering a time-step $\tau > 0$, the evolution inequality (3.10) and the parabolic equation (3.9) with R from (2.3) can be (semi)discretized by the following semi-implicit formula

$$T'\left(\frac{q^k - 2q^{k-1} + q^{k-2}}{\tau^2}\right) + R'\left(\frac{q^k - q^{k-1}}{\tau}\right) + \bar{V}_0'(q^k) + \theta^{k-1}\bar{V}_1'(q^k)$$
$$+ N_Q(q^k) \ni \bar{f}_\tau^k\,, \qquad q^0 = q_0 \text{ and } q^{-1} = q_0 - \tau p_0, \qquad (4.1)$$

$$c\frac{\theta^k - \theta^{k-1}}{\tau} - \mathrm{div}(\kappa\nabla\theta^k) = \alpha\left|\lambda\bullet\frac{\nu^k - \nu^{k-1}}{\tau}\right|^\alpha$$
$$+ (\theta^{k-1} + \vartheta_0)\varphi_1\bullet\frac{\nu^k - \nu^{k-1}}{\tau}\,, \qquad \theta^0 = 0, \qquad (4.2)$$

where $\bar{f}_\tau^k := \frac{1}{\tau}\int_{(k-1)\tau}^{k\tau} \bar{f}(t)\mathrm{d}t$, which represents, for $k = 1, 2, ...$, a recursive variational inequality for $q^k \in Q$ and the linear elliptic boundary-value problem for θ^k if completed by boundary conditions for θ^k. Standardly, we define the approximate solution $(q_\tau, \theta_\tau) \equiv (u_\tau, \nu_\tau, \theta_\tau)$ as piecewise affine interpolation in time. By convexity of T, R, $\bar{V}$ and Q, the inclusion (4.1) for q^k represents just the necessary and sufficient optimality conditions for q_k to solve the following minimization problem:

$$
\left.
\begin{aligned}
\text{Minimize } F^k(q) := \tau^2 T &\left(\frac{q - 2q^{k-1} + q^{k-2}}{\tau^2}\right) \\
+ \tau R &\left(\frac{q - q^{k-1}}{\tau}\right) + \bar{V}_0(q) + \theta^{k-1}\bar{V}_1(q) - \langle \bar{f}_\tau^k, q \rangle \\
\text{subject to } q \equiv (u, \nu) &\in Q \quad \text{defined in (1.8).}
\end{aligned}
\right\} \quad (4.3)
$$

Existence of this approximate solution for any $\tau > 0$ can be proved, similarly as in [44], by compactness arguments together with nonconcentration arguments as in [47].

By testing (4.1) by $q^k - q^{k-1}$ and (4.2) by 1, we get the basic energy balance

$$
\underbrace{T(\dot{q}_\tau(t))}_{\substack{\text{kinetic} \\ \text{energy}}} + \underbrace{\bar{V}_0(q_\tau(t))}_{\substack{\text{elastic} \\ \text{energy}}} + \underbrace{\int_\Omega c(x)\theta_\tau(\cdot, t)\mathrm{d}x}_{\substack{\text{heat} \\ \text{energy}}} - \underbrace{\int_0^t \langle f_\tau, \dot{u}_\tau\rangle\mathrm{d}\xi}_{\substack{\text{work of} \\ \text{external force}}} =
$$

$$
= T(p_0) + \bar{V}_0(q_0) + D_\tau \quad (4.4)
$$

at any $t = k\tau$, where $D_\tau \le 0$ denotes a numerical dissipation, containing the term $-\frac{1}{2}\tau\int_0^t\int_\Omega \varrho|\ddot{u}_\tau|^2\mathrm{d}x\mathrm{d}t$ (with $\ddot{u}_\tau$ denoting $(u^k - 2u^{k-1} + u^{k-2})/\tau^2$) coming from the time discretization of $\varrho\ddot{u}$, the term $-\frac{1}{2}\tau\int_0^t\int_\Gamma \zeta_0(x)|\dot{u}_\tau|^2\mathrm{d}S\mathrm{d}t$ coming from the discretization of $\zeta'(x, u) := \zeta_0(x)u$, and eventually a term created by the presence of the obstacle Q; this last term vanishes whenever $\langle \xi, q^k - q^{k-1}\rangle = 0$ for any $\xi \in N_Q(q^k)$.

The mechanical energy dissipated to the heat within the process during time interval $[0, t]$ is

$$
\int_\Omega c(x)\theta_\tau(\cdot, t)\mathrm{d}x = \int_0^t \langle R'(\dot{q}_\tau), \dot{q}_\tau\rangle\mathrm{d}\xi = \alpha\int_0^t\int_\Omega |\lambda \bullet \dot{\nu}_\tau|^\alpha\mathrm{d}x\mathrm{d}\xi. \quad (4.5)
$$

In particular, for $\alpha = 1$ we have

$$
\text{dissipated energy} = \int_\Omega \operatorname*{Var}_{\xi \in [0, t]} \left[\lambda \bullet \nu_\tau\right](x, \xi)\mathrm{d}x , \quad (4.6)
$$

where "Var" denotes the total variation over the time-interval indicated.

The rigorous analysis of the scheme (4.1)–(4.2) is rather impossible; note that the problem involves for φ_0 quadratic, φ_1, λ linear, and $\alpha = 2$ also the usual thermoelasticity which is still open in the 3D case. For $\alpha = 2$, a partly linearized problem created by considering $(\theta + \vartheta_0) \sim \vartheta_0$ has been analyzed in [21]. Nonlinear R' (i.e. for $\alpha \neq 2$) will probably need a suitable regularization; this analysis is out of the scope of this contribution, however.

5   Computational Experiments

The numerical tests presented in this section are just to verify relevance of the model in the context of expected behaviour of shape memory alloys, proving a good agreement with experiments, cf. e.g. Abeyaratne, Chu, James [1,29], Miyazaki [18], or Huo, Müller, Seelecke [28,40]. For this, we assume

1) $n = 1$ (i.e. we have in mind the situation from Fig. 1),

2) isothermal case (except Fig. 9), i.e. (2.5) with $\bar{V} = \bar{V}(\cdot, \theta + \vartheta_0)$ with θ a constant parameter,

3) slow processes so that kinetic energy can be neglected (by putting $\varrho = 0$),

4) a small specimen so that spatial dependence can be neglected (i.e. we assume ν_x independent of x and hence $u(x)$ affine).

Last two simplifications allows us to suppress all geometrical and dynamical effects and thus to highlight the expected hysteresis behaviour only.

We consider φ composed from three parabolas having minima for the strains $s = \pm 1$ (=two martensitic phases M$^+$ and M$^-$) and $s = 0$ (=one austenitic phase A), namely

$$\varphi(s, \theta) := \min \left(E_m(s \pm 1)^2 + \theta, \, E_a s^2 \right) \tag{5.1}$$

where $E_a = 5$ the Young modulus characterizing elasticity of austenite, $E_m = 7.5$ the Young modulus of martensite; cf. Fig. 2a. Moreover, we count the specimen elastically loaded through a spring, cf. Fig. 1. This was modelled by

$$\zeta(t, \pm l, u) := ku^2, \quad \langle f(t), u \rangle := 2k(u(l) - u(-l))z(t) \tag{5.2}$$

with $z = z(t)$ an external loading and $k = 10$ a constant characterizing the rigidity of the spring. Then $\sigma = 2k(u - z)$ is the macroscopical stress. The function λ is taken according (2.2) with $L^+ = L^- = 1/2$, which means the energy needed for the transformation M$^+\leftrightarrow$A or M$^-\leftrightarrow$A; cf. Fig. 2b.

The initial conditions were chosen as $u_0 = 0$ and $[\nu_0]_x = \frac{1}{2}\delta_{-1} + \frac{1}{2}\delta_1$, where δ_s denotes the Dirac measure supported at the strain s; note that the initial impulse p_0 is not relevant since $T \equiv 0$ is assumed. A response of the strain $s = \nabla u$, the stress σ, and the portion of phases $\lambda \bullet \nu$ within cyclical loading z is displayed on Figs. 3.

The curves on Fig. 3 do not depend on the relative temperature θ if non-positive, i.e. if the austenitic well has greater (or equal) energy than the martensitic ones. The area of the hysteresis loop in the stress/strain diagram

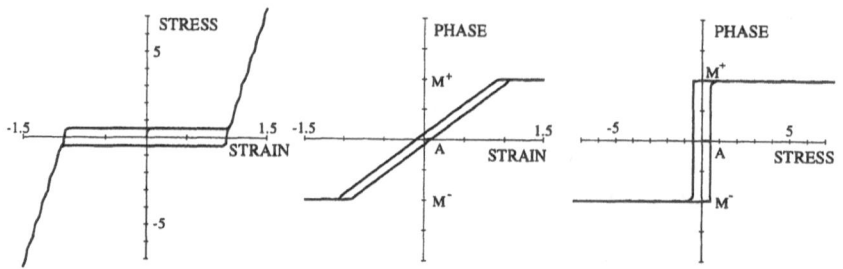

Fig. 3. Quasiplastic hysteresis effects, cyclical loading, $\theta \leq 0$

is precisely $2(L^- + L^+)$, which corresponds to two phase transformations $M^- \to M^+ \to M^-$ undergone within one loop, each of them dissipating the energy $L^- + L^+$, cf. Fig. 2b and formula (4.6).

For positive relative temperature θ, the austenite become dominant at least for unloaded case. For smaller θ, this effect is less markable and quasi-plasticity continuously changes to pseudoelasticity. This transient behaviour is displayed on Fig. 4 (cf. also Müller and Seelecke [40, Fig.7.3]):

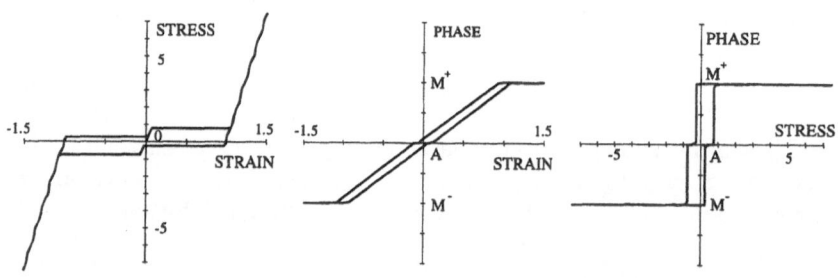

Fig. 4. Transient hysteresis effects, cyclical loading, $\theta = 1/4$

For higher temperature, the austenite become dominant at least for smaller loadings and thus one hysteresis loop is eventually splitted into two loops, which is displayed on Fig. 5.

The area of the hysteresis loops in the stress/strain diagram is precisely $2L^+ = 1$ and $2L^- = 1$, which corresponds to the 2 phase transformations $A \to M^+ \to A$ (right-hand loop) or $A \to M^- \to A$ (left-hand loop) undergone within one loop, each of them dissipating the energy L^+ or L^-, respectively; cf. Fig. 2b.

The higher the temperature, the larger the loading under which no marten-site occurs. This is displayed on Figs. 6 and 7.

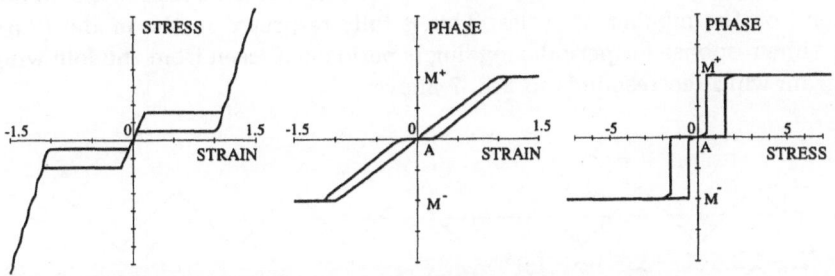

Fig. 5. Pseudoelastic hysteresis effects, cyclical loading, $\theta = 1$

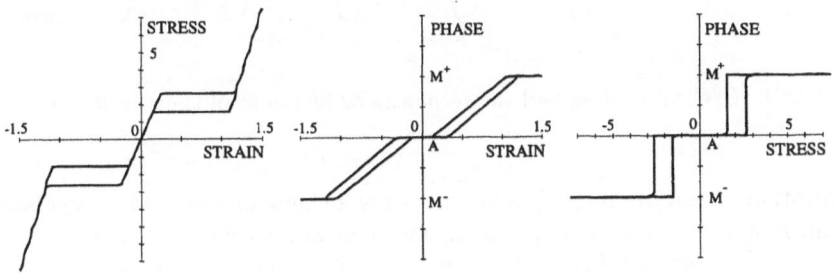

Fig. 6. Pseudoelastic hysteresis effects, cyclical loading, $\theta = 2$

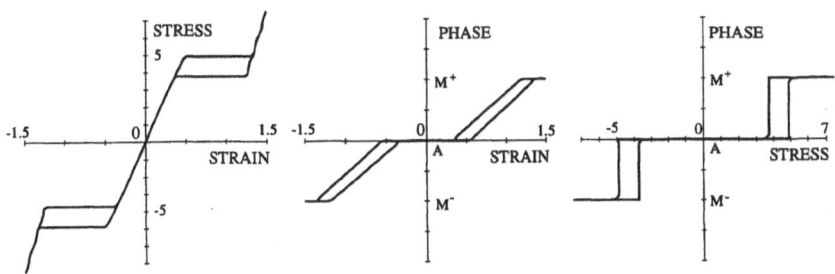

Fig. 7. Pseudoelastic hysteresis effects, cyclical loading, $\theta = 4$

Contrary to classical plasticity, the quasiplasticity as well as pseudoelasticity is a completely reversible process in the sense that the hysteresis loops do not depend on the number of cycles. This is fully respected in the model (2.5). The time response for periodic loading is periodic, as seen from the following diagram which corresponds to Fig. 7 above:

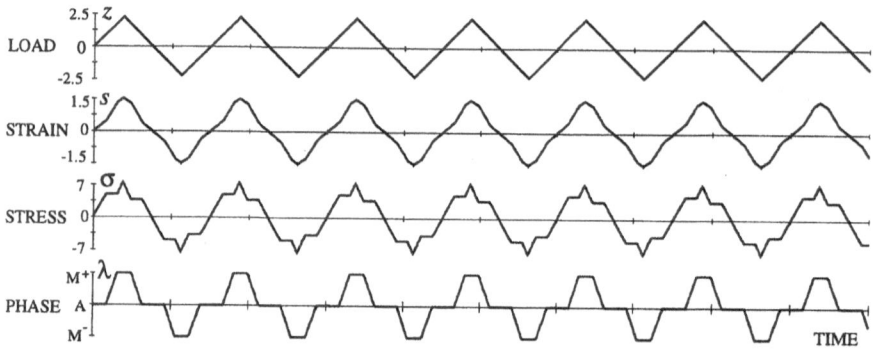

Fig. 8. Cyclical loading and the responses as functions of time for $\theta = 4$

Another standard test of shape-memory behaviour consists in cyclical heating/cooling of a loaded specimen. Let us again consider the situation as in Fig. 1 of a specimen loaded by a spring with z constant in time, cf. (5.2). Namely, the numerical experiment was performed for three values $z = 0.2$, 0.5, and 0.8. Temperature was used as a time-dependent parameter, and varied cyclically within the interval $\theta \in [-1, 5]$, creating again certain hysteresis effects shown in Fig. 9 (cf. also Huo, Müller, Seelecke [28, Fig.1.9] and [40, Fig.3.2] or Miyazaki [18, Fig.34 and 45] for experimental results, or also Hou [26, Fig.9], Rajagopal and Srinivasa [43, Fig.4 and 5] for computational results):

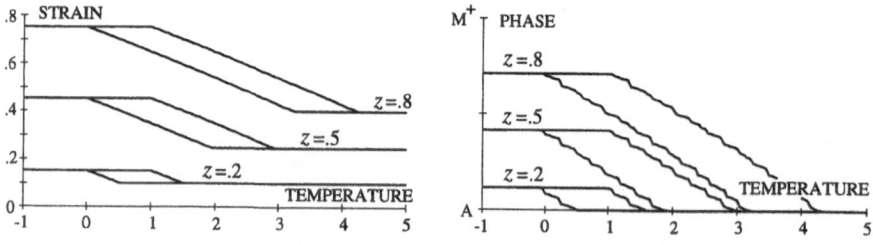

Fig. 9. Hysteresis effects, cyclically varying temperature θ and constant loading z

The choice of initial conditions u_0 and v_0 is not trivial for a loaded case if one wants to start from an equilibrium state. Nevertheless, this choice is not important if one does not care about a short transient event at the beginning of the process, as actually accepted when drawing Fig. 9.

Of course, for all calculations the Young measure v_x^k appearing in each $q^k = (u^k, v_x^k)$ had to be discretized, too. Here we employed the discretization with fixed atoms as in [45,49] which leaded after a suitable transformation (cf. [49]) to a linear-quadratic program to be solved at each time level k. We used 120 fixed atoms supported equidistantly on the considered effective interval $[-1.5, +1.5]$, cf. Fig. 2a. The consequence of this discretization can be seen on small oscillations in the stress/strain diagram of Figs. 3–7 and 9.

6   Concluding Remarks

Typically, the pseudoelastic hysteresis loops has smaller and smaller area within rising temperature, and eventually they disappears if $\theta \geq \theta_C$, where θ_C is the so-called *Curie point*. This can be modelled very simply just by making the energy dissipated within phase transformation dependent on temperature, e.g.

$$L^+ = L^- = L(\theta) = \begin{cases} L_0 \left(1 - \dfrac{\theta}{\theta_C}\right) & \text{for } \theta < \theta_C, \\ 0 & \text{for } \theta \geq \theta_C. \end{cases} \tag{6.1}$$

Then the dissipative potential R would depend also on θ, which would make the analysis of the model still more complicated.

Experiments in Sect. 5 indicates that the isothermal case with the kinetic term neglected gives rise to a hysteresis operator in the standard sense, cf. [7, Chapter 2]. It would be worth specifying explicitly such operator and use it to derive new results.

Moreover, the antiplane shear deformation according Fig. 1 usually cannot be realized, the actual situation being fully vectorial (i.e. both $m > 1$ and $n > 1$) and Q defined in (1.8) is not convex. This makes the formulation and analysis of the problem expectedly much more complicated. For some numerical experiments with a model similar to (2.5) see [48]. Besides, the simple-laminated configuration in Figure 1 is often replaced by higher-order laminates having layers within layers. This makes the evolution extremely complicated even in single crystals, not speaking about polycrystals.

If the energy of the phase transformation $M^- \leftrightarrow M^+$ is not equal to the sum of energies $M^- \leftrightarrow A$ and $M^+ \leftrightarrow A$, then some dissipative potential R more complicated than (2.3) have to be used, involving several λ's with different jumps between the particular phases.

Eventually, it would be very desirable to investigate an attainability of the mesoscopical model presented here by a suitable microscopical-type model like (1.3) but using the dissipative term having the potential, instead of $\frac{1}{2}\int_{\Omega}\mu|\nabla\dot{u}|^2\mathrm{d}x$ used implicitly in (1.3), of some more sophisticated form, e.g.

$$\int_{\Omega}\left|\frac{\partial}{\partial t}\lambda(\nabla u)\right|\mathrm{d}x \qquad (6.2)$$

with λ as above seems a bit more relevant. Besides, a physical mechanism created such sort of dissipation would be worth identifying.

Acknowledgments. The author is thankful for stimulating discussions to M. Brokate, J. Kratochvíl, J. Málek, I. Müller, J. Nečas, V. Novák, K.R Rajagopal, A. Ruffing, P. Šittner, A. Srinivasa, and F. Theil. This research has been partly covered by the USA(NSF)/Czech grant No.2996K1020 during the author's visits at Texas A & M University and by the grant A 107 5707 (Academy of Sciences of the Czech Republic).

References

1. Abeyaratne, R., Chu, C., James, R. D.: Kinetics of Materials with Wiggle Energies: Theory and Application to the Evolution of Twinning Microstructures in a Cu-Al-Ni Shape Memory Alloy. Philosophical Magazine A **73** (1996), 457–497
2. Alt, H. W., Hoffmann, K.-H., Niezgódka, M., Sprekels, J.: A Numerical Study of Structural Phase Transitions in Shape Memory Alloys. Preprint no. 90, Math. Institut, Universität Augsburg, 1985
3. Auricchio, F.: Shape Memory Alloys: Applications, Micromechanics, Macromodelling and Numerical Simulation. PhD-thesis, Univ. of California, Berkeley, 1995
4. Ball, J. M., Holmes, P. J., James, R. D., Pego, R. L., Swart P. J.: On the Dynamics of Fine Structure. J. Nonlinear Science **1** (1991), 17–70
5. Ball, J. M., James, R. D.: Fine Phase Mixtures as Minimizers of Energy. Archive Rat. Mech. Anal. **100** (1988), 13–52
6. Ball, J. M., James, R. D.: Proposed Experimental Tests of a Theory of Fine Microstructure and the Two-Well Problem. Phil. Trans. Royal Soc. London A **338** (1992), 389–450
7. Brokate, M. Sprekels, J.: Hysteresis and Phase Transitions. Springer, New York, 1996
8. Bubner, N.: Landau-Ginzburg Model for a Deformation-Driven Experiment on Shape Memory Alloys. Continuum Mech. Thermodyn. **8** (1996), 293–308
9. Chen, Z., Hoffmann, K.-H.: On a One-Dimensional Nonlinear Thermoviscoelastic Model for Structural Phase Transitions in Shape Memory Alloys. J. Diff. Equations **12** (1994), 325–350
10. Colli, P.: Global Existence for the Three-Dimensional Frémond Model of Shape Memory Alloys. Nonlinear Analysis, Th. Meth. Appl. **24** (1995), 1565–1579
11. Colli, P., Frémond, M., Visintin, A.: Thermo-Mechanical Evolution of Shape Memory Alloys. Quarterly Appl. Math. **48** (1990), 31–47
12. Colli, P., Sprekels, J.: Global Existence for a Three-Dimensional Model for the Thermo-Mechanical Evolution of Shape Memory Alloys. Nonlinear Anal. **18** (1992), 873–888

13. Colli, P., Sprekels, J.: Global Solution to the Full One-Dimensional Frémond Model for Shape Memory Alloys. Math. Meth. Appl. Sci. **18** (1995), 371–385
14. Colli, P., Sprekels, J.: Positivity of Temperature in the General Frémond Model for Shape Memory Alloys. Continuum Mech. Thermodyn. **5** (1993), 255–264
15. Falk, F.: Model Free Energy, Mechanics and Thermodynamics of Shape Memory Alloys. Acta Metallurgica **28** (1980), 1773–1780
16. Falk, F.: Landau Theory and Martensitic Phase Transitions. In: Proc. Int. Conf. on Martensitic Transformations (L.Delaey, M.Chandrasekaran, Eds.), Les Editions de Physique, Les Ulis, 1982
17. Frémond, M.: Matériaux à Mémoire de Forme. C.R. Acad. Sci. Paris **304**, Série II (1987), 239–244
18. Frémond, M., Miyazaki, S.: Shape Memory Alloys. Springer, Wien, 1996
19. Friesecke, G., McLeod, J.B.: Dynamics as a Mechanism Preventing the Formation of Finer and Finer Microstructure. Arch. Rat. Mech. Anal. **133** (1996), 199–247
20. Hoffmann, K.-H., Niezgódka, M., Zheng Songmu: Existence and Uniqueness of Global Solutions to an Extended Model of the Dynamical Development in Shape Memory Alloys. Nonlinear Analysis, Th. Meth. Appl. **15** (1990), 977–990
21. Hoffmann, K.-H., Roubíček, T.: Thermomechanical Evolution of a Microstructure. Quarterly Appl. Math. **52** (1994), 721–737
22. Hoffmann, K.-H., Songmu, Z.: Uniqueness for Structured Phase Transitions in Shape Memory Alloys. Math. Meth. in the Appl. Sci. **10** (1988), 145–151
23. Hoffmann, K.-H., Sprekels, J.: Phase Transitions in Shape Memory Alloys I: Stability and Optimal Control. Numer. Funct. Anal. Optim. **9** (1987), 743–760
24. Hoffmann, K.-H., Zochowski, A.: Existence of Solutions to Some Non-Linear Thermoelastic Systems with Viscosity. Math. Methods in the Applied Sciences **15** (1992), 187–204
25. Hoffmann, K.-H., Zou, J.: Finite Element Approximations of a Landau-Ginzburg's Equation Model for structural Phase Transitions in Shape Memory Alloys. RAIRO Modelisation Math. Anal. Numer. **29** (1995), 629–655
26. Huo, Y.: A mathematical Model for the Hysteresis in Shape Memory Alloys. Continuum Mech. Thermodyn. **1** (1989), 283–303
27. Huo, Y., Müller, I.: Nonequilibrium Thermodynamics of Pseudoelasticity. Continuum Mech. Thermodyn. **5** (1993), 163–204
28. Huo, Y., Müller, I., Seelecke, S.: Quasiplasticity and Pseudoelasticity in Shape Memory Alloys. In: Phase Transitions and Hysteresis. (Eds.: M.Brokate et al.) Lect. Notes in Math. **1584** (1994), 87–146
29. James, R. D.: Hysteresis in Phase Transformations. In: ICIAM 95, Proc. 3rd Int. Congress Indust. Appl. Math., Kirchgässner, K., et al., eds., Akademie Verlag, Berlin, Math. Res. **87** (1996), 135–154
30. Kafka, V.: Shape Memory: A New Concept of Explanation and of Mathematical Modelling. J. Intelligent Material Systems and Structures **5** (1994), 809–824
31. Killough, M. G.: A Diffuse Interface Approach to the Development of a Microstructure in Martensite. PhD-thesis, Dept. of Math., New York University, 1998
32. Klouček, P., Luskin, M.: The Computation of the Dynamics of the Martensitic Transformation. Continuum Mech. Thermodyn. **6** (1994), 209–240

33. Klouček, P., Luskin, M.: Computational Modeling of the Martensitic Transformation with Surface Energy. Math. Comp. Modelling **20** (1994), 101–121

34. Kohn, R. V., Müller, S.: Surface Energy and Microstructure in Coherent Phase Transitions. Comm. Pure Appl. Math. **47** (1994), 405–435

35. Levitas, V. I.: Thermomechanical Theory of Martensitic Phase Transformations in Inelastic Material. Int. J. Solids Structures **35** (1998), 889–940

36. Mielke, A., Theil, F., Levitas, V. I.: Mathematical Formulation of Quasistatic Phase Transformations with Friction Using an Extremum Principle. Preprint No. A8, Sept. 1998, Univ. Hannover

37. Mielke, A., Theil, F.: A Mathematical Model for Rate-Independent Phase Transformations, preprint

38. Morin, P., Spies, R. D.: Convergent Spectral Approximations for the Thermomechanical Processes in Shape Memory Alloys. IMA preprint no.1381, Univ. of Minnesota, Minneapolis, 1996

39. Müller, S.: Variational Models for Microstructure and Phase Transitions. Lect. Notes No. 2, Max-Planck-Institut für Math., Leipzig, 1998

40. Müller, I., Seelecke, S.: Thermodynamic Aspects of Shape Memory Alloys. In: From Microstructure to Macroscopic Properties, Airoldi,, G., ed., Trans. Tech. Publ., Zürich, in print

41. Niezgódka, M., Sprekels, J.: Existence of Solutions for a Mathematical Model of Structural Phase Transitions in Shape Memory Alloys. *Math. Methods in Appl. Sci.* **10** (1988), 197–223

42. Rajagopal, K. R., Srinivasa, A. R.: On the Inelastic Behavior of Solids – Part 1: Twinning. Int. J. Plasticity **11** (1995), 653–678

43. Rajagopal, K. R., Srinivasa, A. R.: On the Thermomechanics of Shape Memory Wires. Z. für angew. Math. u. Physik **50** (1999), 459–496

44. Roubíček, T.: Evolution of a Microstructure: A Convexified Model. Math. Methods in the Applied Sciences **16** (1993), 625–642

45. Roubíček, T.: Finite Element Approximation of a Microstructure Evolution. Math. Methods in the Applied Sciences **17** (1994), 377–393

46. Roubíček, T.: Microstructure Evolution Models. In: Metz Days 1992, Chipot, M., ed., Pitman Res. Notes in Math. **296** (1993), Longmann, Harlow, Essex, pp. 67–73

47. Roubíček, T.: Relaxation in Optimization Theory and Variational Calculus. de Gruyter, Berlin, 1997

48. Roubíček, T., Kružík, M.: Numerical Treatment of Microstructure Evolution Modelling. In: *ENUMATH 97*, Bock, H. G., et al., eds., World Scientific, Singapore, 1998, 532–539

49. Roubíček, T., Srinivasa, A.: An Evolution Model for Martensitic Phase Transformation in Shape-Memory Alloys, in preparation

50. Rybka, P.: Dynamical Modelling of Phase Transitions by Means of Viscoelasticity in Many Dimensions. Proc. Royal Soc. Edinburgh **121A** (1992), 101–138

51. Rybka, P.: The Viscous Damping Prevents Propagation of Singularities in the System of Viscoelasticity. *Proc. Royal Soc. Edinburgh* **127A** (1997), 1067–1074

52. Rybka, P., Hoffmann, K.-H.: Convergence of Solutions to Equation of Viscoelasticity with Capillarity. J. Math. Anal. Appl. **226** (1998), 61–81

53. Sprekels, J.: Global Existence for Thermomechanical Processes with Nonconvex Free Energies of Ginzburg-Landau Form. J. Math. Anal. Appl. **141** (1989), 333–348

54. Srinivasa, A. R., Rajagopal, K. R., Armstrong, R. W.: A Phenomenological Model of Twinning Based on Dual Reference Structures. Acta Metall. (1998), 1–14
55. Swart, P. J., Holmes, P. J.: Energy Minimization and the Formation of Microstructure in Dynamic Anti-Plane Shear. Archive Rat. Mech. Anal. **121** (1992), 37–85
56. Tiihonen, T.: A Numerical Approach to a Shape Memory Model. Report no. 98, Inst. für Math., Universität Augsburg, 1988
57. Theil, F.: Young-Measure Solutions for a Viscoelastically Damped Wave Equation with Nonmonotone Stress-Strain Relation. Arch. Rational Mech. Anal. **144** (1998), 47–78
58. Wineman, A. S., Rajagopal, K. R.: On a Constitutive Theory for Materials Undergoing Microstructural Changes. Arch. Mech. **42** (1990), 53–75
59. Wörsching, G.: Numerical Simulation of the Frémond Model for Shape Memory Alloys. Z. Angew. Math. Mech. **76** (1996), 273–276

A Phase Diagram of Integration and Segregation in a Population of Hawks and Doves

I. Müller

Technische Universität Berlin, Fachgebiet Thermodynamik, Straße des 17. Juni 135, D-10623 Berlin, Germany

Dedicated to Professor Karl-Heinz Hoffmann
on the occasion of his 60th birthday

Abstract. The contest strategy of a population may be used to define a population energy as a function of the composition of the population. The minimum of such an energy corresponds to the composition of the evolutionarily stable strategy which is approached by natural selection.

If, however, the composition is fixed, the population may alter its strategy. If it has two strategies to choose from it will choose the one that is energetically more favourable. This choice can mean an integrated or a segregated population depending on the value of the resource for which different groups of the population compete.

1 Scope

The principal thesis underlying this work is that sociobiology may have something to learn from thermodynamics.

The sociobiologist studies animal contest − a contest for a resource. If an animal wins the resource, it obtains a chance of survival, and survival may provide the possibility for replication. The contest is fought with a strategy − complete with rewards and penalties − and, if the strategy is successful, so that it guarantees survival and the possibility for replication, we call it an evolutionarily stable strategy, an ESS in the jargon of sociobiology.

The thermodynamicist studies molecular interaction, and these studies have a long history. It is now clear that the stability of molecular systems involves criteria such as minimum energy, or maximum entropy, or − in general − minimal free energy. Free energy is a combination of energy and entropy and temperature, where temperature is a weighting factor that determines the relative importance of energy and entropy in the approach to thermodynamic stability and the maintenance of stability.

Therefore, if we wish to apply thermodynamic reasoning to sociobiology, it seems imperative that we should find interpretations of energy and entropy and temperature in terms of contest strategies. In such an extrapolation of thermodynamic concepts it is ironic that the entropy, the most abstract of thermodynamic quantities, is rather easily extrapolated to the new field of sociobiology, whereas the much more suggestive energy is more difficult to generalize. In this work we concentrate on energy – the more difficult task – because energy is *the* quantity that determines integration and separation of groups in a population. And that is the subject of this work: Integration and segregation in a population of hawks and doves.

2   Hawks and Doves – an Evolutionarily Stable Strategy

We consider first a population of hawks and doves, as it was introduced into sociobiology by J. Maynard Smith and G. R. Price [1]. The strategy is succinctly described as follows.[1]

> A hawk fights for the resource; a dove merely engages in a symbolic conflict, posturing and threatening but not actually fighting. Between two hawks there will always be a fight until one is injured. The winner will get the resource worth 50 points, while the loser will get −100 points for being injured. Over many fights each hawk will thus "win" −25 points per fight. If a hawk meets a dove, the hawk will always win the resource and there will be no injury. If two doves meet, they will spend a long time posturing. One will eventually win the resource, but they will both get −10 points for wasted time: In the mean an encounter will thus gain a dove 15 points.

Let x_H and $x_D = 1 - x_H$ be the fractions of hawks and doves in the population. The expectation values E_H and E_D for the number of points won by hawks and doves respectively are

$$E_H = -25 \cdot x_H + 50 \cdot x_D \\ E_D = 0 \cdot x_H + 15 \cdot x_D \, , \tag{2.1}$$

since x_H and x_D are the probabilities that an individual meets hawk or dove, respectively. It is obvious from (2.1) that a few doves in a population of hawks do better than the hawks and that a few hawks in a population of doves do better than the doves. We assume that such advantages translate into an evolutionary payoff so that the fraction of hawks grows when E_H is bigger than E_D. We write

$$\frac{dx_H}{dt} = \alpha \cdot (E_H - E_D) \qquad (\alpha > 0). \tag{2.2}$$

Combining (2.1) and (2.2) we obtain

$$\frac{dx_H}{dt} = -60 \cdot \alpha \cdot \left(x_H - \frac{7}{12} \right). \tag{2.3}$$

---

[1] This is close to a verbatim quote from P. D. Straffin [2].

Thus we conclude that $(x_H, x_D) = \left(\frac{7}{12}, \frac{5}{12}\right)$ is evolutionarily stable. A population with different fractions of hawks and doves will exponentially tend to the stable fractions, see Fig. 1.

On the basis of (2.3) we may construct an energy of the population. Indeed the right-hand side may be written as the gradient of

$$F(x_H) = 30\left(x_H - \frac{7}{12}\right)^2 + C, \qquad (2.4)$$

where C is a constant. This is a convex parabola. We consider $F(x_H)$ as an energy, because obviously the population tends to the minimum of this function, which defines the evolutionarily stable hawk fraction.

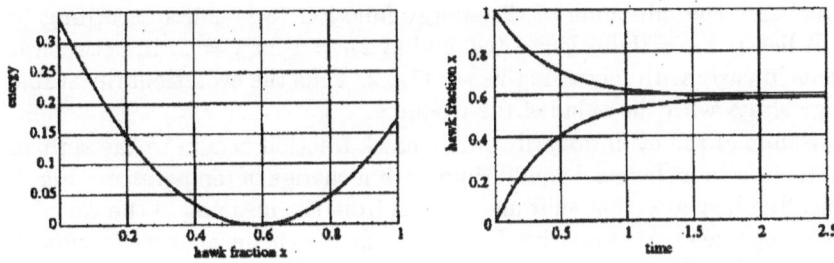

Fig. 1. Left: Energy as function of x_H. Right: Exponential growth and decay of the hawk fraction from initial values $x_H = 0$ and $x_H = 1$, respectively, toward the stable fraction $x_H = \frac{7}{12}$ (α is arbitrarily chosen as $\frac{1}{25}$).

3   Variable Value of Resource – a Chemical Reaction

We proceed to consider the case when the hawk fraction may change in the population as a result of an evolutionary relaxation into an energetic minimum. However, we make the value of the resource variable and we denote it by $50V$. Thus the contest strategy is the same as before, but the penalties and rewards change. We describe the strategy as follows.

A hawk fights for the resource; a dove merely engages in a symbolic conflict, posturing and threatening but not actually fighting. Between two hawks there will always be a fight until one is injured. The winner will get the resource worth 50V points, while the loser will get -100 points for being injured. Over many fights each hawk will thus "win" $\frac{1}{2}(50V-100)$ points per fight. If a hawk meets a dove, the hawk will always win the resource and there will be no injury. If two doves meet, they will spend a long time posturing. One will eventually win the resource, but they will both get -10 points for wasted time: In the mean an encounter will thus gain a dove $\frac{1}{2}(50V-20)$ points.

Note that for $V = 1$ we recover the old Maynard Smith-Price-strategy.

The variable value V of the resource will be reflected in the energy of the population which now reads

$$F(x_H) = 30 \left(x_H - \frac{[25V + 10]}{60} \right)^2 + C; \qquad (3.1)$$

this expression is derived in a manner analogous to the procedure described in Sect. 2.

Obviously the minimum of the energy function (3.1) shifts according to (3.1). It lies at $x_H = 0$ for $V = -0.4$ and at $x_H = 1$ for $V = 2$. In between it proceeds linearly with increasing V, see Fig. 2. Thus the evolutionarily stable strategy shifts with the value of the resource.

The shift of the evolutionarily stable hawk fraction is akin to the shift of concentration in a chemical equilibrium with a change of temperature. Fig. 2 graphically illustrates that shift and – apart from the meaning of the variable x – we recognize great similarity. It is true that the chemical curve is smooth while the sociobiological one has corners. This is due to the fact that we have ignored the entropic effect in our calculation. Entropy tends to smooth out corners.

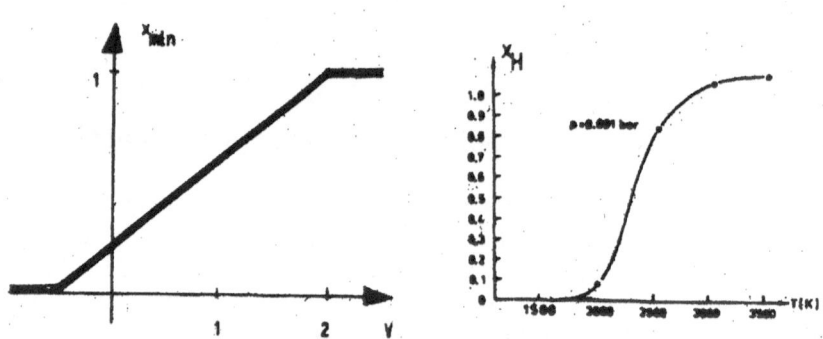

Fig. 2. Left: Evolutionarily stable hawk fraction x_H as a function of value V of resource. Right: Chemical equilibrium concentration x_H in the dissociation $H^2 \rightarrow 2H$ as a function of temperature.

4   Two Strategies – a Phase Transition

Evolution is slow and, if a population starts out with a particular value of x_H, it may take a long time – many generations – before the hawk fraction assumes the value dictated by the evolutionarily stable strategy, i.e. the energetic minimum. Therefore we proceed to discuss the case when x_H is constant in a population and arbitrary, either because the population is newly created or because of a recent immigration of hawks (say).

For anything interesting to occur in the case of a fixed hawk fraction we must have a population which is capable of choosing between two strategies, and which will choose on the basis of which strategy is energetically more favourable.

Therefore we formulate two strategies: A, which is the value-dependent strategy of Sect. 3, and B which is new. The strategies are defined as follows.

A: A hawk fights for the resource; a dove merely engages in a symbolic conflict, posturing and threatening but not actually fighting. Between two hawks there will always be a fight until one is injured. The winner will get the resource worth 50V points, while the loser will get -100 points for being injured. Over many fights each hawk will thus "win" $\frac{1}{2}(50V-100)$ points per fight. If a hawk meets a dove, the hawk will always win the resource and there will be no injury. If two doves meet, they will spend a long time posturing. One will eventually win the resource, but they will both get -10 points for wasted time: In the mean an encounter will thus gain a dove $\frac{1}{2}(50V-20)$ points.

B: A hawk still fights for the sole possession of the resource, again valid 50V points. But the doves will not timidly stand by to let the hawks have their way; they will try to steal the resource from the hawks and we assume that four out of ten times they are successful. But, successful or not, they risk injury with a penalty of -100 points. Therefore a hawk earns $\frac{1}{2}(50V-100)$ points in the mean, if it meets another hawk and $\frac{1}{10}6 \cdot 50V$, if it meets a dove. The contest between doves again consists of posturing and threatening with the same reward and penalty as before. Thus a dove-dove encounter will bring $\frac{1}{2}(50V-20)$ in the mean and a dove-hawk encounter will average $\frac{1}{10}4 \cdot 50V-100$ for the dove.

The value V of the resource may well change much more rapidly than the changes in the population fraction induced by evolution. Indeed, during a single generation – when we may consider x_H to be constant – the value V may vary considerably. We proceed to investigate the effect of such a variation of V on the population.·

The energies associated with the strategies A and B may be calculated as before in Sects. 2 and 3 and they read (for simplicity in writing x will now be the hawk fraction instead of x_H)

$$F^A = 30 \left(x - \frac{[25V + 10]}{60} \right)^2 + C^A \,,$$

$$F^B = -20 \left(x + \frac{[5V + 10]}{40} \right)^2 + C^B \,.$$

(4.1)

The effect of a change of V on the population is determined by the energies F^A and F^B, because — roughly speaking — we expect the population to establish the phase with the lower energy.

On the other hand, we expect the phase A with its timid doves to prevail when the food supply is abundant so that V is small. This means that F^A should be smaller than F^B for low values of V, and we make sure of this by choosing the constants C^A and C^B appropriately, viz.

$$C^A = 0, \qquad C^B = 50.$$

(4.2)

Thus we can now plot the energies F^A and F^B with V as a parameter from (4.1). Fig. 3 shows these curves for some values V between $V = 1$ and $V = 3.5$. The convex curves represent F^A, while the concave ones represent F^B.

Note that for $V = 2.5$ the two curves $F^A(x)$ and $F^B(x)$ begin to intersect each other. Such intersections continue to occur until about the value $V = 3$ whose functions $F^A(x)$ and $F^B(x)$ just barely touch. For $V = 2.75$ there are clear intersections in two points, at about $x \approx 0.19$ and $x \approx 0.92$.

We proceed further on the assumption that a population for a given — and constant — hawk fraction x will assume the strategy A or B depending on the respective energies. Thus it might seem reasonable to assume that, if $F^A < F^B$ holds, the population will employ strategy A, while for $F^B < F^A$ it will employ B.

Thus from inspection of Fig. 3 it would appear that, for $V = 2.75$, the population employs strategy B, if $x \lesssim 0.19$ holds, and again for $x \gtrsim 0.92$. In between it should use strategy A, and in all cases hawks and doves would form a homogeneous mixture, an integrated population.

However, nature has a more interesting alternative. At least the inanimate *solution* of two fluids in the liquid-vapour equilibrium has such an alternative and we assume that this alternative is also available to the *population* of hawks and doves in equilibrium between A and B. The alternative is that — in the language of populations rather than solutions – the population becomes non-homogeneous, i.e. it falls apart into colonies of different hawk fractions.

The reason for the decomposition into colonies is best explained by referring to Fig. 4. That figure shows the graphs of $F^A(x)$ and $F^B(x)$ for $V = 2.70$ which intersect twice. The straight dashed lines in the figure are tangent to $F^A(x)$ and they connect the point of contact of the tangents with the endpoints of $F^B(x)$ at $x = 0$ and $x = 1$. Along these tangents the energy of a

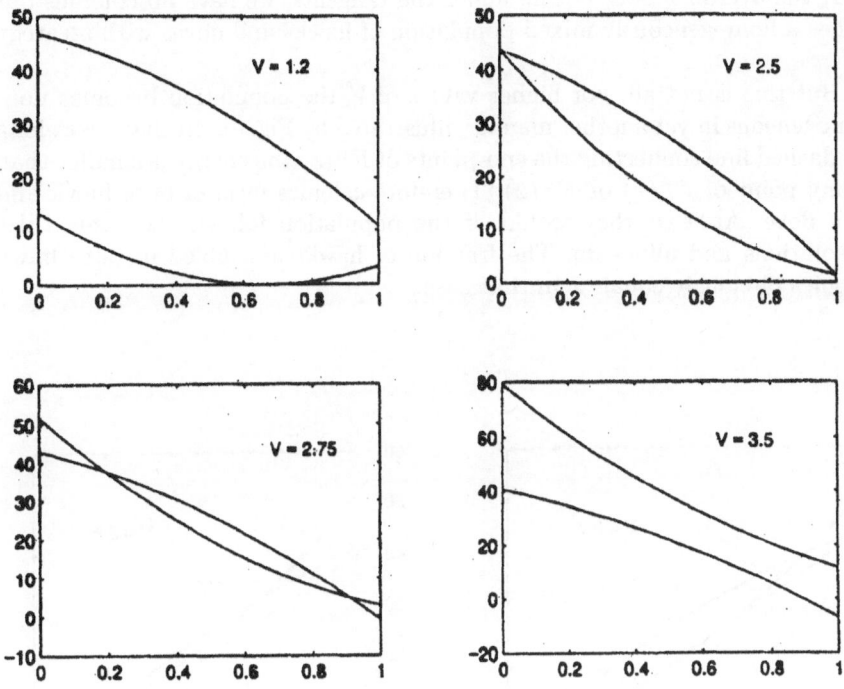

Fig. 3. Energies of the population with strategy A (convex) and strategy B (concave) for different values V

non-homogeneous population is lower than the energy of the homogeneous integrated population with the overall fraction x. Therefore colonies form for those values V for which tangents of the type shown in Fig. 4a occur – with compositions appropriate to the end-points of the tangents. At least that is the assumption; it is motivated by what is indeed observed in solutions and alloys.

In the case where the overall composition x is under the left tangent of Fig. 4a the colonies are pure dove and a hawk-dove mixture with strategy A with hawk fraction x_C^1. The fraction of doves assembled in pure dove colonies is $\frac{x_C^1 - x}{x_C^1}$; obviously it depends on V.

In the case where the overall composition x lies below the right tangent of Fig. 4a the colonies are pure hawk and a hawk-dove mixture with strategy A with the hawk fraction x_C^2. The fraction of hawks assembled in pure hawk colonies is $\frac{x - x_C^2}{1 - x_C^2}$.

If the overall x does not lie under the tangents, we have no colonies but rather a homogeneously mixed population of hawks and doves with strategy A.

But this is not all. For higher values of V the population becomes non-homogeneous in yet another manner, illustrated by Fig. 4b. In that case along the dashed line connecting the end-points of $F^B(x)$ the energy is smaller than in any point of $F^A(x)$ or $F^B(x)$. Therefore colonies form of pure hawk and pure dove. At least they would, if the population follows the same rules as solutions and alloys do. The fraction of hawks assembled in pure hawk colonies obviously equals x in that case.

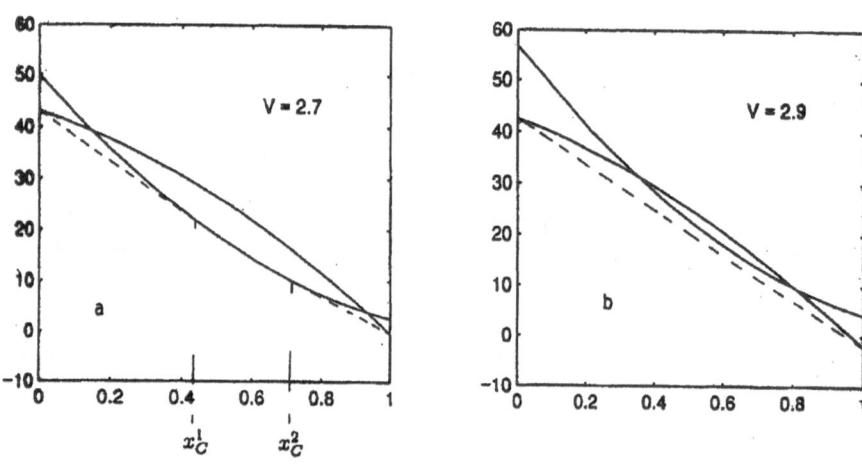

Fig. 4. Energies $F^A(x)$ and $F^B(x)$ and the dashed lines of equilibrium between colonies.   **a.** $V = 2.70$. Colonies consist of pure hawk , or pure dove, and hawk-dove mixture with a hawk fraction x_C **b.** $V = 2.90$. Colonies consist of pure hawk and pure dove.

5   Phase Diagram with Eutectic Point and Miscibility Gap

It is obvious from the construction of the dashed lines in Fig. 4 that the values x_C depend on V. In fact, since the energies $F^A(x)$ and $F^B(x)$ are parabolic, it is easy to calculate the points of contact. Table 5.1 presents the results for some values V in the interesting range.

V	x_C^1	x_C^2
2.48996	–	1
2.49605	0	0.949
2.5	0.066	0.934
2.6	0.341	0.777
2.7	0.481	0.686
2.8	0.593	0.611

Table 5.1 Left and right contact points of tangents
(see Fig. 4a) for different values V

Fig. 5a represents the graphs x_C^1 (V) and x_C^2 (V). They come to an end
where the two tangents coincide at $V_E \approx 0.278$. In analogy to the jargon
of chemical engineering we call V_E the *eutectic* value and the straight line
$V_E = const$ the eutectic line. Above that line no integrated population is
possible, rather the population falls apart into colonies of pure hawks and
pure doves. In chemical engineering the range above $V_E = const$ is called the
miscibility gap.

The word "eutectic" means "easy melting". Indeed the coordinates of
the eutectic point define the hawk fraction of the *highest* value of V that
will "melt" the colonies of pore hawks and pure doves and that creates an
integrated population.

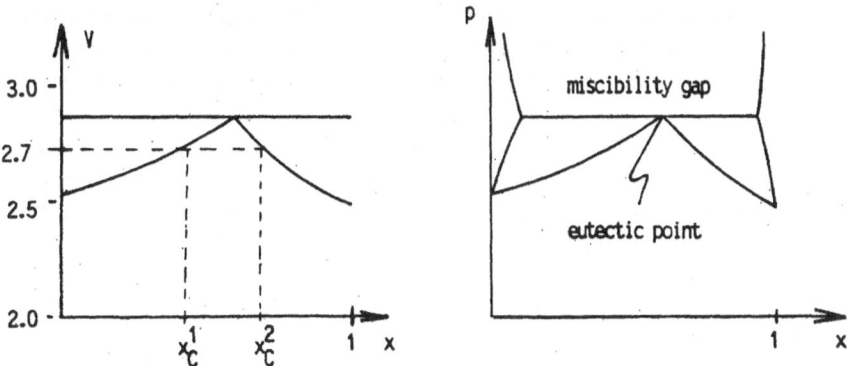

Fig. 5. Left: Phase diagram (V, x) for a population capable of changing strategies.
Right: Phase diagram (p, x) for a solution capable of a liquid-vapor phase change.

Fig. 5b represents a generic phase diagram of a solution[2] with eutectic point and miscibility gap; x is the concentration and p the pressure in that case. There is an obvious similarity between the two diagrams of Fig. 5, but also a noteworthy difference: Indeed the diagram for the solution has narrow lateral strips for $x \gtrsim 0$ and $x \lesssim 1$. In the range of high pressures, which is dominated by the miscibility gap, these strips represent a marginal solubility of one constituent in another one. The strips are missing from the diagram for the population because we have ignored the entropy of mixing $S_{Mix} = x \ln x - (1-x) \ln (1-x)$ in constructing the diagram. Entropic terms will be introduced in future work; their quantitative consideration requires a definition of temperature.

For those readers familiar with metallurgy or chemical engineering the situation described by Fig. 6 will be perfectly clear. To be sure, the variables are different but the phenomena are alike. What is V here, the value of the resource, is the pressure for a liquid solution in equilibrium with its vapour phase. And the hawk fraction of our case corresponds to the concentration, or mol-fraction, of the solution. The low-V-strategy A is the vapour phase in chemical engineering, and the high-V-strategy B is the liquid solution with completely immiscible components.

Still, not everybody may be familiar with thermodynamics of solutions, and for those who are not we describe what happens to a population of a fixed hawk fraction x^0 as the value V goes up, starting from a point in the low-V-, or A-phase, see Fig. 6.

At first nothing much occurs; the strategy A remains stable until the segregation line, denoted by x_C^1, is reached. When that is the case, a small colony of pure doves will precipitate, making the bulk of strategy A a little richer in hawks, so that V can be raised a little higher, before the segregation line is again reached with the same result as before: another pure-dove-colony forms and the hawk fraction increases in the remaining strategy A. This goes on; the state of the pure dove colonies creeps up on the V-axis, and the state of the remaining strategy A with integrated hawks and doves creeps up on the segregation line, as indicated by the arrows. For each value V the fraction of doves in colonies may be determined by the ratio of the lengths $x_C^1 - x^0$ and x_C^1. This goes on until the eutectic point is reached by the remaining phase A. When V is raised further, colonies of pure hawks and pure doves form from the remaining strategy A in the proportion $x_E/(1-x_E)$, where x_E is the abscissa of the eutectic point. The complete segregation of hawks and doves persists for all higher values of V, each group occupying their colonies. Each colony moves up vertically, the doves at $x = 0$ and the hawks at $x = 1$, as indicated by the arrows in Fig. 6.

We conclude that a rise of V will destroy an integrated population and lead to segregation.

---

[2] See any book on thermodynamics, e.g. [3].

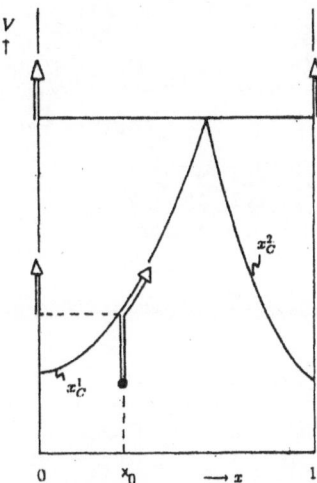

Fig. 6. Increasing V in an integrated population: Precipitation of pure-dove colonies for intermediate values of V, segregation in dove and hawk colonies for high values of V

6   Outlook

The research presented here will continue with the objective of transferring the well-known tenets of the thermodynamic theory of mixtures, solutions, and alloys into sociology and sociobiology. It is hoped that the successful conclusion of this effort will shed a new light on the "mechanisms" of sociobiological phenomena.

Finally we like to say that what has been argued here for a synthetic population of hawks and doves may well be extrapolated by a courageous sociologist to circumstances closer at home, e.g. integration versus segregation of human groups of different religious or ethnic backgrounds. Mimkes [4] is pursuing a similar, albeit more phenomenological path with stimulating and thought-provoking results.

References

1. Maynard-Smith, J., Price, G. R.: The Logic of Animal Conflict, Nature **246** (1973)
2. Straffin, P. D.: Game Theory and Strategy. New Mathematical Library. The Math. Assoc. of America, **36** (1993)
3. Müller, I.: Grundzüge der Thermodynamik. Springer, Heidelberg, 1998
4. Mimkes, J.: Binary Alloys as a Model for the Multicultural Society. Journal of Thermal Analysis **43** (1995)

Existence and Asymptotic Behaviour in Phase-Field Models with Hysteresis

P. Krejčí[1], J. Sprekels[2], and S. Zheng[3]

[1] Mathematical Institute, Academy of Sciences, Žitná 25, CZ–11567 Praha 1,
   Czech Republic
[2] Weierstrass Institute for Applied Mathematics and Stochastics, Mohrenstraße
   39, D–10117 Berlin, Germany
[3] Institute of Mathematics, Fudan University, Shanghai 200433, China

Dedicated to Professor Karl-Heinz Hoffmann
on the occasion of his 60th birthday

Abstract. Phase field systems as mathematical models for phase transitions have drawn increasing attention in recent years. However, while being capable of capturing many of the experimentally observed phenomena, they give only a simplified picture of intrinsic hysteresis effects occurring in phase transition processes. To overcome this shortcoming, the first two authors have recently proposed a new approach in a series of papers which is based on the mathematical theory of hysteresis operators developed in the past fifteen years, and obtained results on existence, uniqueness and regularity of solutions for a class of phase-field systems with hysteresis that includes among others the relaxed Stefan problem and hysteretic analogues of the models due to Caginalp and Penrose–Fife for nonconserved order parameters with zero interfacial energy. Here, we give a brief account of the method, with a special focus on new results obtained by the authors of the present paper and related to the asymptotic behaviour of the system as $t \to +\infty$.

1 Introduction

This paper is devoted to the study of initial-boundary value problems for systems of PDEs of the form

$$\mu(\theta)\, w_t \,+\, f_1[w] \,+\, f_2[w]\,\theta \,=\, 0\,, \tag{1.1}$$

$$(\theta \,+\, F_1[w])_t \,-\, \Delta\theta \,=\, \psi(x,t,\theta)\,, \tag{1.2}$$

in $\Omega \times (0,T)$, where Ω is some open bounded subset of $\mathbb{R}^N$ for some $N \geq 2$ with a Lipschitz boundary $\partial\Omega$, and $T > 0$ is some final time.

Systems of the form $(1.1-2)$ arise as *phase-field equations* from the mathematical study of *phase transitions* and have been studied repeatedly in the literature when f_1, f_2, F_1, ψ are (possibly nonlinear) smooth functions of their respective variables; also cases where (1.1) is replaced by an inclusion as in the so–called relaxed Stefan model have been under continuing study. We refer to the monographs [1] and [10] in this respect.

Models where f_1, f_2, F_1, ψ are *functions* or *graphs* are of restricted value in cases where, due to cycling loads, the phase transition may run in both directions. In such a situation usually hysteresis effects – like undercooling or overheating in a solid-liquid transition – occur. In a series of previous papers [5–7], the first two authors have proposed a new approach to incorporate the occurrence of hysteresis effects into the model by assuming that f_1, f_2, F_1 are hysteresis operators (for the notion of hysteresis operators see the monographs [1,3,4,9]) instead of functions. The possible occurrence of hysteresis effects is not the only reason to consider hysteresis operators in (1.1): in fact, it has been pointed out in [5–7] that already classical models as the relaxed Stefan model can be brought into the form (1.1) with suitable hysteresis operators f_1, f_2, F_1; in this connection, the quantity w can be interpreted as a sort of *time-integrated memory* that the system keeps with respect to changes of the *thermodynamic force* acting on the system. For details, we refer the reader to [5–7]. We also mention the recent paper [2] in which the heat flux has been assumed to contain a thermal memory term in addition to the classical Fourier form. A further direction of recent research has been the study of the asymptotic behaviour of the solutions to the system (1.1 - 2) as $t \to +\infty$. In [8], corresponding results have been established by the authors of the present paper.

In this paper, we will give an overview of the results established in [5–8]. We will address questions of existence, thermodynamic consistency, and asymptotic behaviour, outlining the key ideas of the corresponding proofs. For the details of the proofs, we refer the reader to the original papers [5–8].

2   Statement of the Problem

We consider the system of equations in $\Omega \times (0, T)$

$$\mu(\theta)w_t + f_1[w] + f_2[w]\,\theta = 0\,, \tag{2.1}$$

$$(\theta + F_1[w])_t - \Delta\theta = \psi(x, t, \theta)\,, \tag{2.2}$$

coupled with the initial conditions

$$w(x, 0) = w^0(x)\,, \quad \theta(x, 0) = \theta^0(x)\,, \quad \text{for } x \in \Omega\,, \tag{2.3}$$

and with the Neumann boundary condition

$$\frac{\partial\theta}{\partial n}(x, t) = 0 \quad \text{for } (x, t) \in \partial\Omega \times (0, T)\,, \tag{2.4}$$

where $\partial\theta/\partial n$ denotes the outward normal derivative of θ.

At the first glance, the system (2.1) – (2.4) does not seem to be very difficult from the mathematical point of view. In fact, if f_1, f_2, F_1 were real-valued functions having suitable properties (smoothness, monotonicity, and the like), then this would be true. However, in our case f_1, f_2, F_1 will be hysteresis operators and thus, in particular, non-smooth. Also, when dealing with these operators, we will always have to account for the full history of the inputs which makes the theory less obvious.

We now formulate precisely the assumptions on the mappings μ, f_1, f_2, F_1, ψ.

H1 The initial data are given in such a way that

$$\text{(i)} \quad w^0 \in L^\infty(\Omega), \theta^0 \in W^{1,2}(\Omega) \cap L^\infty(\Omega),$$
$$\text{(ii)} \quad \exists \delta > 0 : \theta^0(x) \geq \delta \quad \text{for a.e. } x \in \Omega. \tag{2.5}$$

H2 The function $\mu : (0, \infty) \to (0, \infty)$ is Lipschitz continuous on compact subsets of $(0, \infty)$, and either

$$\exists \mu_0 > 0 : \quad \mu(\theta) \geq \mu_0 \min\{\theta, 1\} \quad \forall \theta > 0, \tag{2.6}$$

or

$$\exists \mu_0 > 0 : \quad \mu(\theta) \geq \mu_0 \quad \forall \theta > 0. \tag{2.6*}$$

H3 The operators $f_1, f_2 : C[0,T] \to C[0,T]$ are causal, and there exists some $K_1 > 0$ such that

$$w_1, w_2 \in C[0,T] \Rightarrow |f_i[w_1](t) - f_i[w_2](t)| \leq K_1 |w_1 - w_2|_{[0,t]} \tag{2.7}$$
$$\forall t \in [0,T], \ i = 1, 2,$$

where for $z \in C[0,T]$ and $t \in [0,T]$ we denote

$$|z|_{[0,t]} := \max\{|z(\tau)| \ ; \ \tau \in [0,t]\}. \tag{2.8}$$

We moreover assume that either

$\exists \lambda : (0, \infty) \to (0, \infty)$ nondecreasing, with $\limsup\limits_{s \to \infty} \lambda(s)/s = 0$, such that

$$|f_2[w](t)| \leq \lambda(|w|_{[0,t]}) \quad \forall w \in C[0,T], \forall t \in [0,T], \tag{2.9}$$

or

$$\exists K_2 > 0 : \quad |f_i[w](t)| \leq K_2 \quad \forall w \in C[0,T],$$
$$\forall t \in [0,T], \ i = 1, 2. \tag{2.9*}$$

H4 The operator $F_1 : W^{1,1}(0,T) \to W^{1,1}(0,T)$ is causal, and it holds:

$$\exists K_3 > 0 : \quad |F_1[w]_t(t)| \leq K_3|w_t(t)|$$
$$\text{a.e.} \quad \forall w \in W^{1,1}(0,T), \tag{2.10}$$

$$\forall R > 0 \; \exists \Phi_R > 0 : w_1, w_2 \in W^{1,1}(0,T), \; |w_i|_{W^{1,1}(0,T)} \leq R, \; i = 1, 2,$$

$$\Rightarrow |F_1[w_1](t) - F_1[w_2](t)| \leq \Phi_R|w_1 - w_2|_{W^{1,1}(0,t)} \quad \forall t \in [0,T], \tag{2.11}$$

where for $z \in W^{1,1}(0,T)$ and $t \in [0,T]$ we denote

$$|z|_{W^{1,1}(0,t)} := |z(0)| + \int_0^t |\dot{z}(\tau)|d\tau . \tag{2.12}$$

H5 Let $q > r_N^2/(r_N-1)$ be a fixed number, where $r_N := \max\{2, 1+N/2\}$. We assume that $\psi : \Omega \times (0,T) \times \mathbb{R} \to \mathbb{R}$ is a measurable function such that

$$\exists \psi_0 \in L^q(\Omega \times (0,T)) : \quad \theta \leq 0 \Rightarrow \psi(x,t,\theta) = \psi_0(x,t), \tag{2.13}$$

$$\exists K_4 > 0 : \left|\frac{\partial \psi}{\partial \theta}\right| \leq K_4 \quad \text{a.e.}, \tag{2.14}$$

$$\psi_0(x,t) \geq 0 \quad \text{a.e.} \tag{2.15}$$

H6 There exist causal operators $F_2, g : W^{1,1}(0,T) \to W^{1,1}(0,T)$ and a constant $K_5 > 0$ such that for all $w \in W^{1,1}(0,T)$ we have

$$0 \leq g[w]_t \, w_t \leq K_5 \, w_t^2 \quad \text{a.e.}, \tag{2.16}$$

$$F_i[w]_t \leq g[w]_t \, f_i[w] \quad \text{a.e.}, \quad i = 1, 2, \tag{2.17}$$

$$F_1[w](t) \geq 0 \quad \forall t \in [0,T]. \tag{2.18}$$

Remark 2.1 Assumption (2.6) is for instance satisfied if $\mu(\theta) = \mu_0 \theta$, $\mu_0 > 0$ fixed. Then system (2.1) constitutes a hysteretic analogue of the Penrose-Fife model for phase transitions with zero interfacial energy (cf. [6]); on the other hand, (2.6*) is the hysteretic analogue of the *Caginalp* model with zero interfacial energy (see [5]). Note that also the intermediate models $\mu(\theta) = \mu_0 \theta^\alpha$, $0 < \alpha < 1$, $\mu_0 > 0$ are included in (2.6).

Remark 2.2 Under the assumptions above the system (2.1 - 2) is formally thermodynamically consistent provided that for every sufficiently regular solution (w, θ) it holds $\theta > 0$. Indeed, thermodynamic consistency requires in this situation (cf. [5,7]) that

$$\mathcal{U}_t - \theta \, \mathcal{S}_t \leq 0, \tag{2.19}$$

where $\mathcal{U}$ denotes internal energy and $\mathcal{S}$ entropy. However, with the free energy operator

$$\mathcal{F} = \theta\left(1 - \log(\theta)\right) + F_1[w] + F_2[w]\,\theta \qquad (2.20)$$

we obtain

$$\mathcal{U} = \theta + F_1[w]\,, \quad \mathcal{S} = \log(\theta) - F_2[w]\,, \qquad (2.21)$$

and thus, owing to (2.16), (2.17),

$$\mathcal{U}_t - \theta\,\mathcal{S}_t = F_1[w]_t + \theta\,F_2[w]_t$$
$$\leq g[w]_t\,(f_1[w] + \theta\,f_2[w])$$
$$\leq -\mu(\theta)\,g[w]_t\,w_t \leq 0\,.$$

We will see later that the positivity of temperature always holds under the above conditions which will justify the above formal computation.

Remark 2.3 Inequality (2.17), which is the key to verifying the thermodynamic consistency, is satisfied for operators of the general form

$$f_i[w] := \mathcal{P}_i\Big[g[w]\Big]\,, \quad F_i[w] := \mathcal{U}_i\Big[g[w]\Big]\,, \qquad (2.22)$$

where $\mathcal{P}_i$ is a hysteresis operator with a *clockwise admissible* hysteresis potential $\mathcal{U}_i$ in the sense of Sect. 2.5 in [1]. Note that in this case the *dissipation* over a closed cycle is positive and equal to the area enclosed by the traversed hysteresis loop (cf. [5–7]). A typical example is the so-called Prandtl-Ishlinskii operator

$$\mathcal{P}_i[w] := \int_0^\infty h_i(r)s_{Z_r}[w]\,dr\,, \quad \mathcal{U}_i[w] := \frac{1}{2}\int_0^\infty h_i(r)\,s_{Z_r}^2[w]\,dr\,, \qquad (2.23)$$

where s_{Z_r} is the so-called *stop operator* with characteristic set $Z_r = [-r, r]$, and where h_i is a given nonnegative density function. For details, we refer to [1,4] and to [5–7].

3 Main Results

In this section, we will state the main results established in the papers [5–8].

Theorem 3.1 (cf. [7]) *Suppose that the hypotheses* **H1** *to* **H6** *hold with either* (2.6) *and* (2.9*) *or* (2.6*) *and* (2.9)*. Then the system* (2.1 - 4) *has a unique solution* $(w, \theta) \in (L^\infty(\Omega \times (0, T)))^2$ *such that*

$$w_t \in L^\infty(\Omega \times (0, T))\,, \quad \theta_t, \Delta\theta \in L^2(\Omega \times (0, T))\,, \qquad (3.1)$$

and such that (1.1) *and* (1.2) *are satisfied a.e. in* $\Omega \times (0,T)$. $\qquad$ (3.2)

In addition, there exists a constant $\beta > 0$ *such that*

$$\theta(x,t) \geq \delta e^{-\beta t} \quad \text{a.e. in } \Omega \times (0,T). \qquad (3.3)$$

Remark 3.2 Note that (2.9*) requires f_2 to be bounded, which condition is coupled to the situation where $\mu(\theta)$ may tend to zero for $\theta \searrow 0$, while in the case that $\mu(\theta)$ is bounded away from zero a sublinear growth of f_2 with respect to the input w can be admitted.

Proof of Theorem 3.1. We outline the proof in the case when (2.6) and (2.9*) hold. For the detailed proof, the reader is referred to the original paper [7]. Assuming that (2.6) and (2.9*) hold, we introduce the auxiliary functions T_ε, $\mu_\varepsilon : \mathbb{R} \to \mathbb{R}^+$ by

$$T_\varepsilon(s) := \max\{\varepsilon, |s|\}, \quad \mu_\varepsilon(s) := \mu(T_\varepsilon(s)), \qquad (3.4)$$

for $s \in \mathbb{R}$. Let γ_ε be the operator

$$\gamma_\varepsilon[w,\theta] := -\frac{1}{\mu_\varepsilon(\theta)} \left(f_1[w] + T_\varepsilon(\theta) f_2[w] \right). \qquad (3.5)$$

We then consider the "cut off"-system

$$w_t = \gamma_\varepsilon[w,\theta], \qquad (3.6)$$

$$(\theta + F_1[w])_t - \Delta\theta = \psi(x,t,\theta). \qquad (3.7)$$

Using the general existence result of Theorem 3.1 in [7] for initial-boundary value problems for systems of the form (3.6 - 7), we can conclude that (3.6 - 7) admits a unique solution $(w^\varepsilon, \theta^\varepsilon)$ satisfying the initial and boundary conditions (2.3 - 4) such that $\theta^\varepsilon, w^\varepsilon, w_t^\varepsilon \in L^\infty(\Omega \times (0,T))$, as well as

$$\theta_t^\varepsilon, \ \Delta\theta^\varepsilon \in L^2(\Omega \times (0,T)).$$

We now test (3.7) with an arbitrary function $p \in W^{1,2}(\Omega \times (0,T))$ such that $p \leq 0$ almost everywhere. We then get, for a.e. $t \in (0,T)$, that

$$\int_\Omega \left(p\,\theta_t^\varepsilon + \langle \nabla p, \nabla\theta^\varepsilon \rangle \right)(x,t)\,dx$$

$$= \int_\Omega p\Big(\psi_0(x,t) + \psi(x,t,\theta^\varepsilon) - \psi(x,t,0) \Big)\,dx + \int_\Omega \left(|p|\, F_1[w^\varepsilon]_t \right)(x,t)\,dx$$

$$\leq K_4 \int_\Omega \left(|p||\theta^\varepsilon| \right)(x,t)\,dx + \int_\Omega \left(|p|\, g[w^\varepsilon]_t\, f_1[w^\varepsilon] \right)(x,t)\,dx, \qquad (3.8)$$

where

$$\int_\Omega \left(|p| \, g[w^\varepsilon]_t \, f_1[w^\varepsilon] \right)(x, t) \, dx$$

$$= - \int_\Omega \left(|p| \, \frac{g[w^\varepsilon]_t}{w_t^\varepsilon} \, \frac{f_1[w^\varepsilon]}{\mu_\varepsilon(\theta^\varepsilon)} \right) \left(f_1[w^\varepsilon] + T_\varepsilon(\theta^\varepsilon) \, f_2[w^\varepsilon] \right)(x, t) \, dx \,.$$

(3.9)

To estimate the last integral, we first notice that for every $a, b, r \in \mathbb{R}$ we have

$$- a^2 - rab \leq \frac{1}{2} \left(\sqrt{1 + r^2} - 1 \right) \left(a^2 + b^2 \right)$$

$$\leq \frac{|r|}{2} \min \left\{ 1, |r| \right\} \left(a^2 + b^2 \right) \,.$$

(3.10)

Hence, by assumptions (2.16) and (2.6),

$$- \frac{g[w^\varepsilon]_t}{w_t^\varepsilon} \, \frac{f_1[w^\varepsilon]}{\mu_\varepsilon(\theta^\varepsilon)} \left(f_1[w^\varepsilon] + T_\varepsilon(\theta^\varepsilon) \, f_2[w^\varepsilon] \right)$$

$$\leq \frac{K_5}{2\mu_0} \left((f_1[w^\varepsilon])^2 + (f_2[w^\varepsilon])^2 \right) T_\varepsilon(\theta^\varepsilon) \,.$$

(3.11)

Combining inequalities (3.8), (3.9) and (3.11) with assumption (2.9*), we obtain that

$$\int_\Omega \left(p\,\theta_t^\varepsilon + \langle \nabla p, \nabla \theta^\varepsilon \rangle \right)(x, t) \, dx$$

$$\leq \left(K_4 + \frac{K_5 K_2^2}{\mu_0} \right) \int_\Omega \left(|p| \, T_\varepsilon(\theta^\varepsilon) \right)(x, t) \, dx \,.$$

(3.12)

Put $\beta := K_4 + K_5 \, K_2^2 / \mu_0$, $\varepsilon := \delta e^{-\beta T}$, and

$$p(x, t) := - \left(\delta e^{-\beta t} - \theta^\varepsilon(x, t) \right)^+ \quad \text{for } (x, t) \in \Omega \times (0, T). \quad (3.13)$$

Then it follows from inequality (3.12) that

$$\int_\Omega \left(p \, (p + \delta e^{-\beta t})_t + |\nabla p|^2 \right)(x, t) \, dx \leq$$

$$\beta \int_\Omega |p| \, (|p| + \delta e^{-\beta t})(x, t) \, dx \,.$$

(3.14)

This yields, in particular,

$$\frac{1}{2} \frac{d}{dt} \int_\Omega p^2(x, t) dx + \int_\Omega |\nabla p|^2(x, t) dx \leq \beta \int_\Omega p^2(x, t) dx \,, \quad (3.15)$$

whence, by Gronwall's lemma, $p \equiv 0$. Hence $\theta^\varepsilon(x, t) \geq \delta \, e^{-\beta t} > \varepsilon$ a.e., and, in particular, $T_\varepsilon(\theta^\varepsilon) = \theta^\varepsilon$, $\mu_\varepsilon(\theta^\varepsilon) = \mu(\theta^\varepsilon)$. We thus have proved that

$(w, \theta) = (w^\varepsilon, \theta^\varepsilon)$ is a solution satisfying the conditions of Theorem 3.1. Uniqueness follows from Theorem 3.1 in [7].   □

We now focus our attention on the asymptotic behaviour of the solution to the system (2.1 - 4) as $t \to +\infty$. To this end, we introduce the notations

$$C_{\text{loc}} := \{u : [0, \infty) \to \mathbb{R} ; \; u|_{[0,T]} \in C[0,T] \quad \forall T > 0\},$$

$$W_{\text{loc}}^{1,1} := \{u : [0, \infty) \to \mathbb{R} ; \; u|_{[0,T]} \in W^{1,1}(0,T) \quad \forall T > 0\},$$

$$L_{\Omega,\text{loc}}^p := \{u : \Omega \times (0, \infty) \to \mathbb{R} ; \; u|_{\Omega \times [0,T]} \in L^p(\Omega \times (0,T)) \quad \forall T > 0\}$$

$$\text{for } 1 \le p \le +\infty.$$

In addition, we have to impose somewhat more restrictive hypotheses on the quantities defining our system; we consider the hypotheses:

H2* The function $\mu : [0, \infty) \to (0, \infty)$ satisfies (2.6*) and is continuously differentiable on $[0, \infty)$.

H3* The operators f_1, f_2 map C_{loc} into C_{loc}, as well as $W_{\text{loc}}^{1,1}$ into $W_{\text{loc}}^{1,1}$, and there exist $K_1 > 0, K_2 > 0$ such that for $i = 1, 2$ and a.e. $t > 0$ it holds:

$$|f_i[w_1](t) - f_i[w_2](t)| \le K_1 |w_1 - w_2|_{[0,t]} \quad \forall w_1, w_2 \in C_{\text{loc}}, \quad (3.16)$$

$$\left| \frac{d}{dt} f_i[w] \right| \le K_1 \left| \frac{dw}{dt}(t) \right| \quad \forall w \in W_{\text{loc}}^{1,1}, \quad (3.17)$$

$$|f_i[w](t)| \le K_2. \quad (3.18)$$

H4* The operator F_1 maps $W_{\text{loc}}^{1,1}$ into itself and is causal, and the conditions (2.10) and (2.11) hold for every $T > 0$.

H5* It holds $\psi \equiv 0$.

H6* Let s denote the *stop operator* with characteristic $[0, 1]$ (for the definition of s see, for instance, [5–8] or [1,3,4,9]). We assume that there exists a causal operator $F_2 : W_{\text{loc}}^{1,1} \to W_{\text{loc}}^{1,1}$ such that for all $w \in W_{\text{loc}}^{1,1}$ we have

$$F_i[w]_t \le f_i[w] s[w]_t \quad \text{a.e.}, \quad i = 1, 2, \quad (3.19)$$

$$F_i[w] \ge 0 \quad \text{a.e.}, \quad i = 1, 2. \quad (3.20)$$

After these preliminaries, we can now state the main result concerning the asymptotic behaviour. To this end, note that under the conditions **H1** and **H2*** - **H6*** the system (2.1 - 4) admits a unique solution $(w, \theta) \in$

$L^\infty_{\Omega,\text{loc}} \times L^\infty_{\Omega,\text{loc}}$ such that (2.1) and (2.2) are satisfied almost everywhere on $\Omega \times (0,\infty)$ and such that $w_t \in L^\infty_{\Omega,\text{loc}}, \theta_t, \Delta\theta \in L^2_{\Omega,\text{loc}}$, as well as $\theta(x,t) \geq \delta e^{-\beta t}$ with some $\beta > 0$ which does not depend on t. Denoting by $\|\cdot\|$ the $L^2(\Omega)$-norm, we have the following result.

Theorem 3.3 (cf. [8]) *Suppose the hypotheses* **H1** *and* **H2*** *to* **H6*** *are satisfied. Then there is some constant* $\hat{C} > 0$ *such that*

$$\|\theta(t)\|_{L^\infty(\Omega)} \leq \hat{C}, \quad \|w_t(t)\|_{L^\infty(\Omega)} \leq \hat{C} \quad \text{for a. e. } t > 0. \tag{3.21}$$

In addition, for $t > 0$ *put*

$$E(t) := \frac{1}{2}(\|\nabla\theta(t)\|^2 + \|s[w]_t(t)\|^2). \tag{3.22}$$

Then we have

$$\int_0^\infty E(t)\, dt \leq \hat{C}, \tag{3.23}$$

and there exists a function $E_* : [0,\infty) \to [0,\infty)$ *such that*

$$E(t) = E_*(t) \quad \text{a. e.}, \quad \lim_{t\to\infty} E_*(t) = 0. \tag{3.24}$$

In particular, the function E *satisfies the condition*

$$\lim_{t\to\infty} \operatorname{sup\,ess}\{E(s)\, ; s > t\} = 0. \tag{3.25}$$

Proof. We only sketch the proof. For the details the reader is referred to the original paper [8]. In that follows, we denote by C_i, $i \in \mathbb{N}$, positive constants that may depend on the data of the problem but neither on x nor on t. At first, we integrate (2.2) over Ω to obtain that

$$\int_\Omega \theta(x,t)\, dx + \int_\Omega F_1[w](x,t)\, dx \leq C_1, \tag{3.26}$$

while (2.1) implies that

$$|w_t(x,t)| \leq C_2\,(1 + \theta(x,t)) \quad \text{a.e. in } \Omega \times (0,\infty). \tag{3.27}$$

Therefore,

$$|F_1[w]_t(x,t)| \leq C_3\,(1 + \theta(x,t)) \quad \text{a.e. in } \Omega \times (0,\infty). \tag{3.28}$$

We now use the following general result (cf. [8]):

Lemma 3.4 *Suppose that* $u^0 \in L^\infty(\Omega)$ *is given and that*

$$u \in L^\infty(0,T; L^\infty(\Omega)) \cap L^2(0,T; H^1(\Omega))$$

for any $T > 0$ is a solution to the problem

$$u_t - \Delta u = \Phi[u] \qquad in \; \Omega \times (0, \infty), \qquad (3.29)$$

$$u(x, 0) = u^0(x) \qquad in \; \Omega, \qquad (3.30)$$

$$\frac{\partial u}{\partial n} = 0 \qquad on \; \partial \Omega \times (0, \infty), \qquad (3.31)$$

where $\Phi : L^\infty_{\Omega, \mathrm{loc}} \to L^\infty_{\Omega, \mathrm{loc}}$ satisfies

$$|\Phi[u](x, t)| \leq A(1 + |u(x, t)|) \qquad a.e. \; in \; \Omega \times (0, \infty) \qquad (3.32)$$

for some constant $A > 0$. In addition, let

$$\int_\Omega |u(x, t)| \, dx \leq B \qquad for \; all \; t \geq 0 \qquad (3.33)$$

with some constant $B > 0$. Then there is some constant $R > 0$ such that

$$\|u(t)\|_{L^\infty(\Omega)} \leq R \qquad for \; a. \, e. \; t > 0. \qquad (3.34)$$

Using the above lemma, we conclude that (3.21) holds. Next, observe that $\lambda(x, t) := \log \theta(x, t)$ satisfies

$$\lambda_t - \Delta\lambda = \frac{1}{\theta}(\theta_t - \Delta\theta) + \left|\frac{\nabla\theta}{\theta}\right|^2, \qquad (3.35)$$

where, owing to (3.19) and (2.1),

$$\theta_t - \Delta\theta = -F_1[w]_t \geq -f_1[w] \, s[w]_t$$
$$\geq \mu(\theta) \, w_t \, s[w]_t + \theta \, f_2[w] \, s[w]_t. \qquad (3.36)$$

Now recall that for the stop operator it holds (cf. [1,3,4,9])

$$w_t \, s[w]_t = (s[w])_t^2 \qquad a.e. \qquad (3.37)$$

Hence, a.e. in $\Omega \times (0, \infty)$,

$$\theta_t - \Delta\theta \geq \mu(\theta) \, |s[w]_t|^2 + \theta \, F_2[w]_t, \qquad (3.38)$$

whence,

$$\lambda_t - \Delta\lambda \geq F_2[w]_t + \frac{\mu(\theta)}{\theta} |s[w]_t|^2 + \left|\frac{\nabla\theta}{\theta}\right|^2. \qquad (3.39)$$

Integrating (3.39) over $\Omega \times (0, t)$ for $t > 0$, and using (3.20), (3.21) and **H1**, we conclude that

$$\int_0^t \int_\Omega \left(\frac{\mu(\theta)}{\theta} |s[w]_t|^2 + \left|\frac{\nabla\theta}{\theta}\right|^2\right) dx \, d\tau \leq C_4. \qquad (3.40)$$

Therefore, invoking (2.6*) and the already proved inequality (3.21), we can infer that (3.23) holds.

Next, we multiply (2.2) by θ_t, integrate over Ω, and use (3.19), (3.37) and (3.21) to arrive at the estimate

$$||\theta_t(t)||^2 + \frac{d}{dt} ||\nabla\theta(t)||^2 \le C_5 \quad \text{a.e. in } (0,\infty). \tag{3.41}$$

Now, we differentiate the equation

$$w_t + \frac{1}{\mu(\theta)} f_1[w] + \frac{\theta}{\mu(\theta)} f_2[w] = 0$$

with respect to t, and then test with $s[w]_t$, to obtain that a.e. in $\Omega \times (0,\infty)$ we have

$$(w_{tt} \, s[w]_t)(x,t) \le C_6 \left(1 + |\theta_t(x,t)|\right), \tag{3.42}$$

whence, a.e. in $(0,\infty)$,

$$\int_\Omega (w_{tt} \, s[w]_t)(x,t) \, dx \le \frac{1}{2} ||\theta_t(t)||^2 + C_7. \tag{3.43}$$

Combining (3.41) with (3.43), we see that

$$\int_\Omega (w_{tt} \, s[w]_t)(x,t) \, dx + \frac{1}{2} \frac{d}{dt} ||\nabla\theta(t)||^2 \le C_8, \tag{3.44}$$

$$\text{for a.e. } t \in (0,\infty).$$

We notice (for a proof, see [8], where a more general case has been discussed) that the function

$$P(t) := \int_0^t \int_\Omega (w_{tt} \, s[w]_t)(x,\tau) \, dx \, d\tau - \frac{1}{2} ||s[w]_t(t)||^2$$

equals a nondecreasing function almost everywhere. From (3.44) it follows that the function E defined by (3.22) satisfies the inequality

$$P(t) - P(s) + E(t) - E(s) \le C_8 (t - s) \quad \text{for a.e. } t > s > 0, \tag{3.45}$$

hence there exists a nondecreasing function q_* such that $q_*(t) := C_8 \, t - E(t)$ a.e. Putting

$$E_*(t) := C_8 \, t - q_*(t) \qquad \forall t \ge 0, \tag{3.46}$$

we obtain from (3.23) that

$$\int_0^\infty E_*(t) \, dt \le \hat{C}. \tag{3.47}$$

By a lemma established in [8], we derive from (3.47), (3.46) that

$$\lim_{t\to\infty} E_*(t) = 0\,,$$

which concludes the proof of the assertion.   □

References

1. Brokate, M., Sprekels, J.: Hysteresis and Phase Transitions. Appl. Math. Sci. Vol. **121**, Springer-Verlag, New York, 1996
2. Gilardi, G., Krejčí, P., Sprekels, J.: Hysteresis in Phase-Field Models with Thermal Memory. Preprint WIAS No. **497** (1999), submitted
3. Krasnoselskii, M. A., Pokrovskii, A. V.: "Systems with Hysteresis". Springer-Verlag, Heidelberg, 1989
4. Krejčí, P.: Hysteresis, Convexity and Dissipation in Hyperbolic Equations. Gakuto Int. Series Math. Sci. & Appl. Vol. **8**, Gakkōtosho, Tokyo, 1996
5. Krejčí, P., Sprekels, J.: A Hysteresis Approach to Phase-Field Models. To appear in Nonlinear Anal.
6. Krejčí, P., Sprekels, J.: Hysteresis Operators in Phase-Field Models of Penrose-Fife Type. Appl. Math. **43** (1998), 207-222
7. Krejčí, P., Sprekels, J.: Phase-Field Models with Hysteresis. Preprint WIAS No. **458** (1998), submitted
8. Krejčí, P., Sprekels, J., Zheng, S.: Asymptotic Behaviour for a Phase-Field System with Hysteresis. In preparation
9. Visintin, A.: Differential Models of Hysteresis. Appl. Math. Sci. Vol. **111**, Springer-Verlag, New York, 1994
10. Visintin, A.: Models of Phase Transitions. Progress in Nonlinear Differential Equations and their Applications Vol. **28**, Birkhäuser, Basel, 1996

On a New Class of Nonlocal Unilateral Problems in Thermomechanics

J.-F. Rodrigues*

CMAF/Universidade de Lisboa, Av. Prof. Gama Pinto, 2, P-1649-003 Lisboa, Portugal

Dedicated to Professor Karl-Heinz Hoffmann
on the occasion of his 60th birthday

1 Introduction

In this study we consider the equilibrium of an elastic membrane over a rigid obstacle subjected to a temperature field strongly depending on the contact with the obstacle. This problem can be formulated as an unilateral problem coupled with a heat diffusion equation with a discontinuous function depending on the contact region. It corresponds to an interior free boundary problem in thermoelasticity of different type with respect to the boundary unilateral contact problem, or the thermal Signorini problem, considered in [2] with mollification of the discontinuous heat source.

Contrary to the thermal boundary obstacle problem, which is still an open problem, the interior one can be solved using, as a new application, the continuous dependence properties of the characteristic function of the coincidence set in the obstacle problem. This requires appropriate nondegeneracy conditions, as indicated in [7], in order to obtain existence and, in more particular situations, also uniqueness results.

After the formulation in the first section of the model problem for an elastic membrane with a thermal obstacle, we show in Sects. 2 and 3 that the known theory can yield the existence of at least one solution of the coupled problem, and provides a uniqueness criterion.

Considering the asymptotic situation corresponding to the arbitrary high conductivity of the heat in the membrane, which yields a uniformly constant temperature field, we are led to a new class of nonlocal unilateral problem in a domain $\Omega \subset \mathbb{R}^2$ of the following type

$$(*) \quad \begin{cases} u \geq \psi \,, \\ -\Delta u \geq f\left(\text{meas}\{u = \psi\}\right) \,, \\ (u - \psi)\left[\Delta u + f\left(\text{meas}\{u = \psi\}\right)\right] = 0 \,. \end{cases}$$

* This work was partially supported by the project PRAXIS/2/2.1/MAT/125/94.

In this case, the loading force depends on a nonlocal a priori unknown quantity – the area of the contact set between the membrane and the obstacle. This problem is considered in a slightly more general case in Sects. 5 and 6, where it can be obtained as a limit problem in which the force or the elasticity coefficient may depend on the temperature field, respectively. Similar results as those arising in reaction-diffusion systems as in [6] can also be obtained for time dependent problems by extending the techniques of [4].

2   An Elastic Membrane with a Thermal Obstacle

The classical obstacle problem consists of finding, at each point $x \in \Omega \subset \mathbb{R}^2$ of an open bounded domain of the plane, the vertical displacement $u = u(x)$ of a membrane loaded by a force $f = f(x)$, constrained to lie above a body, representing an obstacle $\psi = \psi(x)$, and with prescribed height g on the smooth boundary $\partial \Omega$.

The equilibrium position u, by the principle of minimum potential energy, will minimize the functional

$$E(u) = \frac{1}{2} \int_\Omega \alpha |\nabla u|^2 \, dx - \int_\Omega fu \, dx \qquad (2.1)$$

in the convex set of admissible displacements

$$\mathbb{K} = \left\{ v \in H^1(\Omega) \colon \; v \geq \psi \; \text{in} \; \Omega, \; v = g \; \text{on} \; \partial \Omega \right\}, \qquad (2.2)$$

where in (2.1) $\alpha > 0$ is an elastic coefficient and in (2.2) the Sobolev space $H^1(\Omega)$ means that the displacement, when it exists, has finite energy.

If the constraining body has a temperature $\tau = \tau(x)$ supposed to be different of the temperature of the air, normalized to be zero, the temperature of the membrane $\vartheta = \vartheta(x)$ is assumed to satisfy

$$\kappa \, \Delta \vartheta = \vartheta - \tau \, \chi_{\{u=\psi\}} \qquad \text{in} \; \Omega . \qquad (2.3)$$

Here the constant $\kappa > 0$ is proportional to the conductivity of the membrane and $\chi_{\{u=\psi\}}$ denotes the characteristic function of the coincidence set $\{x \in \Omega \colon u(x) = \psi(x)\}$:

$$\chi_{\{u=\psi\}}(x) = \begin{cases} 1 & \text{if} \; u(x) = \psi(x) , \\ 0 & \text{if} \; u(x) > \psi(x) . \end{cases} \qquad (2.4)$$

Then (2.3) means that the heat diffusion across the membrane is proportional to the difference of its temperature and the temperature at the exterior of the obstacle in the contact zone or of the air in its complement. For simplicity, we shall assume that $\psi < g$ on $\partial \Omega$ and there is no heat flux at the boundary, so that

$$\frac{\partial \vartheta}{\partial n} = 0 \quad \text{on} \; \partial \Omega . \qquad (2.5)$$

If we assume in the energy functional $\alpha = 1$ but the force may depend, in some controlled way, on the temperature field, i.e.

$$f = f(x, \vartheta(x)) , \qquad (2.6)$$

we obtain then a coupled free boundary problem for (u, ϑ). Indeed, it is well-known (see [5] or [7], for instance) that the minimum of (2.1) over (2.2) corresponds to a variational inequality which is formally equivalent to find u and the free boundary $\Phi = \partial\{u = \psi\} \subset \Omega$, such that

$$-\Delta u = f(\vartheta) \quad \text{in } \{u > \psi\} \subset \Omega , \qquad (2.7)$$

$$u = \psi \text{ and } \frac{\partial u}{\partial n} = \frac{\partial \psi}{\partial n} \quad \text{on } \Phi , \qquad (2.8)$$

$$u = g \quad \text{on } \partial\Omega . \qquad (2.9)$$

This problem is then coupled with the thermal problem (2.3)–(2.5). We observe that the right hand side term in (2.3) is related to the displacement u through the characteristic function (2.4) of the contact set. The continuous dependence of $\chi_{\{u=\psi\}}$ on the temperature is given through (2.7) and, in the general case, may be a delicate problem, since it may be ill-posed.

Under appropriate compatibility and non-degeneracy assumptions we shall consider first the membrane thermal problem with obstacle in the following variational form.

Find a pair $(u, \vartheta) \in \mathbb{K} \times H^1(\Omega)$, such that

$$\int_\Omega \nabla u \cdot \nabla(v - u)\, dx \geq \int_\Omega f(\vartheta)\,(v - u)\, dx , \qquad \forall v \in \mathbb{K} , \qquad (2.10)$$

$$\kappa \int_\Omega \nabla \vartheta \cdot \nabla \zeta\, dx + \int_\Omega \vartheta \zeta\, dx = \int_\Omega \tau \chi_{\{u=\psi\}} \zeta\, dx , \qquad \forall \zeta \in H^1(\Omega) . \qquad (2.11)$$

We shall require the following assumptions:

$$\kappa > 0 , \ \tau \in L^\infty(\Omega), \ m \leq \tau(x) \leq M \ \text{ for a.e. } x \in \Omega , \qquad (2.12)$$

$$\psi \in W^{2,p}(\Omega), \ g \in W^{2-1/p,p}(\partial\Omega), \ \psi(x) < g(x), \ \forall x \in \partial\Omega , \qquad (2.13)$$

for some $p > 1$, where $W^{2-1/p,p}(\partial\Omega)$ denotes the trace space for the Sobolev space $W^{2,p}(\Omega)$. Notice that $W^{2,p}(\Omega) \subset C^{r-1}(\bar\Omega)$ for $p > r = 1, 2$ when the space dimension is two. We also assume $m < M$.

We assume $f(x, \theta) : \Omega \times \mathbb{R} \to \mathbb{R}$ is a Carathéodory function, i.e., continuous in the variable θ for a.e. $x \in \Omega$, measurable in x for all θ and we require

$$\sup_{\theta \in [m_-, M_+]} |f(x, \theta)| \leq C \quad \text{a.e. } x \in \Omega , \qquad (2.14)$$

where $m_- = 0$ if $m \geq 0$, $m_- = m$ if $m < 0$, $M_+ = 0$ if $M \leq 0$ and $M_+ = M$ if $M > 0$.

For the existence result we assume the following weak nondegeneracy condition

$$f(x, \theta) + \Delta \psi(x) \neq 0 \quad \text{a.e.} \quad x \in \Omega, \quad \forall \theta \in [m_-, M_+] . \tag{2.15}$$

Theorem 2.1. *Under the above assumptions, namely (2.12)–(2.15), there exists at least one solution (u, ϑ) to the coupled thermal obstacle problem (2.10)–(2.11), such that*

$$\begin{aligned} &u \in W^{2,p}(\Omega) \cap C^0(\bar{\Omega}), \ p > 1, \quad and \\ &\vartheta \in W^{2,q}(\Omega) \cap C^1(\bar{\Omega}), \ \forall q < \infty . \end{aligned} \tag{2.16}$$

To obtain a uniqueness result, we shall require a stronger nondegeneracy assumption, supposing the existence of a constant $\lambda > 0$, such that

$$f(x, \theta) + \Delta \psi(x) \leq -\lambda < 0 \quad \text{a.e.} \quad x \in \Omega, \quad \forall \theta \in [m_-, M_+] \tag{2.17}$$

and that the nonlinearity of f in θ is locally Lipschitz continuous, i.e.

$$\big| f(x, \theta_1) - f(x, \theta_2) \big| \leq L |\theta_1 - \theta_2| \tag{2.18}$$

for a.e. $x \in \Omega$ and for all $\theta_1, \theta_2 \in [m_-, M_+]$, with $L > 0$.

Theorem 2.2. *If, in addition to the assumptions of Theorem 2.1, we assume (2.17), (2.18) with the restriction*

$$\mu L < \lambda \tag{2.19}$$

where $\mu = \max(M_+, -m_-) = \|\tau\|_{L^\infty(\Omega)} > 0$, then the solution (u, ϑ) to (2.10)–(2.11) is unique.

3   Existence of a Solution (Proof of Theorem 2.1)

The proof of the existence result may be done with an appropriate construction of a nonlinear continuous and compact mapping $\mathcal{S}$, for instance, in the following closed convex subset of $L^2(\Omega)$:

$$\mathcal{C} = \Big\{ \sigma \in L^2(\Omega): \ m_- \leq \sigma(x) \leq M_+ \ \text{a.e.} \ x \in \Omega \Big\}, \tag{3.1}$$

where $m_- \leq 0 \leq M_+$ are the constants introduced in (2.14).

In fact, for any $\sigma \in \mathcal{C}$, we solve the obstacle problem

$$u_\sigma \in \mathbb{K}: \int_\Omega \nabla u_\sigma \cdot \nabla (v - u_\sigma) \, dx \geq \int_\Omega f_\sigma (v - u) \, dx, \quad \forall v \in \mathbb{K}, \tag{3.2}$$

with $f_\sigma = f(\sigma) \in L^\infty(\Omega)$, for which the general theory is well established (see, for instance, [5], [3] or [7]). In particular, by the assumptions (2.13) and (2.14), the mapping

$$\mathcal{C} \ni \sigma \longmapsto u_\sigma \in \mathbb{K} \cap W^{2,p}(\Omega)$$

is continuous for the strong topologies (see Theorem 5:4.5 of [7]).

However, this is not enough for our purposes and we need a stronger continuous dependence property. In fact, we have, for every $q < \infty$:

$$L^2(\Omega) \supset \mathcal{C} \ni \sigma \longmapsto \chi_{\{u_\sigma=\psi\}} \in L^q(\Omega) , \qquad (3.3)$$

is a continuous mapping for the strong topologies, which is a direct consequence of the following auxiliary result.

Lemma 3.1. *(Continuous dependence of the characteristic functions). Let* $\chi_j = \chi_{\{u_j=\psi\}}$ *be given by (2.4) for the solution* u_j *of (3.2) corresponding to a sequence of forces* $f_j \in L^p(\Omega)$, $p > 1$. *Then*

$$f_j \underset{j}{\to} f \quad in \ L^p(\Omega) \quad implies \quad \chi_j \underset{j}{\to} \chi \quad in \ L^q(\Omega), \ \forall q < \infty , \qquad (3.4)$$

provided the limit force f *is such that*

$$f(x) + \Delta\psi(x) \neq 0 \quad a.e. \ in \ \Omega . \qquad (3.5)$$

Proof. We simplify the argument of Theorem 6:6.1 of [7], since by the regularity properties of u_j and u, the solutions of (3.2) corresponding to f_j and f, respectively, satisfy the equations (see Theorem 5:4.3 of [7])

$$- \Delta u_j + (f_j + \Delta\psi) \chi_j = f_j \quad a.e. \ in \ \Omega \qquad (3.6)$$

and

$$- \Delta u + (f + \Delta\psi) \chi = f \quad a.e. \ in \ \Omega . \qquad (3.7)$$

Since $0 \leq \chi_j \leq 1$, there exists a subsequence and a function η, $0 \leq \eta \leq 1$, a.e. in Ω, such that

$$\chi_j \underset{j}{\rightharpoonup} \eta \quad in \ L^q(\Omega)\text{-weak}, \ \forall q < \infty .$$

Taking this limit in the sense of distributions in (3.6), we obtain

$$-\Delta u + (f + \Delta\psi)\eta = f \quad a.e. \ in \ \Omega ,$$

which compared with (3.7), yields

$$(f + \Delta\psi)(\eta - \chi) = 0 \quad a.e. \ in \ \Omega$$

and, by the assumption (3.5), implies

$$\eta = \chi \quad \text{a.e. in } \Omega .$$

Then, since χ_j and χ are characteristic functions

$$\int_\Omega \chi_j^q \, dx = \int_\Omega \chi_j \, dx \xrightarrow{j} \int_\Omega \chi \, dx = \int_\Omega \chi^q \, dx$$

what implies their strong convergence in $L^q(\Omega)$, proving (3.4). $\qquad\square$

Now we may consider the linear Neumann problem

$$\theta \in H^1(\Omega): \kappa \int_\Omega \nabla\theta \cdot \nabla\xi \, dx + \int_\Omega \theta\,\xi \, dx = \int_\Omega \tau\chi\xi \, dx \ , \quad \forall \xi \in H^1(\Omega), \qquad (3.8)$$

with $\chi = \chi_{\{u_\sigma = \psi\}}$, for which the following lemma holds in general.

Lemma 3.2. *Let (2.12) hold and $m_- < M_+$ be the constants introduced after (2.14). For any $\chi \in L^\infty(\Omega)$, such that $0 \le \chi \le 1$ a.e. in Ω, there exists a unique solution θ of (3.8). Moreover θ satisfies the estimates*

$$\|\theta\|_{L^2(\Omega)} \le \|\tau\|_{L^\infty(\Omega)} \|\chi\|_{L^2(\Omega)} , \qquad (3.9)$$

$$\|\theta\|_{W^{2,q}(\Omega)} \le C_{\kappa,q} \|\chi\|_{L^q(\Omega)}, \quad \forall q < \infty \qquad (3.10)$$

$$m_- \le \theta(x) \le M_+ , \quad \forall x \in \bar\Omega . \qquad (3.11)$$

Proof. This lemma is an immediate consequence of the linear theory of elliptic equations. Indeed (3.9) follows by taking $\xi = \theta$ in (3.8) and using Schwartz inequality, while (3.10) is a consequence of the well-known L^q regularity. Finally (3.11) is a consequence of the weak maximum principle: since (2.12) implies

$$m_- \le m\chi \le \tau\chi \le M\chi \le M_+ \quad \text{a.e. in } \Omega ,$$

taking $\xi = (\theta - M_+)^+ = \max(\theta - M_+, 0)$ in (3.8), we obtain

$$\int_{\{\theta > M_+\}} (\tau\chi - \theta)\,(\theta - M_+)\, dx = \kappa \int_\Omega \nabla\theta \cdot \nabla(\theta - M_+)^+ \, dx$$

$$= \kappa \int_\Omega \left|\nabla(\theta - M_+)^+\right|^2 dx$$

which implies meas$\{\theta > M_+\} = 0$, i.e. $\theta \le M_+$; analogously by taking $\xi = (\theta - m_-)^-$ in (3.8) we conclude $\theta \ge m_-$. $\qquad\square$

In order to conclude the proof of Theorem 2.1 is sufficient to consider θ_σ, the solution of (3.8) corresponding to $\chi = \chi_{\{u_\sigma = \psi\}}$ and observe that the so constructed mapping

$$\mathcal{S}: \mathcal{C} \to L^q(\Omega) \to \mathcal{C} \cap W^{2,q}(\Omega)$$

$$\sigma \mapsto \chi_{\{u_\sigma = \psi\}} \to \theta_\sigma$$

is not only continuous but also compact as a mapping of $\mathcal{C} \subset L^2(\Omega)$ into itself, due to the compactness of $W^{2,q}(\Omega) \subset L^2(\Omega)$.

Hence by the Schauder fixed point theorem, there exists at least a fixed point ϑ, i.e.

$$\vartheta = S(\vartheta) \in \mathcal{C} \cap W^{2,q}(\Omega)$$

and the pair (u_ϑ, ϑ) solves the problem (2.10)–(2.11).

4   Proof of the Uniqueness Theorem 2.2

We start by recalling the following property of the linear Neumann problem (3.8):

$$\|\theta\|_{L^1(\Omega)} \leq \mu \, \|\chi\|_{L^1(\Omega)} \tag{4.1}$$

where we recall $\mu = \|\tau\|_{L^\infty(\Omega)} > 0$.

In order to show this property we consider the function

$$s_\delta(w) = \begin{cases} 1 & \text{if } w \geq \delta , \\ \dfrac{w}{\delta} & \text{if } |w| \leq \delta , \\ -1 & \text{if } w \leq -\delta , \end{cases}$$

for $\delta > 0$, and take $\xi = s_\delta(\theta)$ in (3.8). Since

$$\int_\Omega \nabla\theta \cdot \nabla s_\delta(\theta) \, dx = \frac{1}{\delta} \int_{\{|\theta|<\delta\}} |\nabla\theta|^2 \, dx \geq 0 ,$$

we obtain

$$\int_\Omega \theta \, s_\delta(\theta) \, dx \leq \int_\Omega \tau \chi \, s_\delta(\theta) \, dx \leq \mu \, \|\chi\|_{L^1(\Omega)} . \tag{4.2}$$

Letting $\delta \to 0$, since $\theta \, s_\delta(\theta) \to |\theta|$ a.e., by Lebesgue theorem, we have

$$\int_\Omega \theta \, s_\delta(\theta) \, dx \longrightarrow \int_\Omega |\theta| \, dx$$

and (4.2) yields the estimate (4.1).

From Chapter 5 of [7], we recall, in an appropriate form to this problem, the strong continuous dependence globally in $L^1(\Omega)$ of the characteristic functions of the coincidence sets, under the more restrictive condition for two different forces

$$f_i(x) + \Delta\psi(x) \leq -\lambda < 0 \quad \text{a.e. in } x \in \Omega, \quad i = 1, 2 , \tag{4.3}$$

for the same obstacle ψ and the same Dirichlet boundary condition satisfying the assumption (2.13).

Lemma 4.1. *Let (2.13) and (4.3) hold. If* $\chi_i = \chi_{\{u_i=\psi\}}$, $i = 1, 2$, *denote the characteristic function associated to the solution* u_i *of the variational inequality (3.2) corresponding to* $f_i \in L^p(\Omega)$, *then the following estimate holds*

$$\|\chi_1 - \chi_2\|_{L^1(\Omega)} \leq \frac{1}{\lambda} \|f_1 - f_2\|_{L^1(\Omega)} . \tag{4.4}$$

Proof. Since $u_i \in W^{2,p}(\Omega)$, by the regularity theory for the obstacle problem (see Theorem 5:4.3 of [7]), u_i is such that

$$- \Delta u_i + \zeta_i = f_i \quad \text{a.e. in } \Omega \tag{4.5}$$

with

$$\zeta_i = -(f_i + \Delta\psi)^- \chi_i = (f_i + \Delta\psi)\chi_i \in L^p(\Omega) . \tag{4.6}$$

Let $\text{sign}(w)$ denote the sign function ($\text{sign}(0) = 0$) and Σ denote the monotone graph corresponding to it, i.e.

$$\Sigma(w) = \begin{cases} 1 & \text{if } w > 0 , \\ [-1, 1] & \text{if } w = 0 , \\ -1 & \text{if } w < 0 . \end{cases}$$

Since u_1 and u_2 take the same values on $\partial\Omega$ it is not difficult to show that

$$0 \geq \int_\Omega \text{sign}(u_1 - u_2) \, \Delta(u_1 - u_2) \, dx = \int_\Omega s \, \Delta(u_1 - u_2) \, dx \tag{4.7}$$

for any measurable function $s = s(x)$ such that $s \in \Sigma(u_1 - u_2)$ a.e. $x \in \Omega$. In particular, we may choose (see [7] and its references)

$$s(x) = \begin{cases} -1 & \text{on } \{u_1 < u_2\} \cup \{\zeta_1 < \zeta_2\} , \\ 0 & \text{on } \{u_1 = u_2\} \cap \{\zeta_1 = \zeta_2\} \\ 1 & \text{on } \{u_1 > u_2\} \cup \{\zeta_1 > \zeta_2\} \end{cases}$$

and, recalling (4.5), we obtain

$$\begin{aligned} \int_\Omega |\zeta_1 - \zeta_2| \, dx &= \int_\Omega (\zeta_1 - \zeta_2) \, s \, dx \\ &= \int_\Omega [(f_1 - f_2) + \Delta(u_1 - u_2)] \, s \, dx \\ &\leq \int_\Omega |f_1 - f_2| \, dx . \end{aligned} \tag{4.8}$$

On the other hand, by (4.6) and the assumptions (4.3), we have the pointwise inequality

$$|\zeta_1 - \zeta_2| = |(f_1 + \Delta\psi)\chi_1 - (f_2 + \Delta\psi)\chi_2| \geq \lambda|\chi_1 - \chi_2| \quad \text{a.e. in } \Omega,$$

which combined with (4.8) yields the conclusion (4.4).   □

Now we observe that, if (u_1, ϑ_1) and (u_2, ϑ_2) denote two solutions to (2.10)–(2.11), by setting $\chi_1 = \chi_{\{u_1=\psi\}}$ and $\chi_2 = \chi_{\{u_2=\psi\}}$, combining the estimate (4.1) with (4.4) and recalling the hypothesis (2.18) we obtain

$$\|\vartheta_1 - \vartheta_2\|_{L^1(\Omega)} \leq \mu\|\chi_1 - \chi_2\|_{L^1(\Omega)}$$

$$\leq \frac{\mu}{\lambda}\|f(\vartheta_1) - f(\vartheta_2)\|_{L^1(\Omega)}$$

$$\leq \frac{\mu L}{\lambda}\|\vartheta_1 - \vartheta_2\|_{L^1(\Omega)}.$$

But this is impossible if (2.19) holds, unless $\vartheta_1 = \vartheta_2$. This proves the uniqueness of the solution.

5   The Nonlocal Problem with High Conductivity: $\kappa \to \infty$

In the sequel we denote by $(u_\kappa, \vartheta_\kappa)$ any solution to (2.10)–(2.11) corresponding to a fixed κ, $0 < \kappa < \infty$. In order to analyse the asymptotic behaviour as $\kappa \to \infty$, we observe that, by the maximum principle (see Lemma 3.2), we have

$$m_- \leq \vartheta_\kappa(x) \leq M_+, \quad \forall x \in \bar{\Omega}. \tag{5.1}$$

Taking $\xi = \vartheta_\kappa$ in (2.11), we easily obtain

$$\kappa\int_\Omega |\nabla\vartheta_\kappa|^2\,dx + \frac{1}{2}\int_\Omega |\vartheta_\kappa|^2\,dx \leq \frac{1}{2}\int_\Omega \tau^2\,\chi_{\{u_\kappa=\psi\}}\,dx,$$

which implies the estimate

$$\|\nabla\vartheta_\kappa\|_{L^2(\Omega)} \leq \frac{1}{\sqrt{2\kappa}}\|\tau\|_{L^2(\Omega)}. \tag{5.2}$$

Denoting the average in Ω by

$$\fint_\Omega v = \frac{1}{\text{meas}(\Omega)}\int_\Omega v(x)\,dx,$$

by choosing $\xi = 1$ in (2.11), we obtain

$$\fint_\Omega \vartheta_\kappa = \fint_\Omega \tau\,\chi_{\{u_\kappa=\psi\}}. \tag{5.3}$$

Applying the well-known Poincaré type inequality

$$\left\| \vartheta_\kappa - \int_\Omega \vartheta_\kappa \right\|_{L^2(\Omega)} \leq C \left\| \nabla \vartheta_\kappa \right\|_{L^2(\Omega)} ,$$

using (5.2), as $\kappa \to \infty$ we immediately conclude that

$$\vartheta_\kappa \underset{\kappa}{\to} \Theta \quad \text{in } H^1(\Omega) , \tag{5.4}$$

$$\int_\Omega \vartheta_k \underset{\kappa}{\to} \Theta \quad \text{in } \mathbb{R} , \tag{5.5}$$

for some constant $\Theta \in [m_-, M_+]$.

On the other hand, by (5.1) and the assumption (2.14) we conclude that the solutions u_κ to (2.10) are bounded in $H^1(\Omega) \cap W^{2,p}(\Omega)$ independently of κ. Hence we may extract a subsequence of $\kappa \to \infty$ such that

$$u_\kappa \underset{\kappa}{\rightharpoonup} w \quad \text{in } W^{2,p}(\Omega)\text{-weak and } H^1(\Omega)\text{-strong} . \tag{5.6}$$

Since $f(\vartheta_\kappa) \underset{\kappa}{\to} f(\Theta)$ in $L^q(\Omega)$, $\forall q < \infty$, we see that this limit w is the unique solution to the variational inequality

$$w \in \mathbb{K}: \int_\Omega \nabla w \cdot \nabla (v - w) \, dx \geq \int_\Omega f(\Theta) \, (v - w) \, dx , \quad \forall v \in \mathbb{K}, \tag{5.7}$$

and, in fact, the convergence (5.6) also holds strongly in $W^{2,p}(\Omega)$.

Applying Lemma 3.1 to this sequence, under the assumption (2.15), we conclude that

$$\chi_{\{u_\kappa = \psi\}} \underset{\kappa}{\to} \chi_{\{w = \psi\}} \quad \text{in } L^q(\Omega), \quad \forall q < \infty . \tag{5.8}$$

Recalling (5.3) and (5.5), we also find that

$$\int_\Omega \vartheta_\kappa \underset{\kappa}{\to} \int_\Omega \tau \chi_{\{w = \psi\}} = \Theta . \tag{5.9}$$

We introduce the following operator for measurable functions v defined in Ω

$$\mathcal{M}\{v = \psi\} = \int_\Omega \tau \chi_{\{v = \psi\}} , \tag{5.10}$$

where $\chi_{\{v = \psi\}}$ is defined, in general, by (2.4). Then the property (5.9) yields that, in fact, w is a solution to the nonlocal obstacle problem

$$u \in \mathbb{K}: \int_\Omega \nabla u \cdot \nabla (v - u) \, dx \geq \int_\Omega f(\mathcal{M}\{u = \psi\}) \, (v - u) \, dx ,$$
$$\forall v \in \mathbb{K}, \tag{5.11}$$

and we have the following result:

Theorem 5.1. *Under the assumptions (2.13), (2.14) and (2.15), there exists at least one solution u to the nonlocal variational inequality (5.11), which can be obtained as limits when $\kappa \to \infty$*

$$u_\kappa \underset{\kappa}{\to} u \quad in \;\; W^{2,p}(\Omega) \,, \tag{5.12}$$

$$\vartheta_\kappa \underset{\kappa}{\to} \mathcal{M}(\{u = \psi\}) \quad in \;\; H^1(\Omega) \,, \tag{5.13}$$

where $(u_\kappa, \vartheta_\kappa)$ are solutions to (2.10)–(2.11).

Remark 1. Adapting the fixed point argument as described in Sect. 3, by solving (5.7) with an arbitrary real parameter $\Theta \in [m_-, M_+]$ and setting $S(\Theta) = \mathcal{M}\{w_\Theta = \psi\} \in [m_-, M_+]$, since S is a continuous real function, we can obtain directly the existence of at least a solution to (5.11), without considering the approximating coupled problems (2.10)–(2.11). However, in general, it is not clear if any solution of (5.11) can be approximated by solutions of the coupled problem, unless its uniqueness is assured as in the special case below. See [1] for other nonlocal unilateral problems.

Theorem 5.2. *Under the assumptions (2.13)–(2.14) and (2.17)–(2.18) with the restriction ($|\Omega| = \mathrm{meas}(\Omega)$)*

$$\mu L < \lambda |\Omega|$$

there exists a unique solution to the variational inequality (5.11).

Proof. This is an immediate consequence of Sect. 4 and the Banach fixed point theorem for strict contractions, since we have

$$\left\| \chi_{\{u_1 = \psi\}} - \chi_{\{u_2 = \psi\}} \right\|_{L^1(\Omega)} \le \frac{1}{\lambda} \left\| f(\mathcal{M}\{u_1 = \psi\}) - f(\mathcal{M}\{u_2 = \psi\}) \right\|_{L^1(\Omega)}$$

$$\le \frac{L}{\lambda} \left\| \mathcal{M}\{u_1 = \psi\} - \mathcal{M}\{u_2 = \psi\} \right\|_{L^1(\Omega)}$$

$$\le \frac{\mu L}{\lambda |\Omega|} \left\| \chi_{\{u_1 = \psi\}} - \chi_{\{u_2 = \psi\}} \right\|_{L^1(\Omega)} \,.$$

$\square$

Remark 2. If we set $\tau = |\Omega|$, then $\mathcal{M}\{u = \psi\} = \mathrm{meas}\{u = \psi\}$ and Theorems 5.1 and 5.2 (with $L < \lambda$) apply to the introduction problem $(*)$.

6   The Case with Temperature Dependent Elastic Coefficient

In this section we analyse the more delicate situation when in (2.1) the elastic coefficient $\alpha = \alpha(\vartheta)$ may depend on the temperature ϑ of the membrane. For simplicity, we assume now $f = \varphi(x)$ independent of ϑ, since otherwise it can be dealt as in the previous sections. We shall consider then the coupled obstacle problem:

$$u \in \mathbb{K}: \quad \int_\Omega \alpha(\vartheta) \, \nabla u \cdot \nabla(v - u) \, dx \geq \int_\Omega \varphi(v - u) \, dx \, , \quad \forall v \in \mathbb{K} \, , \qquad (6.1)$$

$$\vartheta \in H^1(\Omega): \quad \kappa \int_\Omega \nabla \vartheta \cdot \nabla \zeta \, dx + \int_\Omega \vartheta \, \zeta \, dx = \int_\Omega \tau \, \chi_{\{u=\psi\}} \, \zeta \, dx \, , \qquad (6.2)$$
$$\forall \zeta \in H^1(\Omega) \, .$$

We shall assume the function $\alpha \colon \mathbb{R} \to \mathbb{R}_+$ of class C^1 such that

$$\alpha(\theta) \geq \alpha_* > 0, \quad \forall \theta \in \mathbb{R} \qquad (6.3)$$

and for any $\theta \in C^1(\bar{\Omega})$, we shall consider the elliptic operator

$$A_\theta w = \nabla \cdot \left(\alpha(\theta) \, \nabla w \right) = \alpha(\theta) \, \Delta w + \alpha'(\theta) \, \nabla \theta \cdot \nabla w \, . \qquad (6.4)$$

For a given $\varphi \in L^\infty(\Omega)$ and an obstacle as in (2.13), we require now the nondegeneracy assumption in the form

$$\varphi(x) + A_{\theta(x)} \, \psi(x) \neq 0 \quad \text{a.e.} \ \ x \in \Omega \qquad (6.5)$$

for any function θ belonging to the convex subset

$$\mathcal{D} = \left\{ \sigma \in C^1(\bar{\Omega}): \ m_- \leq \sigma(x) \leq M_+ \ \text{a.e. in } \Omega \right\} , \qquad (6.6)$$

where the constants $m_- < M_+$ are determined from the assumption (2.12) as in Sect. 2.

For the existence result, we can now extend the argument of Sect. 3, by applying the Schauder fixed point theorem in $\mathcal{D} = \mathcal{C} \cap C^1(\bar{\Omega})$ and constructing the nonlinear continuous and compact mapping $\mathcal{P}: \mathcal{D} \to \mathcal{D}$ as follows:

$$\mathcal{P}: \quad \mathcal{D} \to \mathbb{K} \cap W^{2,p}(\Omega) \to \ L^q(\Omega) \ \to \mathcal{D} \cap W^{2,q}(\Omega)$$
$$\sigma \mapsto \qquad u_\sigma \qquad \mapsto \chi_{\{u_\sigma = \psi\}} \mapsto \qquad \theta_\sigma \ .$$

Here u_σ will be the unique solution of the variational inequality (6.1) with θ replaced by an arbitrary $\sigma \in \mathcal{D}$. The continuous dependence result $\sigma \mapsto \chi_{\{u_\sigma = \psi\}}$ is now more delicate, since it is necessary to extend Lemma 3.1 to elliptic operators with variable coefficients. However, as in Theorem 6:6.1 of [7], this can be done exactly as in Sect. 3. Indeed, if $\sigma_j \underset{j}{\to} \sigma$, for instance in $H^1(\Omega)$, then well-known results imply $u_{\sigma_j} \underset{j}{\to} u_\sigma$ in $H^1(\Omega)$ and we can pass to the limit in j in the equation

$$A_{\sigma_j} u_j + (\varphi + A_{\sigma_j} \psi) \chi_j = \varphi \quad \text{a.e. in } \Omega \, ,$$

since we have $A_{\sigma_j} u_j \underset{j}{\to} A_\sigma u_\sigma$ and $A_{\sigma_j} \psi \to A_\sigma \psi$ in $L^1(\Omega)$-strong. Hence, the embedding $W^{2,q}(\Omega) \subset C^1(\bar{\Omega})$ being compact for $q > 2$, clearly the fixed point $\vartheta = \mathcal{P}\vartheta$ with $u = u_\vartheta$ solves the problem (6.1)–(6.2) and we have proven the following existence result.

Theorem 6.1. *Let the preceding assumptions hold, in particular (6.3) and (6.5). Then there exists at least one solution $(u, \vartheta) \in [\mathbb{K} \cap W^{2,p}(\Omega)] \times W^{2,q}(\Omega)$, $\forall q < \infty$, to the coupled problem (6.1)–(6.2).*   □

As in Sect. 5, we can also consider the limit case when $\kappa \to +\infty$, which corresponds to the nonlocal problem

$$u \in \mathbb{K}: \quad \alpha\big(\mathcal{M}\{u = \psi\}\big) \int_\Omega \nabla u \cdot \nabla(v - u)\, dx \geq \int_\Omega \varphi(v - u)\, dx,$$

$$\forall v \in \mathbb{K} . \quad (6.7)$$

Due to the assumption (6.3) this problem is exactly the same as in (5.13) with $f(x, \mathcal{M}\{u = \psi\}) = \varphi(x)/\alpha(\mathcal{M}\{u = \varphi\})$ and we also have the following asymptotic convergence.

Theorem 6.2. *If $(u_\kappa, \vartheta_\kappa)$ denote a solution to (6.1)–(6.2) under the assumptions of Theorem 6.1, when $\kappa \to \infty$ we have*

$$u_\kappa \underset{\kappa}{\to} u \quad in \ H^1(\Omega)$$

$$\vartheta_\kappa \underset{\kappa}{\to} \mathcal{M}\{u = \psi\} \quad in \ H^1(\Omega)$$

where u is a solution to (6.7) and $\mathcal{M}\{u = \psi\}$ is given by (5.10).

Proof. As in the proof of Theorem 5.1, we show first that $\vartheta_\kappa \underset{\kappa}{\to} \Theta$ in $H^1(\Omega)$. This is sufficient to pass to the limit in (6.1), obtaining the convergence $u_\kappa \underset{\kappa}{\to} u$ in $H^1(\Omega)$ and allowing the identification of Θ as the constant $\mathcal{M}\{u = \psi\}$.   □

Remark 3. As in Theorem 5.1, it is not necessary to require $\alpha \in C^1(\mathbb{R})$ to solve (6.7) and a continuity property is sufficient for this purpose. However, the uniqueness argument of Theorem 5.2 is not applicable to (6.1)–(6.2), and uniqueness seems to be an open question.

References

1. Chipot, M., Rodrigues, J.F.: On a Class of Nonlocal Nonlinear Elliptic Problems. M^{2}AN - Math. Mod. Num. Anal. **26** (1992), 447–468
2. Duvaut, G.: Free Boundary Problem Connected with Thermoelasticity and Unilateral Contact. In "Free Boundary Problems", Proc. Pavia Seminar, Ist N. Alt. Mat., Roma (1980), Vol. I, 217–236
3. Friedman, A.: Variational Principles and Free-Boundary Problems. J. Wiley, New York, 1982
4. Hilhorst, D., Rodrigues, J.F.: On a Nonlocal Diffusion Equation with Discontinuous Reaction. Adv. Diff. Equations (in press)
5. Kinderlehrer, D., Stampacchia, G.: An Introduction to Variational Inequalities and their Applications. Academic Press, New York, 1980
6. Nishiura, Y.: Global Structure of Bifurcating Solutions of Some Reaction-Diffusion Systems. SIAM J. Math. Anal. **13** (1982), 555–593
7. Rodrigues, J.F.: Obstacle Problems in Mathematical Physics. North-Holland, Amsterdam, 1987

Dynamics of Diffusive Phase Transitions Driven by Coupled Mechanisms

M. Niezgódka

ICM, Warsaw University, Pawińskiego 5a, 02-106 Warsaw, Poland

Dedicated to Professor Karl-Heinz Hoffmann
on the occasion of his 60th birthday

Abstract. The paper gives an overview of basic set-ups for modelling dynamic phase separation phenomena in binary systems governed by various driving mechanisms coupled with diffusion. In particular, non-isothermal situations are treated and two-scale systems with conserved macroscopic order parameters while including non-conserved mesoscopic components are considered.

1   Introduction

Mathematical studies of phase change and a broad class of diffusion-driven phase transition phenomena are typically focused on a macroscopic scale of resolution, with continuum description of the medium involved. As well exercised over quite long period of time, such approaches are often neither capable to reproduce even basic features of the reportedly modelled phenomena nor exhibit desired mathematical correctness on finite time intervals.

First upon re-examining seemingly fundamental concepts of diffusion as viewed on continuum level [1,4,10,12], with thermodynamical well-posedness gained also in case of complex multi-component systems subject to phase transitions a speed-up could be seen in the related mathematical studies.

Still, only rather limited advances have been achieved so far in acknowledging the following aspects of high relevance for validating the models:

- strong coupling of various physical fields governing the dynamic developments,
- necessity to reflect mutual interactions of driving mechanisms that act within different scales of resolution (from macro- via meso-scale up to microscopic effects).

This refers not only to the theoretical advances in the mathematical treatment, but also to computational approaches. Coherent results have been achieved in a few rather special classes of models. It is the objective of this paper to indicate the status and outline some of challenging problems.

2   A Few Specific Models for Diffusion-Driven Phase Transitions

The coupling of dynamic diffusive developments with accounting for their non-isothermal nature contributes to serious troubles in finding a thermo-dynamically consistent description that, at the same time, would be mathematically well-posed. In all approaches so far either the dynamic components of the coupling between these two fields have been simplified to a kind of prototype set-up or no results on the uniqueness of solutions are available [1,2,5,8,13].

Driven by diffusive forces, non-isothermal dynamic developments in multi-component systems give rise to nonlinear parabolic systems when modelled within phenomenological mean-field continuum approximation approach [7,12]. When limited to thermal and diffusive phenomena, whole bulk of electro-magnetic actions, mechanical deformations and chemical reactions put aside, the resulting model is constituted of the dynamic energy and mass balances, possibly subject to constraints imposed so as to remind of other phenomena involved. The way in which the evolution terms enter into the energy balance equation is critical for constitution of the system. The subtle interplay of en-thalpic and entropic components there proves to play a key role for analysis of the model.

The way in which the singularities that enter into the evolution term in energy balance equation are handled becomes of special significance in systems with non-smooth terms due to constraints imposed on state variables. The non-differentiable contributions may reflect external forcing, either in the form of an applied control action or an impact of some external fields (decoupled due to essentially different time scale, in particular).

In the sequel, two classes of models will be discussed,

- a model for non-isothermal diffusive phase separation in continuum,
- a two-scale model for diffusive phase separation (macroscopic scale), coupled with meso-scale model for kinetic ordering transformations.

To avoid the necessity of treating an additional level of complexity, only two-component (binary) systems will be accounted for.

2.1   Non-Isothermal Diffusive Phase Separation (Continuum Model)

Within an extended Ginzburg-Landau approach a reference form of a system governing non-isothermal phase separation can be reduced to the parabolic system of the equations for energy and mass balances [7–9,12],

$$\rho(u)_t + \lambda(w)_t - \Delta u = f(t,x) \qquad \text{in } Q_T := (0,+\infty) \times \Omega, \qquad (2.1)$$

$$\frac{\partial u}{\partial n} + n_o u = h(t, x) \qquad \text{on } \Sigma_T := (0, +\infty) \times \Gamma, \tag{2.2}$$

$$u(0, \cdot) = u_o \qquad \text{in } \Omega, \tag{2.3}$$

$$w_t - \Delta(-\kappa \Delta w + \xi + g(w) - \lambda'(w)u) = 0 \qquad \text{in } Q, \tag{2.4}$$

$$\xi \in \beta(w) \qquad \text{in } \in Q_T, \tag{2.5}$$

$$\frac{\partial w}{\partial n} = 0 \qquad \text{on } \Sigma_T := (0, +\infty) \times \partial \Omega, \tag{2.6}$$

$$\frac{\partial}{\partial n}\{-\kappa \Delta w + \xi + g(w) - \lambda'(w)u\} = 0 \qquad \text{on } \Sigma_T, \tag{2.7}$$

$$w(0, \cdot) = w_o \qquad \text{in } \Omega. \tag{2.8}$$

In the above system, u is related to the temperature field θ and w is an order parameter (representing local phase ratio or the number of phase particles, e.g.). Here Ω is a bounded domain in $\mathbf{R}^N$, $1 \leq N \leq 3$, with smooth boundary Γ, g and λ smooth functions defined on $\mathbf{R}$, λ' the derivative of λ; $n_o > 0$ and $\kappa > 0$ constants, f, h, u_o, w_o prescribed data. In this system, to reflect possible constraints, β is a maximal monotone graph in $\mathbf{R} \times \mathbf{R}$. The function ρ in (2.1) is increasing, with open domain $D(\rho)$ and range $R(\rho)$ in $\mathbf{R}$, locally bi-Lipschitz.

Since it may happen in situations which are physically relevant that $D(\rho) \neq \mathbf{R}$, this contributes to a singular form of the evolution term in (2.1). In particular, in the models of non-isothermal phase separation developed on the level of the non-equilibrium thermodynamics formalism [1,4,11]),

$$\rho(u) = -\frac{1}{u}, \qquad -\infty < u < 0$$

is (up to scaling factor) the Kelvin temperature.

Let us note that for the case of $\beta = \partial I_{[\sigma_*, \sigma^*]}$, subdifferential of the indicator function of a finite segment, this choice may reflect limited resolution of the model around pure phase structures (inability to measure the level of local impurities), hence treating a structure to be in pure phase state below a certain level of admixtures. To give an example, such a treatment is well-justified for *non-polynomial* Flory-Huggins models of phase separation in polymeric materials [1].

There are quite a few studies of special cases of the system (2.1)-(2.8), without the non-smooth part of ξ and without the term $\lambda'(w)u$ representing nonlinear coupling of energy and mass transfer, as a rule also postulating temperature to be constant.

For general systems of the form (2.1)-(2.8), results on the existence, uniqueness of solutions, their large-time behaviour, and qualitative analysis covering a spectrum of stability questions were given in [7–9], provided $\rho(u)$ an increasing, globally bi-Lipschitz on **R**. Note that the latter condition is not fulfilled in the case of $\rho(u) = -\frac{1}{u}$, $-\infty < u < 0$, where no regularity u in t sufficient for uniqueness of the solution is available in multidimesional situations.

2.2   2.2 Two-Scale Model of Diffusive Phase Separation Coupled with Kinetic Structural Transformation

In standard diffusion models for developments in binary systems each of the components was considered physically homogeneous. For the processes in crystallic materials an additional differentiation may prove significant up to becoming critical in specific circumstances. The possibility of alternating between a number of structurally different phases may heavily affect physico-chemical properties of the material.

For such situations, a prototype model has been introduced by Cahn and Novick-Cohen [3]. The approach they have proposed assembles:

- diffusion driven by equations of Cahn-Hilliard type (continuum component, macroscopic scale),
- kinetic structural transformation governed by equations of Allen-Cahn type (discrete mesoscopic scale).

In a reference form, the system can be given the following structure, with w standing for conserved macroscopic order parameter and v representing a non-conserved mesoscopic order parameter (for processes in solid binary alloys v characterizes types of system's crystalline structure):

$$w_t - \nabla.\, Q\nabla\{\xi + f(w,v) + \alpha w - \eta\Delta w\} = 0 \qquad \text{in } Q_T = (0,+\infty) \times \Omega, \tag{2.9}$$

$$v_t - \eta\Delta v + g(w,v) + \beta v = 0 \qquad \text{in } Q_T, \tag{2.10}$$

$$\xi \in \beta(w) \qquad \text{in } \in Q_T, \tag{2.11}$$

$$\frac{\partial w}{\partial n} = \frac{\partial(\Delta w)}{\partial n} = \frac{\partial v}{\partial n} = 0 \qquad \text{on } \Sigma_T, \tag{2.12}$$

$$v(0, \cdot) = v_o, \quad w(0, \cdot) = w_o \quad \text{in } \Omega. \tag{2.13}$$

For coupled two-scale models of the form (2.9)-(2.13), results on existence, uniqueness, large-time behaviour of solutions and existence of global attractor have been obtained recently [6].

The way in which the dynamic coupling of conserved variables (Cahn-Hilliard type component of the model) with non-conserved ones (Allen-Cahn type component) can be handled in more general set-up, accounting in particular for stochastic nature of meso-scale part of the model, seems to represent a challenging direction for future studies.

References

1. Alt, H. W., Pawlow, I.: Existence of Solutions for Non-Isothermal Phase Separation. Adv. Math. Sci. Appl. 1 (1992), 319-409
2. Blowey, F. E., Elliott, C. M.: The Cahn-Hilliard Gradient Theory for Phase Separation with Non-Smooth Free Energy, Part I: Mathematical Analysis. European J. Appl. Math. 2 (1991), 233-280
3. Cahn, J. W., Novick-Cohen, A.: Evolution Equations for Phase Separation and Ordering in Binary Alloys. J. Stat. Phys. 76 (1994), 877-909
4. DeGroot, S. R., Mazur, P.: Non-Equilibrium Thermodynamics, Dover Publ., New York (1984)
5. Elliott, C. M., Zheng S.: On the Cahn-Hilliard Equation. Arch. Rat. Mech. Anal. 96 (1986), 339-357
6. Gokieli, M.: An Extended Model for Phase Separation and Ordering, to appear (1999)
7. Kenmochi, N., Niezgódka, M.: Non-Linear System for Non-Isothermal Diffusive Phase Separation. J. Math. Anal. Appl. 188 (1994), 651-679
8. Kenmochi, N., Niezgódka, M.: Viscosity Approach to Modelling Non-Isothermal Diffusive Phase Separation. Japan J. Ind. Appl. Math. 13 (1996), 135-169
9. Kenmochi, N., Niezgódka, M., Zheng, S.: Global Attractor of a Non-Isothermal Model for Phase Separation.In: A. Damlamian et al., eds., Curvature Flows and Related Topics, GAKUTO Int. Ser., Math. Sci. Appl., Tokyo vol. 5 (1995), 129-143
10. Luckhaus, S., Visintin, A.: Phase Transition in Multicomponent Systems. Manuscripta Math. 43 (1983), 261-288
11. Oono, Y., Puri, S.: Study of the Phase Separation Dynamics by Use of Cell Dynamical Systems, I. Modelling. Phys. Rev. A 38 (1988), 434-453
12. Penrose, O., Fife, P. C.: Thermodynamically Consistent Models of Phase-Field Type for the Kinetics of Phase Transitions. Physica D, 43 (1990), 44-62
13. Zheng, S.: Asymptotic Behaviour of the Solution to the Cahn-Hilliard Equation. Applicable Anal. 23 (1986), 165-184

Free Boundary Problems and their Stabilisation

J. R. Ockendon

Oxford Centre for Industrial and Applied Mathematics, University of Oxford,
24-29 St Giles, Oxford OX1 3LB, UK.

Dedicated to Professor Karl-Heinz Hoffmann
on the occasion of his 60th birthday

1 Introduction

Free boundary problems (FBPs) are those in which some partial differential
equations have to be solved in a region with a boundary whose position is to
be found as part of the problem. This article will briefly review some of the
theory and folklore concerning such problems and their stabilisation with the
aim of highlighting the major open problems that currently confront mathematicians working in the area. It is a subject with which Prof. Hoffmann
has been intimately involved over the past three decades and it is a pleasure
to acknowledge the contributions he has made to our understanding of problems ranging from the Stefan problem, which created so much excitement in
the 1970's, to FBPs in superconductivity which are currently "all the rage".
Of course I must also include Prof. Hoffmann's involvement with the more
political applications of "les frontières libres".

Sect. 2 contains some general remarks about the stability of problems in
which the free boundary has one dimension fewer than that of the space of
independent variables; assuming the free boundary is orientable, such problems can be *one-phase* when the solution of the field equations is trivial on
one side of the FB. This is followed in Sect. 3 by a catalogue of some of
the regularisations that have been proposed for these problems. Sects. 4 and
5 contain corresponding remarks about the less-intensively studied class of
"codimension-two" FBPs.

2 The Stability of Codimension-One FBPs

The stability of this, the most frequently-occurring class of FBPs, has been
studied extensively. Unfortunately the applications are so ubiquitous in applied science that the results are scattered in many different journals and
books. The easiest stability analyses are for one-dimensional problems in
which the free boundary is a curve $x = s(t)$ in the space-time domain
(x, t), and in which only one-dimensional perturbations are considered so

that $x \sim s_0(t) + \varepsilon s_1(t) + \ldots$, where s_0 is known and ε is a prescribed small parameter. Such analyses are not usually very interesting except when the field equations are hyperbolic conservation laws, in which case they soon lead into fascinating problems concerning existence, uniqueness and numerical approximation. The work of researchers such as Lax and Oleinik has resolved many of these problems for scalar equations, but there remain many open questions about systems.

Even when we exclude hyperbolic problems, but generalise to two space dimensions so that the free boundary is, say, $t = \omega(x, y)$, an astonishing number of possibilities open up. Some of the best known examples are listed below. In each case we cite the results of a linear stability analysis in which the free boundary is nearly straight so that $\omega(x, y) \simeq x/V$ where V is its velocity. In all cases the field equations are linear, with constant coefficients, so that we can expand the dependent variable in exponentials in space and time and also write the free boundary as

$$x \sim Vt + e^{\sigma t} \sin ky + \ldots,$$

say, where $k > 0$, and derive a dispersion relation for $\sigma(k)$. All the results are for perturbations that are local to the free boundary and they only apply to the simplest of the models that have been proposed: the derivations, with appropriate references, can be found in [1].

2.1   Inviscid, Irrotational Surface Gravity Waves

Taking x to be vertical and V to be zero, so that the liquid is basically at rest, we find that

$$\sigma = \pm i(k/F)^{\frac{1}{2}}, \tag{2.1}$$

where F is the Froude number, which measures the importance of the fluid inertia relative to gravity. This is an example of the Rayleigh-Taylor instability and it reflects the fact that instability results if the liquid is above the free boundary ($F < 0$) and neutral stability if it is below ($F > 0$).

2.2   Vortex Sheets in Inviscid Hydrodynamics

Taking x to be normal to the undisturbed sheet and again setting $V = 0$, we find that

$$\sigma = ik/2(U_+ + U \pm i(U_+ - U_-)), \tag{2.2}$$

where $U_\pm$ are the tangential fluid velocities on either side of the sheet. This is an example of the Kelvin-Helmholtz instability, which always occurs when $U_+ \neq U_-$.

2.3   Hele-Shaw Flow, Porous Medium Flow, Phase Changes in Type-I Superconductors, Stefan Problems

The one-phase Hele-Shaw model is the simplest special case of many models for FBPs in all these areas and its dispersion relation is

$$\sigma = kV, \tag{2.3}$$

where V is the velocity of the free boundary into the active phase, i.e. the liquid. Hence, Hele-Shaw flow is stable when the liquid region expands and unstable when it contracts. The two-phase generalisation to porous medium flow (the Muskat problem) is that, in the absence of gravity, instability occurs when a region saturated with a less viscous fluid expands into a region saturated with a more viscous one: however, the introduction of gravity can, effectively, change the sign of V. For phase changes modelled by Stefan problems, the generalisation is that instability occurs whenever either phase is superheated or supercooled; however, because the field equation is now parabolic rather than elliptic, this result is only strictly true when k is large, and long wavelength disturbances may not suffer such instabilities. Finally, type-I superconductors can sometimes be modelled as one-phase Stefan problems and hence they are stable when the superconducting region contracts and unstable when it expands.

   This list could be prolonged almost indefinitely. However, the results quoted illustrate the remarkable fact that even though the field equations for each of the problems, i.e. Laplace's equation or the heat conduction equation, are usually associated with well-posed mathematical models, the presence of the free boundary can readily lead to a local linearisation being either well-posed or ill-posed. In many cases, a simple reversal in the direction of motion of the boundary will bring about a switch from well- to ill-posed. When such ill-posedness does occur, the temporal growth of the solution may necessitate the introduction of nonlinearity into the model after some finite time. Note however that the growth rate associated with (2.1) may be such that, for sufficiently well-behaved initial conditions, blow-up may not occur in finite time; indeed none of the examples (2.1-2.3) are as dramatic as the dispersion relation $\sigma = k^2$ for the backward heat equation, for which finite time blow-up occurs even for analytic initial data.

   All the above results apply to very simple models in which apparently small effects such as surface energy and dissipation are neglected. The incorporation of such effects offers immediate possibilities for regularisation, as we now describe.

3   Regularisation of Codimension-One FPBs

3.1   Introduction of Higher Derivatives

From either the mathematical or physical viewpoint, the most obvious way to regularise any of the unstable models listed above is to introduce some new surface energy or relaxation effects into the free boundary conditions, thereby increasing the order of the derivatives that occur in these conditions. Hence, results like 2.3 become

$$\sigma = kV - \delta k^3,$$

for some positive parameter δ. Unfortunately when parameters such as δ are small, the resulting singular perturbation problems are notoriously difficult to analyse, mathematically or numerically. They typically involve "exponential asymptotics" so that perturbations of amplitude say $O(e^{-1/\delta})$ can have a dramatic effect on the solution for small positive δ. The effect of such regularisations has only been described systematically for certain simple configurations, such as travelling waves, in the past decade [2], while accurate numerical results for evolution problems are only just beginning to appear [3]. It is noteworthy that both these studies involve the Hele-Shaw problem, whose properties under conformal transformations give us far more global knowledge of its solution structure than we have for the other models of Sect. 2.

The general picture that seems to be emerging is that regularisation via the introduction of higher derivatives can be extremely powerful as a pattern selector[1] . Not only do the extra terms prevent the development of many kinds of singularities, but they also select preferred free boundary morphologies over times greater than the time to blow-up in the unregularised problem. However, a simple cautionary example concerns the introduction of a surface tension into the two-dimensional Hele-Shaw problem

$$\Delta p = 0 \tag{3.1}$$

in a closed, simply connected domain $\Omega(t)$, with a "sink" at $r^2 = x^2 + y^2 = 0$. The free boundary conditions are

$$p = \delta\kappa, \; \frac{\partial p}{\partial n} = -\nu_n \text{ on } \partial\Omega(t), \tag{3.2}$$

where κ is the appropriately-signed curvature and ν_n the velocity of $\partial\Omega$, and

$$p \sim Q \log r$$

as $r \to 0$. Even with the regularising surface tension δ, the centre of mass of Ω cannot move because

$$\frac{d}{dt} \iint_\Omega x \, dA = \int_{\partial\Omega} x\nu_n \, ds = -\int_{\partial\Omega} x\frac{\partial p}{\partial n} \, ds = \int_{\partial\Omega} p\frac{\partial x}{\partial n} \, ds = \int_{\partial\Omega} \delta\frac{d\theta}{ds} \cos\theta \, ds \,,$$

where $dx/ds = \cos\theta$. This quantity is zero and so all the fluid cannot be extracted from $\Omega(0)$ without a topology change in $\partial\Omega$, unless $\Omega(0)$ is a circle centred at the origin. This leads us to our next regularisation.

---

[1] However, this view has recently been challenged in [4]

3.2   Topology Change

Even well-posed problems of the type listed above can easily develop singularities in finite time. For example, it has been demonstrated in [5] that, in the regularised Hele-Shaw problem for a simply-connected fluid domain containing a "sink", singularities can occur via loss of univalency of the map from the flow domain to the unit circle just as easily when the free boundary "touches itself" with finite speed (allowing analytical continuation) as when an infinite-speed cusp develops. In the former case, a relatively simple regularisation is possible in which the topology is allowed to change. In fact, such topology changes are intrinsic in the complex variable models proposed in [6].

3.3   Smoothing Transformations

Mathematically, perhaps the most interesting class of regularisations are smoothing transformations of the classical models, either by direct integration or via a weak formulation. One of the best known is the Baiocchi transformation of the following model for steady one-phase porous medium flow in which the pressure is $p(x,y)$, y is vertical and $0 < y < h(x)$:

$$\nabla^2 p = 0$$

with $p = \frac{\partial}{\partial n}(p + y) = 0$ on the free boundary $y = h(x)$,

and
$$\frac{\partial p}{\partial y} = 0 \;\; \text{on} \;\; y = 0,$$

and suitable conditions on vertical walls $x = $ const.

Defining
$$u(x,y) = \int_y^h p(x,y')dy',$$

we find
$$\nabla^2 u = 1 , \tag{3.3}$$

with
$$u = \frac{\partial u}{\partial n} = 0 \;\; \text{on} \;\; y = h . \tag{3.4}$$

Hence u can be continued across the free boundary as a C^1 function and our smoothing integration has led to a problem that is equivalent to a variational inequality [7].

This type of smoothing is even more interesting when gravity is neglected but the flow is allowed to evolve in a region Ω . Then (3.1,3.2) becomes the Hele-Shaw model in which the free boundary conditions for $p(x,y,t)$ are

$$p = 0 = \frac{\partial p}{\partial t} - |\nabla p|^2 \;\; \text{on} \;\; t = \omega(x,y);$$

the Baiocchi transformation is

$$u = \begin{cases} \int_0^t p(x,y,\tau)\,d\tau & \text{for}(x,y) \in \Omega_0, \text{the region occupied by fluid,} \\ \int_{\omega(x,y)}^t p(x,y,\tau)\,d\tau & \text{for}(x.y) \in \Omega \setminus \Omega_0. \end{cases}$$

When Ω is expanding, another formal calculation leads to (3.3,3.4), with t appearing as a parameter ; we can either write

$$\nabla^2 u = \chi(x)H(u),$$

where χ is the characteristic function of Ω , or the linear complementarity problem

$$u \geq 0, \ \chi - \nabla^2 u \geq 0, \ (\chi - \nabla^2 u)u = 0.$$

In either case, our smoothing integration has led to a helpful weak formulation of the problem. However there is a dramatic irreversibility because, when Ω contracts, we cannot even determine $u(x,y,0)$ without solving the ill-posed Cauchy problem

$$\nabla^2 u = 1 \ \text{in} \ \Omega_0 \ \text{with} \ u = \frac{\partial u}{\partial n} = 0 \ \text{on} \ \partial\Omega_0.$$

In exactly the same spirit, we can formally establish a relationship between the one-phase Stefan model

$$\frac{\partial T}{\partial t} = \nabla^2 T \ \text{with} \ T = 0, \ \frac{\partial T}{\partial n} = -\nu_n \ \text{on} \ \partial\Omega \qquad (3.5)$$

where ν_n is again the normal velocity of $\partial\Omega$ and the Crank-Gupta model for u defined as $\int_\omega^t T(x,y,\tau)\,d\tau$ in $\Omega \setminus \Omega_0$:

$$\nabla^2 u = \frac{\partial u}{\partial t} + 1, \ u = \frac{\partial u}{\partial n} = 0 \ \text{on} \ \partial\Omega . \qquad (3.6)$$

Now (3.5) is formally the time derivative of (3.6) as long as no other constraints are imposed on u, and this statement is true whether or not $\partial\Omega$ is expanding or contracting, i.e. whether or not the free boundary is stable or unstable. However, (3.6) is a well-posed model for oxygen-diffusion in a tissue in either case [8]. The explanation for this apparent dichotomy is the positivity of u in the model. This allows the contracting free boundary problem to stabilise itself by nucleating new components of the free boundary at points at which u falls to zero (see [9] for a discussion of this situation in one space dimension).

A related, and equally famous regularisation also concerns (3.6) but with the free boundary conditions $\frac{\partial u}{\partial n} = -\nu_n$, $u = 0$ on $\partial\Omega$. This is the "resistance welding problem" and is such that if the initial temperature $u(x,0)$ is negative and the unit volumetric heating is sufficient to raise the temperature to the melting temperature $u = 0$ at some point, then a superheated region is formed locally with an unstable free boundary. However an "enthalpy" smoothing in which (3.6) is replaced by

$$\frac{\partial h}{\partial t} = \nabla^2 u(h) + 1$$

with
$$u = \begin{cases} h - 1 & , \ h > 1 \\ 0 & , \ 1 > h > 0 \\ h & , \ 0 > h \end{cases}$$

taken in the sense of distributions, legislates against superheating by producing a mushy region in which $u \equiv 0$; at the mush boundary $u = \frac{\partial u}{\partial n} = 0$ so locally the enthalpy regularisation has retrieved the Crank-Gupta model..

In summary, there seem to be few interesting questions concerning the regularisation of co-dimension-one free boundary problems that are linearly well-posed, but this is far from being the case for ill-posed problems. The principal hurdle to be overcome is the understanding of what happens at a blow-up time or point, when the unregularised free boundary develops a singularity beyond which no obvious classical continuation is possible. Despite intensive analytical and numerical studies of physically-regularised singular perturbation models (see e.g. [3,5] for cusp formation in the Hele-Shaw problem, [10,11] for singularity formation in the Kelvin-Helmholtz and Rayleigh-Taylor problems), what happens after such a singularity remains largely a matter of conjecture; however, it is known that, in two dimensions, vortex sheets cannot roll up into a concentrated vortex. One speculation concerning the Hele-Shaw model is that the subsequent free boundary morphology takes the form of "cracks" whose tips move in such a way that the map from the physical domain to some prescribed domain be extremal and univalent. However, this variational formulation admits too many possible crack paths, and as discussed in [5] the tip dynamics problem remains as challenging as it is in the classical theory of brittle fracture. Such thoughts lead us into another class of free boundary problems.

4 Regularisations Leading to Codimension-Two FBP's

There are two commonly-occurring types of codimension-one FBP that can collapse into a more classical problem in partial differential equations.

The first is in the happy circumstance that the problem is one-phase, and the free boundary is close to a fixed one, with the active region lying between the two. Then, using a suitable asymptotic approximation, it is usually possible to write down a partial differential equation for the thickness of the active region. This happens in many thin films and shallow water flows and, even for two-phase problems can lead to an integro-differential equation such as the Benjamin-Ono equation [1].

Another such situation arises when a two-dimensional free boundary is not near a fixed boundary but is nonetheless nearly one-dimensional.

4.1   Vortices and Dislocations

An obvious configuration to study is one in which the codimension-one free boundary is in the shape of a thin tube. Examples include the boundary between the rotational and irrotational regions of a high-Reynolds-number fluid vortex, or the phase boundary between the normal and superconducting regions in a Type II superconducting vortex. In such cases it is simplest to start by modelling the vortex as a distribution of singularities along a curve Γ. In the fluid case, this results in the famous Biot-Savart Law

$$\mathbf{u}(\mathbf{r}) = -\frac{\omega_0}{4\pi} \int_\Gamma \mathbf{r} \wedge d\mathbf{r}' / |\mathbf{r} - \mathbf{r}'|^3 \qquad (4.1)$$

for the velocity $\mathbf{u}$, which satisfies $\mathbf{curl\,u} = \mathbf{0}$ except on Γ and $\mathbf{div\,u} = 0$; in the superconducting case we find that the current $\mathbf{j}$ satisfies

$$\mathbf{j}(\mathbf{r}) = \frac{1}{2} \int_\Gamma \nabla \left(e^{-|\mathbf{r}-\mathbf{r}'|}/|\mathbf{r}-\mathbf{r}'| \right) \wedge d\mathbf{r}' . \qquad (4.2)$$

where $\mathbf{j} = \mathbf{curl\,H}$ and $\mathbf{curl}^2\mathbf{H} + \mathbf{H} = \mathbf{0}$ (the London equation) except on Γ, and $\mathbf{div\,H} = 0$. Here $\omega_0/2\pi$ is the circulation around the fluid vortex, where ω_0 is arbitrary, but the superconducting vortex is quantised. In both situations, we have prescriptions for the field variables once the position of Γ is known but, to obtain an FBP for the evolution of Γ, it is necessary to introduce some new dynamical considerations into the modelling of the "core" of the vortex. Indeed, such a philosophy can be applied to the derivation of codimension-one free boundary problems but the solutions of the field equations are usually smooth enough for the relevant free boundary conditions to be derivable from a simple conservation argument. Here, on the other hand, the behaviour of (4.1,4.2) as the distance $\rho = |\mathbf{r} - \mathbf{r}'|$ from $\mathbf{r}$ to a point $\mathbf{r}'$ on Γ tends to zero contains two singular terms:
(i) a two dimensional "vortex" contribution in which $\mathbf{u}$ or $\mathbf{j}$ is of $O(\rho^{-1})$
(ii) a term proportional to $\log \rho^{-1}$.
Hence a singular perturbation analysis is needed before we can make any progress towards an evolution model, and, to do this, a regularisation must be introduced as will be described below.

An analogous situation arises in the classical Volterra model for dislocations in crystals, where it is the incompatibility (roughly, the curl of the strain tensor) that suffers as singularity along Γ. As discussed in [12], this means that the curl of the elastic distortion tensor $\boldsymbol{\beta}$ (taken row by row) satisfies

$$\mathbf{curl}\,\beta = \boldsymbol{\delta}_\Gamma \otimes \boldsymbol{b}$$

where $\boldsymbol{b}$ is the Burgers vector of the dislocation and $\boldsymbol{\delta}_\Gamma(\mathbf{r}) = \int_\Gamma \delta(\mathbf{r} - \mathbf{r}')\,d\mathbf{r}'$.

The inversion formula for the stress tensor is too complicated to quote here (see [13]) but again its asymptotic expansion near the core of the dislocation yields two singular terms, the second being logarithmically infinite. It is traditional in elasticity to regularise the relevant integral by introducing a suitable "cut-off" i.e. by excising a region $\rho \leq \varepsilon$ from the curvilinear integrals [13,14].

A quite different type of codimension-two problem can arise as follows.

4.2   Water Entry

A quite different situation arises when we consider the complement of the problems mentioned at the beginning of this section. A paradigm is a limiting case of the famous two-dimensional water-entry problem [15]. This is a model for a "shallow" rigid body $y = -t + \varepsilon f(x)$, where f is convex and even with $f(0) = 0$, which penetrates an inviscid half space $y < 0$ at time $t = 0$. The impact velocity is assumed so great that surface tension and gravity can be neglected. If ε was not small, we would have a codimension-one FBP but, for ε small, it has been proposed in [16] that the impacting body can be modelled as an expanding flat plate $-d(t) < x < d(t)$, $y = 0$, where the velocity potential φ satisfies

$$\nabla^2\varphi = 0\,,\ y < 0\,,\ |\nabla\varphi| \to 0 \ \text{ as } \ x^2 + y^2 \to \infty \tag{4.3}$$

with

$$\frac{\partial\varphi}{\partial y} = -1\,,\ \frac{\partial\varphi}{\partial t} < 0 \ \text{ on } \ y = 0 \ |x| < d \tag{4.4}$$

and

$$\varphi = 0\,,\ \frac{\partial\varphi}{\partial y} = \frac{\partial h}{\partial t}\,,\ h < f - t \ \text{ on } \ y = 0,\ |x| > d \tag{4.5}$$

and

$$\varphi = h = 0 \ \text{ at } \ t = 0\,; \tag{4.6}$$

here $h(x,t)$ is the codimension-one free boundary displacement from $y = 0$ and the inequalities are based on the physical assumptions of no cavitation or interpenetration respectively. We are thus led to a codimension-two FBP for the free boundary $x = \pm d(t)$, $y = 0$, but, as in our models for vortices and dislocations, it needs more information before it has any chance of being well posed.

In all these situations, we need to examine the singularity at the codimension-two boundary more carefully, both physically and mathematically. The upshot will be a local regularisation which both removes the singularity at the codimension-two free boundary and also gives the information necessary to close the model.

4.3   Local Behaviour Near Codimension-Two FBPs

We first discuss the regularisations that are needed to close our codimension-two models for vortices and dislocations. The easiest case is for superconducting vortices for which the London equation leading to (4.2) is in fact the leading approximation to the higher-order Ginzburg-Landau model. Hence the latter can be used to regularise the core singularity inherent in (4.2). As described in [12], a matching between (4.2) away from the core and the Ginzburg-Landau solution (which is locally axisymmetric) near the core yields the solvability condition that the vortex velocity is approximately proportional to $\kappa \mathbf{n}$, where κ is the curvature of Γ and $\mathbf{n}$ the principal normal. The positive constant of proportionality tends to infinity like $\log(\varepsilon^{-1})$ where ε is the core radius (or "cut off" length) compared to the radius of curvature of Γ .

Thus the dynamics of an isolated superconducting vortex is reduced to the problem of curvature flow is three-dimensions. This looks innocuous enough and, indeed the regularisation procedure can even be used to model the interaction of vortices with each other (or with a boundary) when the separation distance is of the order of the core radius. However the law of motion has undesirable consequences in some situations, as will be mentioned in the next section.

We can adopt the same approach for fluid vortices in two ways, each of which yields the same result. We can either appeal to the Navier-Stokes equations to again regularise the core by introducing higher derivatives or we may use physical insight to allocate a sufficiently smooth vorticity distribution in an inviscid core. In either case the matching procedure yields that the vortex velocity is, to leading order proportional to $\kappa \mathbf{b}$, where $\mathbf{b}$ is the binormal to Γ. The constant of proportionality is subject to the same remarks as in the case of superconducting vortices.

The case of dislocations in an elastic crystal is fundamentally different from that of vortices because the stress near the core turns out to be of order $|\mathbf{b}|/\rho$ as $\rho \to 0$. Hence the core radius is comparable to the length of the Burgers vector and is thus of atomic dimensions. This leads to a completely new class of "discrete" regularisations which are only at all well understood in the physically irrelevant case of one space dimension. All that can be stated with any confidence is that a virtual work argument, which invokes a line tension in the dislocation as its "regularisation", yields that the self-induced dislocation velocity is in the direction of the normal. However, crystallographic considerations often demand that Γ always lies in a prescribed plane [13].

We emphasise that all the above results are only true to lowest order and only apply to isolated codimension-two free boundaries. If as is often the case, this free boundary is the union of millions of components, and if these components are sufficiently close together, then the results must be reassessed.

Let us now return to the water entry problem (4.3-4.6), for which the local solution near the codimension-two boundary $x = \pm d$, $y = 0$ does not require us to invoke a new physical mechanism, but rather the retention of the codimension-one FB in a region sufficiently close to, say, $x = d$, $y = 0$. As shown in [17], to lowest order the "inner" problem is a well-known Helmholtz flow, whose solution contains no singularities and can be written down explicitly. Matching the solution of this inner problem with that of (4.3-4.6), which only applies sufficiently far from $x = d$, $y = 0$, leads us to the condition that the codimension-one free boundary effectively meets the impacting body at $x = d$, $y = 0$. Thus

$$h(d(t), t) = f(d(t)) - t \ ,$$

and this is the required extra information that emerges from the regularised inner solution. It is believed that, with this information, this model is well posed as a complementarity problem and the local analysis requires that $|\nabla \varphi| = O((x - d)^2 + y^2)^{-\frac{1}{2}}$ as the codimension-two free boundary is approached. However the interplay between the inequalities in (4.4,4.5) and the singularity behaviour has never been established rigorously.

In any event, an explicit solution is possible and it yields

$$f(d(t)) = \int_0^t \frac{d(t)d\tau}{\sqrt{(d(t))^2 - (d(\tau))^2}}$$

which has the following solution for $d(t)$:

$$d^{-1}(x) = \frac{2}{\pi} \int_0^x \frac{f(s)\,ds}{\sqrt{x^2 - s^2}}$$

It is noteworthy that the water exit problem in which the rigid body is $y = Vt + \varepsilon f(x)$, with $V > 0$ yields an under-determined Abel equation for $d(t)$. Worse still, a local linear stability analysis in which the codimension-two free boundary is the curve

$$x = d(t) + \varepsilon \, e^{\sigma t} \sin kz \ , \ y = 0$$

yields that

$$\sigma \leq kV$$

is a condition for stability [17]. Hence entry problems are stable and exit problems unstable on the basis of linearised theory.

A similar approach can be used to describe the initial stages of a wide variety of contact problems between liquids and solids, the most famous being the elastic contact problem of Hertz. We will not describe the details here because they can be found in many texts both mathematical [7] and engineering, but it does appear that the process of unloading in a Hertz problem is quite different from the time-reversal of the loading problem [18]. Since the same kind of irreversibility appears in other codimension-two problems in porous medium flow and electrical painting, it has been conjectured that these problems are just as prone to instability as are their codimension-one progenitors.

We must also mention that the even more important problem of crack propagation, brittle or ductile, is effectively a codimension-two FBP. In the traditional static stress analysis of brittle cracks, the crack geometry is prescribed and the stress concentration near the tip needs to be computed. However, this stress concentration invalidates linear elasticity theory locally near the crack tip. There is no space here to discuss the different kinds of regularisations that have been proposed, but it is almost certain that the identification of the physically relevant regularisation is the key to the evolution problem (see [20,21] for background and [22] for a recent mathematical approach). Note the contrast with contact problems, where the "tip" of the contact region has the freedom to adjust its position so as to ensure that the stress remains everywhere bounded.

In view of the evident instability of water exit problems, and also of the instability of superconducting vortices in suitably aligned external magnetic fields [23], it would be pleasant to be able to report results concerning the regularisation of the second class of problems described in sec. 4 in a framework similar to that of sec. 3. Unfortunately the only regularisation of which the author is aware is the "mushy region" model proposed in [24] for spray regions in water waves. Such a smoothing of the density between its values for water and air may be appropriate to regularise the water exit problem.

5   Conclusion

We conclude by mentioning a spectre that hangs over many of the models cited above, especially those that appear to be ill-posed on the basis of linear stability theory. This is that both observations and the theory of exponential asymptotics shows that the free boundary usually becomes exceedingly irregular as time evolves, famous examples being the dendritic solidification of supercooled melts and fringing patterns in porous medium flow. In such situations it becomes imperative to study configurations in which many components of the free boundary interact with each other. In the physical sciences, many heuristic models have been proposed for such mushy regions and their validation is currently one of the most important challenges facing the applied mathematical community. This remark applies both to "codimension-one" mushes such as microstructure in solids, and "codimension-two" mushes such as vortex tangles as models for turbulence, or three-dimensional type II superconductors in appropriate magnetic fields.

References

1. Ockendon, J. R., Howison, S. D., Lacey, A. A., Movchan, A. B.: Applied Partial Differential Equations. Oxford University Press, 1999
2. Kessler, D. A., Levine, H.: Theory of the Saffman-Taylor Finger Pattern. Phys. Rev. **33** (1986), 2621–2639
3. Hou, T. Y., Si, H.: Numerical Study of Hele-Shaw Flow with Suction. Preprint, 1998
4. Mineev-Weinstein, M.: Selection of the Saffman-Taylor Finger Width in the Absence of Surface Tension. Phys. Rev. Lett. **80** (1998), 2113
5. Hohlov, Y. E., Howison, S. D. Huntingford, C., Ockendon, J.R., Lacey, A. A.: A Model for Nonsmooth Free Boundaries in Hele-Shaw Flows. Q. J. Mech. App. Math. **47** (1998), 107–128
6. Richardson, S.: Some Hele-Shaw Flows with Time-Dependent Free Boundaries. J. Fluid Mech. **102** (1981), 263–278
7. Elliott, C. M., Ockendon, J. R.: Weak and Variational Methods for Moving Boundary Problems. Pitman, London, 1982
8. Crank, J., Gupta. R. S.: A Moving Boundary Problem Arising in the Diffusion of Oxygen. J. Inst. Math. Applic. **10** (1972), 19–44
9. Fasano, A., Primicerio, M., Howison, S. D., Ockendon, J. R.: Some Remarks on the Regularisation of One-Phase Stefan Problems. Quart. App. Math. **48** (1990), 153–168
10. Moore, D. W.: The Spontaneous Appearance of a Singularity in an Evolving Vortex Sheet. Proc. R. Soc. London. **A365** (1979), 105–119
11. Caflisch, R. E., Siegel, M.: Singularity Formation during Rayleigh-Taylor Instability. J. Fluid Mech. **252** (1993), 51–78
12. Carpio, A., Chapman, S. J., Howison, S. D.: Dynamics of Line Singularities. Phil. Trans. R. Soc. London **355** (1997), 2013–2024
13. Head, A. K., Howison, S. D., Ockendon, J. R., Tighe, S. P.: An Equilibrium Theory of Dislocation Continua. SIAM Review **35** (1993), 580–609
14. Hirsch, J. P., Lothe, J.: Theory of Dislocations. Wiley, 1982
15. Garabedian, P. R.: Oblique Water Entry of a Wedge. Commun. Pure Appl. Maths. **6** (1953), 157–165
16. Wagner, H. : Über Stoß- und Gleitvorgänge an der Oberfläche von Flüssigkeiten. Z. Angew. Math. Mech. **12** (1932), 193–215
17. Howison, S. D., Ockendon, J. R., Wilson, S. K.: Incompressible Water-Entry Problems at Small Deadrise Angles. J. Fluid. Mech. **222** (1990), 215
18. Turner, J. R. : The Frictional Unloading Problem in a Linear Elastic Half-Space. J. Inst. Math. Applics. **24** (1979), 439–470
19. Howison, S. D., Morgan. J., Ockendon, J. R.: A Class of Codimension-Two Free Boundary Problems. SIAM Rev. **39** (1996), 187–220
20. Freund, L. B.: Dynamic Fracture Mechanics. Cambridge University Press, 1990
21. Barenblatt, G. I.: The Mathematical Theory of Equilibrium Cracks in Brittle Fracture. Adv. Appl. Mech. **7** (1962), 55–129
22. Friedman, A., Bei Hu Velazquez, J. J.: Propagation of Cracks in Elastic Media. Preprint, 1999
23. Richardson, G., Stoth, B.: Ill-Posedness of the Mean-Field Model of Superconducting Vortices and the Regularisation thereof. Euro. J. Appl. Math., (2000), to appear

24. Rogers, J. C. W., Szymezak, W. G.: Computation of Violent Surface Motions: Comparisons with Theory and Experiment. Phil. Trans. R. Soc. London. **A355** (1997), 649–664

A New Model for the Dynamics of Dispersions in a Batch Reactor[*]

A. Fasano and F. Rosso

University of Firenze, Department of Mathematics *Ulisse Dini*, Viale Morgagni, 67/a, 50134 Firenze, Italy

Dedicated to Professor Karl-Heinz Hoffmann
on the occasion of his 60th birthday

Abstract. A new model for coalescence and breakage of liquid-liquid dispersion is presented. The main features are: (i) the introduction of an efficiency factor which controls the time rate of the various processes affecting the size distribution function of droplets, (ii) a new effect - that we call volume scattering - which is consistent with the experimentally observed circumstance of the existence of a top size limit for droplets depending on the general dynamical conditions. The model is proved to be mathematically and physically correct by proving existence and uniqueness of a regular solution to the Cauchy problem.

1 Introduction

A liquid-liquid dispersion is a mechanical system formed by a population of droplets of variable size of a given component A, finely distributed into another component B, immiscible with A. We will study here the evolution of such a system due to purely mechanical effects, although chemical and thermal actions can generally be important. These systems are rather common in important industrial and environmental applications such as colloid chemistry, crude oil extraction and pipelining, photographic processes, meteorology. We remark that the mathematical research about droplets interactions dates back to the beginning of the century (see [20] for example) and our reference list is far from being complete.

Rational design of dispersed phase reactors requires a knowledge of the evolution of the size distribution function, which is the object of several recent papers [1,2,4,5,7–9,16,17]. The literature about coalescence-fragmentation processes has grown considerably during recent years also because such phenomena are studied in connection with the dynamics of polymers chains (see, for example, [9]). Our work is partially inspired by those of Melzak [11], and Valentas *at al.* [18,19].

In this paper we present a new model for the evolution equation in which we introduce some new features in order to take into account some important experimentally observed facts. Some preliminary results had been sketched in [6]. First

[*] This work was partially supported by the G.N.F.M. Strategic Project "Metodi Matematici in Fluidodinamica e Dinamica Molecolare" and by the C.N.R. contract # 98.01027.CT01

of all we assume that there exists a finite top size limit v_o of the droplet volume[1], a fact usually disregarded in most of the mathematical research and which requires the presence of a new interaction mechanism as we shall see in a moment. If, on one side, letting the droplet volume to go to infinity simplifies the statement of the problem, on the other side the resulting mathematical complications (e. g. summability questions) are far from being trivial. Secondly we notice that, under certain conditions, the breakage action may be ineffective below a (positive) critical droplet volume v_{crit}, so that pre-existing (at time $t = 0$) droplets in $[0, v_{crit}]$ can only disappear because of coalescence. Finally we introduce an efficiency factor linked to some average property of the dispersion.

Differently from all the models considered in the literature, we add a *volume scattering term* not affecting the total number of droplets during interaction, and an *efficiency function* depending on both the total number of droplets $\mathfrak{N}(t)$ and the total interfacial area $\mathfrak{S}(t)$, which controls the time rate of all kinds of interactions . The action of the scattering term is to prevent the appearance of "too large" droplets, without introducing artificial cut-off in the coalescence kernel. In other words, volume scattering allows the formation of a *virtual* large droplet that does not survive and decays into a pair of droplets, each within the admissible volume range. Even if one introduces a singularity in the breakage kernel at $v = v_o$, coalescence would naturally lead to the formation of droplets of volume larger than v_o; therefore volume scattering seems to be the only mechanism consistent with the experimental evidence of a finite top size v_o.

To be more precise, let $f(v, t)$ (v = volume, t = time) be the distribution function of droplet size (per unit volume of dispersion). We assume droplets to be uniformly distributed in the reactor so that f does not depend on spatial coordinates. This is quite reasonable when the imposed shear rate is sufficiently high and the viscosity of the dispersion sufficiently low[2]. We also assume that the system is isolated, so that there is no heat or mass exchange effects.

It is worth noting that the theory illustrated in the present paper can be extended to the case of *multi-component* dispersions.

We formulate the following evolution equation and then we will comment the meaning of each term.

---

[1] It is quite obvious that there cannot exist droplets of arbitrarily large volume and some Authors introduce an artificial upper bound; however it should be possible to relate v_o to actual physical quantities such as the imposed shear rate, reactor (blade-impeller) geometry, percentage of dispersed phase (hold-up), temperature, and the chemical and physical characteristics of the two liquid phases, including surface tension. Although an explicit law for v_o as a function of all the above parameters is not available, experiments show very clearly that in a batch reactor v_o decreases when the impeller velocity is increased (see Figs.4, 4).

[2] When these conditions are not satisfied, droplets generally show observable convective motions due to buoyancy and gravity combined. However most of research on dispersions focusses upon the final droplet distribution *under perfect mixing condition*.

$$\frac{\partial f}{\partial t} = \phi(t) \left(\int_0^{v/2} \tau_c(w, v - w) f(w, t) f(v - w, t) \, dw \right.$$

$$- f(v, t) \int_0^{v_0 - v} \tau_c(v, w) f(w, t) \, dw$$

$$+ \int_v^{v_0} \alpha(w) \beta(w, v) f(w, t) \, dw - \alpha(v) f(v, t)$$

$$+ \int_{v_0 - v}^{v_0} dw \int_{v + w - v_0}^{(v+w)/2} \tau_s(u, v + w - u; v, w) f(u, t)$$

$$\times f(v + w - u, t) \, du$$

$$\left. - f(v, t) \int_{v_0 - v}^{v_0} f(w, t) \, dw \int_{v + w - v_0}^{(v+w)/2} \tau_s(v, w; u, v + w - u) \, du \right) .$$

(1.1)

Equation (1.1) models the three binary interaction processes that make f evolve with time. Let us describe their action in detail. Functions appearing in (1.1) have the following meaning:

(a) **Efficiency factor.** We defined $\phi(t) = \Phi\left[\mathfrak{N}(t), \mathfrak{S}(t)\right]$, where

$$\mathfrak{N}(t) = \int_0^{v_0} f(v, t) \, dv, \quad \mathfrak{S}(t) = \int_0^{v_0} v^{(2/3)} f(v, t) \, dv ,$$

(1.2)

represent respectively the number of droplets and the interfacial area per unit volume, and $\Phi(\mathfrak{N}, \mathfrak{S})$ is a given continuous function. We call $\phi(t)$ the *efficiency factor*: the underlying idea is that all interactions are driven by the internal power dissipation which in turn is related to both $\mathfrak{N}$ and $\mathfrak{S}$. The role of the factor Φ is to enhance or depress the dynamics, while the mechanical structure of the interactions is described by the respective kernels. However our model allows a high degree of freedom in the choice of Φ. This is a novelty with respect to similar models within this topic and it brings additional mathematical complications particularly in the existence proof of the Cauchy problem.

(b) **Coalescence kernel.** Let $\tau_c(v, w)$ represent the (binary) *coalescence kernel*. This is proportional to the probability that two colliding droplets of respective volumes v and w coalesce to form a unique droplet of volume $v + w$. A natural choice is to assume that the probability of coalescence is proportional to the total cross sectional area, that is

$$\tau_c \simeq \left(v^{1/3} + w^{1/3}\right)^2 .$$

(1.3)

Clearly $\tau_c(v, w)$ has to be symmetric:

$$\tau_c(v, w) = \tau_c(w, v) .$$

(1.4)

In this connection we notice that binary coalescence is the rule in most experimental observations: indeed coalescence appears to be a multi-step process and there is a very little probability that it involves more than two drops simultaneously.

In the first two integral terms the arguments of τ_c vary in the interval $(0, v_o)$, consistently with our assumption, but τ_c can be continued as in (1.3) in a larger domain. In the first term, which represents the gain rate at level volume v, the arguments of τ_c have the ordering $w < v - w$, otherwise coalescence events, giving rise to a droplet of volume v, would be counted twice. Instead, in the second term, representing loss rate from the same cell in the volume space, we must consider coalescence of droplets of volume v with droplets of any other size w such that $v + w \leq v_o$. We define the formal coalescence operator[3]

$$L_c f(v, t) = \int_0^{v/2} \tau_c(w, v - w) f(w, t) f(v - w, t)\, dw$$

$$- f(v, t) \int_0^{v_o - v} \tau_c(v, w) f(w, t)\, dw \ . \tag{1.5}$$

(c) **Breakage kernel.** Let

$$\tau_b(w, v) = \alpha(w)\beta(w, v), \quad (w > v) \ , \tag{1.6}$$

represent the *breakage kernel.* Here $\beta(w, u)\, du$ is the probability that a droplet of volume in the interval $(u, u + du)$ is generated by breakage of a droplet of volume w. We assume, for simplicity, that a single breakage event cannot produce more than two droplets (*binary* breakage). In this case u and $w - u$ have the same probability, that is

$$\beta(w, u) = \beta(w, w - u) \ , \tag{1.7}$$

and

$$\beta(w, u) = 0, \quad \text{if } w \leq u \ . \tag{1.8}$$

Accordingly $\beta(w, u)$ is normalized as follows

$$\int_0^{w/2} \beta(w, v)\, dv = 1 \ . \tag{1.9}$$

The third and fourth terms in (1.1) give the gain and loss rate, respectively, due to breakage.

Unlike coalescence, experiments show that *multiple* breakage may not be exceptional; however the case of multiple breakage adds further complications just to the definition of relevant kernels, not to the essential mathematical difficulties of the whole problem. For this reason we decided to confine ourselves to the binary case.

The continuous increasing function $\alpha(w)$ is the breakage rate. The integral

$$T_B^{-1} = \frac{1}{v_o} \int_0^{v_o} \alpha(v)\, dv \ , \tag{1.10}$$

---

[3] The functional space in which we choose f will be specified later.

is the *overall breakage frequency* (we call T_B *characteristic breakage time*).

We define the breakage operator by

$$L_\mathrm{b} f(v,t) = \int_v^{v_0} \alpha(w)\beta(w,v)f(w,t)\,dw - \alpha(v)f(v,t) \ . \tag{1.11}$$

If $\alpha(w) = 0$ only for $w = 0$, we say that we have *complete* breakage. The case of *limited* breakage corresponds to the existence of a lower positive limit $v_\mathrm{crit}^{(1)}$ to droplet volume so that α vanishes in $\left[0, v_\mathrm{crit}^{(1)}\right]$. The limit $v_\mathrm{crit}^{(1)}$ depends on several peculiar parameters[4].

We may also assume that droplets with volume less than $v_\mathrm{crit}^{(2)}$ are not *produced* by rupture. This is obtained by taking

$$\beta(w,u) = 0, \quad \text{for } w \in \left[0, 2v_\mathrm{crit}^{(2)}\right) \quad \text{and for } u \in \left[0, v_\mathrm{crit}^{(2)}\right) \ , \tag{1.12}$$

which clearly forbids that one of the two daughters of w, either u or $w - u$, with volume less that $v_\mathrm{crit}^{(2)}$ be produced by breakage. However, the cut-off limits for α and β are not necessarily related. In order to simplify our exposition and reduce the number of free parameters, we have decided to identify the two cut-off limits by setting $v_\mathrm{crit}^{(1)} = 2v_\mathrm{crit}^{(2)} := 2v_\mathrm{b}$. The factor 2 is due for consistency: indeed there are no chances to see droplets with volume $u < v_\mathrm{b}$ (resulting from the rupture of larger ones) and droplets with volume $u < 2v_\mathrm{b}$ do not break into smaller ones (one of which would necessarily have volume below v_b). Consequently, unless $v_\mathrm{b} = 0$, droplets with volume in the range $(0, v_\mathrm{b}]$ are those pre-existing at the initial instant and can only disappear, by coalescence, into larger droplets within the admissible range. The function $\beta(w,v)$ needs an extension for $w > v_0$ too, that will be defined below.

(d) Scattering kernel. Let $\tau_s(v, w; v + w, u)$ represent the *scattering* kernel regulating interactions of pairs of droplets whose cumulative volume exceeds v_0. This term is proportional to the frequency of collisions resulting in the following volume re-distribution: a pair (v, w) such that $v + w > v_0$ produces by coalescence a droplet which immediately decays into a pair $(u, v + w - u)$. Physically this amounts to say that, due to τ_s, large droplets (with volume greater than v_0) produced by coalescence are unstable. We call this process *volume scattering* for its analogy with the collision term in Boltzmann equation. Following this interpretation we are lead to the choice

$$\tau_s(v, w; v + w, u) = \tau_c(v, w)\beta(v + w, u) \ . \tag{1.13}$$

The function $\beta(s, u)$ for $s \in (v_0, 2v_0)$ preserves the symmetry property (1.7) and is normalized so that

$$\int_{s-v_0}^{s/2} \beta(s, u)\,du = 1 \ . \tag{1.14}$$

---

[4] A widely used (empirical) law, called *Weber relation*, defines

$$v_\mathrm{crit}^{(1)} = 10^{-4}\pi\,(\sigma/\varrho)^{1.8}\left(\omega^2 D^{4/3}\right)^{-1.8} \ ,$$

where σ and ϱ are, respectively, the surface tension and the density of the dispersed phase, ω is the angular velocity of the impeller and D is the impeller diameter.

At this point, by assuming (1.13) and (1.14), we can write the gain and loss rate terms by scattering in (1.1) by defining the following *scattering operator*

$$
L_s f(v, t) = \int_{v_o - v}^{v_o} dw \int_{v+w-v_o}^{(v+w)/2} \tau_c(u, v + w - u)\beta(v + w, v)f(u, t)
$$

$$
\times f(v + w - u, t)\, du - f(v, t) \int_{v_o - v}^{v_o} \tau_c(v, w)f(w, t)\, dw \ . \tag{1.15}
$$

The last term is nothing but the continuation of the loss term in the coalescence operator and is obtained by performing the integration w. r. t. the variable u, thanks to (1.14). The integration intervals are consistent with the range of v, w and the requirement that the outcome of scattering is a pair in the admissible range.

Equation (1.1) has to be completed with an initial condition $f(v, 0) = f_0(v)$ on $[0, v_o]$ and solved in the region $[0, v_o] \times [0, T]$, where f is required to be continuous.

For the reduced equation

$$
\frac{\partial f}{\partial t} = L_c f + L_b f \ , \tag{1.16}
$$

the available theory is rather complete. In [11] Melzak proved a global existence and uniqueness theorem of classical solution for sufficiently regular initial data. Other interesting results in this direction were obtained by Spouge in [14–16] and in [17] where he considered also a polydispersed case (with coalescence interaction only), by McLeod in [10] and, more recently, by Friedman et al. in [1,7]. Papers by Spouge [15,16] treat discrete cases and are more oriented towards numerical integration of (1.16). In [7] the difficulty of the problem is complicated by the assumption that each droplet carries a chemical species which is uniformly distributed within it, the concentration c of the chemical species being a variable quantity. This implies that the unknown function f depends on c too. This version of the model is particularly suitable for studying photographic emulsions. In both papers [1,7] the asymptotic behaviour of solutions for $t \to \infty$ is also investigated.

In the present paper we study existence and uniqueness to the Cauchy problem for the more general equation

$$
\frac{\partial f}{\partial t} = \phi(t)\, (L_c f + L_b f + L_s f) \ . \tag{1.17}
$$

2   Some Qualitative Properties

In this Section we prove some fairly intuitive qualitative properties, assuming that the kernels are continuous functions satisfying the hypotheses listed so far, that are simply suggested by the physics of the problem, namely:

(i) $\tau_c(v, w)$ is a positive continuous symmetric function on $[0, v_o] \times [0, v_o]$; we put
$$
\max_{[0, v_o] \times [0, v_o]} \tau_c = \tilde{\tau}_c \ (< +\infty).
$$

(ii) $\beta(w, v)$ for $w \in (0, v_o]$ is a non-negative continuous function such that

$$
\beta(w, v) = 0, \qquad \text{for } v \in [w, v_o] \text{ and } v \in [0, v_b] \ ,
$$
$$
\beta(w, v) = \beta(w, w - v), \quad \text{for } v \in [0, w] \ ,
$$

where $v_b \geq 0$ and $v_b/v_o \ll 1$. For $w \in (0, v_o]$, the function $\beta(w, v)$ is normalized according to (1.9)

(iii) $\beta(s, v)$ for $s \in (v_o, 2v_o]$ has the same properties as above, with the normalization described by (1.14)

(iv) $\alpha(w)$ is a continuous function in $[0, v_o]$ which vanishes in $[0, 2v_b]$ and is increasing for $w > 2v_b$; we put [5] $\max\limits_{[0,v_o]} \alpha = \tilde{\alpha}\,(< +\infty)$.

(v) $\tilde{\Phi} = \sup\limits_{\mathbb{R}^2_+} \Phi < +\infty$, and $\quad \hat{\Phi} = \inf\limits_{\mathbb{R}^2_+} \Phi > 0$.

The next two propositions state expected physical properties of solutions; their proofs are based on the properties of kernels and the usage of some appropriate changes of variables.

Proposition 2.1. *Let $f(v,t)$ be a non-negative solution of equation (1.17). Then*

(a) $\displaystyle\int_0^{v_o} L_c f(v,t)\, dv \leq 0,\quad$ *that is coalescence tends to decrease the number of droplets per unit volume.*

(b) $\displaystyle\int_0^{v_o} L_b f(v,t)\, dv \geq 0,\quad$ *that is the number of droplets per unit volume tends to increase because of (binary) breakage.*

(c) $\displaystyle\int_0^{v_o} L_s f(v,t)\, dv = 0,\quad$ *that is scattering (with binary exit) does not alter the number of droplets.*

Moreover, if for a given $t \geq 0$, $\operatorname{supp} f(\cdot, t) \neq \emptyset$, then (a) holds in a strict sense, while the integral in (b) is strictly positive, provided that $\operatorname{supp} \alpha(\cdot) \cap \operatorname{supp} f(\cdot, t) \neq \emptyset$.

Proof. Define

$$D = \{(w, v) \in \mathbb{R}^2 / w \in (0, v),\quad v \in (0, v_o)\}\ ,$$

$$E = \{(w, v) \in \mathbb{R}^2 / v \in (0, v_o),\quad w \in (0, v_o - v)\}\ ,$$

and consider the (measure preserving) change of variables $\mathcal{C}_1 : \mathbb{R}^2 \to \mathbb{R}^2$

$$\begin{cases} \sigma = v - w\ , \\ \xi = w\ . \end{cases} \tag{2.1}$$

---

[5] Taking α bounded in $[0, v_o]$ is just a matter of simplicity. With some additional work a singularity for $w \to v_o$ could be allowed.

Since $C_1(D) = E$, we have

$$\int_0^{v_o} L_c f(v,t)\, dv = \frac{1}{2} \iint_D \tau_c(w, v-w) f(w,t) f(v-w,t)\, dv dw$$

$$- \iint_E \tau_c(w,v) f(w,t) f(v,t)\, dv dw$$

$$= \frac{1}{2} \iint_{C_1(D)} \tau_c(\xi,\sigma) f(\xi,t) f(\sigma,t)\, d\xi d\sigma \qquad (2.2)$$

$$- \iint_E \tau_c(w,v) f(w,t) f(v,t)\, dv dw$$

$$= -\frac{1}{2} \iint_E \tau_c(w,v) f(w,t) f(v,t)\, dv dw \ .$$

which, for $f(\cdot,t) \geq 0$, implies (a). If supp $f(\cdot,t) \neq \emptyset$ then necessarily

$$(\text{supp } f(\cdot,t))^2 \cap E \neq \emptyset$$

and then the above integral is negative. Similarly we have

$$\int_0^{v_o} L_b f(v,t)\, dv = \int_0^{v_o} \left\{ \int_v^{v_o} \alpha(w)\beta(w,v) f(w,t)\, dw \right\} dv$$

$$- \int_0^{v_o} \alpha(v) f(v,t)\, dv \qquad (2.3)$$

$$= \iint_A \alpha(w)\beta(w,v) f(w,t)\, dw dv - \int_0^{v_o} \alpha(v) f(v,t)\, dv \ ,$$

where

$$A = \{(w,v) \in \mathbb{R}^2 / w \in (v, v_o),\quad v \in (0, v_o)\}\ ;$$

however we also have

$$A = \{(w,v) \in \mathbb{R}^2 / w \in (0, v_o),\quad v \in (0, w)\}\ ,$$

so that

$$\int_0^{v_o} L_b f(v,t)\, dv = \int_0^{v_o} \left\{ \int_0^w \beta(w,v)\, dv \right\} \alpha(w) f(w,t)\, dw$$

$$- \int_0^{v_o} \alpha(v) f(v,t)\, dv \ . \qquad (2.4)$$

Because of (1.9), we get

$$\int_0^{v_o} L f(v,t)\, dv = \int_0^{v_o} \alpha(w) f(w,t)\, dw \ , \qquad (2.5)$$

and (b) is proved. Notice that the inequality (b) is strict if

$$\text{supp } \alpha(\cdot) \cap \text{supp } f(\cdot,t) \neq \emptyset \ .$$

In order to prove (c) we set $s = v + w$ and rewrite the drops producing integral in the form

$$\int_{v_o}^{v_o+v} ds \int_{s-v_o}^{s/2} \tau_c(u, s - u)\beta(s, v) f(u, t) f(s - u, t) \, du \ . \qquad (2.6)$$

Next we integrate in the variable v, obtaining

$$\int_0^{v_o} dv \int_{v_o}^{v_o+v} ds \ \beta(s, v) \int_{s-v_o}^{s/2} \tau_c(u, s - u) f(u, t) f(s - u, t) \, du \ . \qquad (2.7)$$

Now we interchange the integral order of the first two integrals. Along this way we can factorize

$$\int_{s-v_o}^{v_o} \beta(s, v) \, dv = 2 \ ,$$

and with the change of variables

$$\begin{cases} \bar{u} = u \ , \\ \bar{w} = s - u \ . \end{cases} \qquad (2.8)$$

we reduce the above integral to the form

$$2 \iint_{T_1} \tau_c\,(\bar{u}, \bar{w})\, f\,(\bar{u}, t)\, f\,(\bar{w}, t) \ d\bar{u} d\bar{w} \ ,$$

where T_1 is the triangle in the $(\bar{u}, \bar{w})$-plane with vertices $(0, v_o)$, $(v_o/2, v_o/2)$, (v_o, v_o). The loss term has instead the form

$$\iint_{T_1 \cup T_2} \tau_c(u, w) f(u, t) f(w, t) \, du dw \ ,$$

where we have written u instead of v and $T_1 \cup T_2$ is the triangle in the (u, w)-plane with vertices $(0, v_o)$, (v_o, v_o), $(v_o, 0)$. Since the integral over T_2 is obtained from the integral over T_1 by exchanging the roles of u and w, we conclude that

$$\iint_{T_1 \cup T_2} \cdots = 2 \iint_{T_1} \cdots$$

because of the symmetry of τ_c. Thus the loss term exactly compensates the gain term and the proof is complete.

$\square$

It is physically evident that each of the processes of coalescence, breakage and scattering are volume preserving and therefore we expect that the total volume of the dispersed phase is constant. Indeed we can prove the following

Proposition 2.2. Let $f(v, t)$ be a solution of equation (1.17). Then the global volume of droplets does not change with time, i.e.

$$\mathcal{V}(t) = \int_0^{v_o} v f(v, t) \, dv = \int_0^{v_o} v f_o(v) \, dv = \mathcal{V}(0) \ .$$

Proof. We first check that

$$\int_0^{v_o} v L_c f(v, t)\, dv = 0 \ . \tag{2.9}$$

Proceeding as in the proof of Proposition 2.1, we get

$$\iint_D v\tau_c(w, v - w) f(w, t) f(v - w), t)\, dw dv$$

$$= \iint_E (\sigma + \xi)\tau_c(\xi, \sigma) f(\xi, t) f(\sigma, t)\, d\xi d\sigma = 2 \iint_E \tau \tau_c(\xi, \sigma) f(\xi, t) f(\sigma, t)\, d\xi d\sigma \ ;$$

Recalling (2.2) and the symmetry of τ_c, (2.9) is proved.

Notice now that, as in the proof of Proposition 2.1

$$\int_0^{v_o} v \left\{ \int_v^{v_o} \alpha(w)\beta(w, v) f(w, t)\, dw \right\} dv$$

$$= \int_0^{v_o} \left\{ \int_0^w v\beta(w, v)\, dv \right\} \alpha(w) f(w, t)\, dw \ . \tag{2.10}$$

Because of the symmetry of β, it is easy to check that the average volume

$$\frac{1}{2} \int_0^w v\beta(w, v)\, dv = \frac{w}{2} \ .$$

Therefore

$$\int_0^{v_o} v L_b f(v, t)\, dv = \int_0^{v_o} w\alpha(w) f(w, t)\, dw - \int_0^{v_o} v\alpha(v) f(v, t)\, dv = 0 \ .$$

We finally prove that

$$\int_0^{v_o} v L_s f(v, t)\, dv = 0 \ . \tag{2.11}$$

First we operate on the volume producing term

$$\int_0^{v_o} v\, dv \int_{v_o-v}^{v_o} dw \int_{v+w-v_o}^{(v+w)/2} \tau_c(u, v + w - u)\beta(v + w, v) \tag{2.12}$$
$$f(u, t) f(v + w - u, t)\, du \ ,$$

in the same way as we did on (2.6), i. e. we set $v + w = s$ and interchange the order of the integrals in v and in s. This allows to isolate the integral

$$\int_{s-v_o}^{v_o} v\beta(s, v)\, dv \ ,$$

which is twice the averaged volume of the outgoing droplets, i. e. s. Thus we have reduced (2.12) to the form

$$\int_{v_o}^{2v_o} s\, ds \int_{s-v_o}^{s/2} \tau_c(u, s - u) f(u, t) f(s - u, t)\, du \ , \tag{2.13}$$

which, by means of transformation (2.8) reduces to

$$\iint_{T_1} (\bar{u} + \bar{w}) \, \tau_c \, (\bar{u}, \bar{w}) \, f \, (\bar{u}, t) \, f \, (\bar{w}, t) \; d\bar{u} d\bar{w} \; .$$

This has to be compared with the loss term, that can be written as

$$\iint_{T_1 \cup T_2} u \tau_c (u, w) f(u, t) f(w, t) \, du dw \; .$$

It is now immediate to check that the two contributions cancel each other.

$\square$

3   Global Existence and Uniqueness

We take some further assumptions on the kernels, to be added to (i)–(v) of the previous Section.

(vi) The functions $\tau_c(w, v)$ and $\alpha(v)$ are continuously differentiable[6];

(vii) $\beta(w, v)$ is piecewise continuously differentiable for $w \in (2v_b, 2v_o)$ and for all $v \in (v_b, w)$ and in addition:

    (a) $\alpha(w) \left| \dfrac{\partial \beta(w, v)}{\partial v} \right|$ has at most an integrable singularity as $w \to 2v_b$;

    (b) for w in a neighbourhood of $2v_o$, e.g. $w \in (3v_o/2, 2v_o)$, $\beta(w, v)$ is independent of v.

The latter assumption guarantees that $\dfrac{\partial \beta(w, v)}{\partial v}$ is identically zero when β approaches the behaviour of a Dirac distribution. Although it may look artificial, this is physically meaningful and may be refined provided that the integral of $\left| \dfrac{\partial \beta(w, v)}{\partial v} \right|$ on the support of β remains bounded. To be more specific, we can define $\beta(w, v)$ as follows:

$$\beta(w, v) = 12 \frac{(v - v_b)_+ \, (w - v - v_b)_+}{(w - 2v_b)_+^3}, \quad \text{for } w \in (2v_b, v_o + v_b) \; , \qquad (3.1)$$

$$\beta(s, v) = -\frac{c(s)}{2} \left(v - \frac{s}{2} \right)^2 + d(s), \quad \text{for } s \in (v_o + v_b, 2v_o) \; , \qquad (3.2)$$

where $c(s)$ can be chosen as a linear decreasing continuous function which in $s = v_o + v_b$ equals the curvature of $\beta(w, v_o)$ for $w \to (v_o + v_b)^-$ and is zero in $\left[\dfrac{3v_o}{2}, v_o \right]$, while $d(s)$ is determined by the the normalization request (1.14).

---

[6] Recall however footnote 5

Remark 3.1. According to (3.1), assumption (a) above requires

$$\alpha(w) \approx (w - 2v_b)^{1+\nu} \ ,$$

with $\nu > 0$. We observe that this implies that the integral

$$\int_{v+v_b}^{v_o} \alpha(w)\beta(w,v)\,dw$$

remains bounded as $v \to v_b$

Remark 3.2. It is important to stress that, for $s \in \left(\dfrac{3}{2}v_o, 2v_o\right)$, we have $\beta(s,v) = (v_o - s/2)^{-1}$ for $v \in (s - v_o, s/2)$ and consequently, for v in the vicinity of v_o

$$\int_{v_o}^{v_o+v} \beta(s,v)\,ds = \int_{2v}^{v_o+v} \frac{1}{v_o - s/2}\,ds = 2\ln 2 \ .$$

We finally need some assumption on the initial distribution $f_o(v)$

(viii) $f_o(v) \in C^1\left([0, v_o(v)]\right)$, $f_o(v) \ge 0$, with $f_o(0) = 0$.

We want to prove the following

Main Theorem. *The Cauchy problem given by (1.1) with the initial condition $f(v,0) = f_o(v)$ has a unique solution which exists for all times. The solution is non-negative, continuously differentiable in t and Lipschitz continuous in v.*

The proof follows from the collection of Theorems 3.1, 3.2, 3.3 below. The classical and elegant proof of local existence for the reduced equation (1.16) given by Melzak in [11] cannot be duplicated in this case. This *is not* due to the presence of the additional scattering term, since the key point in Melzak's proof is that the r.h.s. of the integro-differential evolution equation is bilinear. The main complication here is caused by the factor $\phi(t)$ which is a functional of f. However, we can see that Melzak's procedure of representing the solution as a power series still works provided we work step by step along a suitably defined infinite sequence of time intervals. This approach requires nontrivial modifications of the proof given in [11]: in particular, we have to delay the argument in $\phi(t)$, to choose conveniently the time intervals, and to obtain an estimate of both the first derivatives of the elements of the approximating sequence, so to get compactness. For the sake of brevity we will not report this modified proof here, because we want to follow a different procedure which has the remarkable advantage of being an effective shortcut in the proof of positivity.

In the first step of our reasoning, we introduce the following modified Cauchy problem (3.3): find a function $\psi(v,t)$, continuously differentiable in t and Lipschitz continuous in v such that

$$\begin{cases} \dfrac{\partial \psi}{\partial t} = \phi(t) \left(L_c^+ \psi + L_b^+ \psi + L_s^+ \psi \right) , \\[2mm] \psi(v,0) = f_o(v) \geq 0 , \end{cases} \tag{3.3}$$

where the operators L_λ^+ $(\lambda = c, b, s)$ are defined as follows

$$L_c^+ \psi = \int_0^{v/2} \tau_c(w, v - w) \psi_+(w, t) \psi_+(v - w, t)\, dw$$
$$-\psi(v, t) \int_0^{v_o - v} \tau_c(v, w) |\psi(w, t)|\, dw , \tag{3.4}$$

$$L_b^+ \psi = \int_v^{v_o} \alpha(w) \beta(w, v) \psi_+(w, t)\, dw - \alpha(v) \psi(v, t) , \tag{3.5}$$

$$L_s^+ \psi = \int_{v_o - v}^{v_o} dw \int_{v + w - v_o}^{(v+w)/2} \tau_c(u, v + w - u) \beta(v + w, v) \psi_+(u, t) \psi_+(v + w - u, t)\, du$$
$$-\psi(v, t) \int_{v_o - v}^{v_o} \tau_c(v, w) \left| \psi(w, t) \right| dw . \tag{3.6}$$

Theorem 3.1. *(Positivity) Under the assumptions (i)-(viii) listed above all solutions of (3.3) are non-negative.*

Proof. The proof follows immediately from inequality

$$\frac{\partial \psi}{\partial t} \geq -P(v, t) \psi(v, t), \quad \psi(v, 0) \geq 0 , \tag{3.7}$$

in which

$$P(v, t) = \int_0^{v_o - v} \tau_c(v, w) \left| \psi(w, t) \right| dw + \alpha(v) + \int_{v_o - v}^{v_o} \tau_c(v, w) \left| \psi(w, t) \right| dw ,$$

is non-negative and bounded for bounded ψ and in particular is strictly positive for a given v preventing the existence of a negative minimum of ψ. □

Notice now that, if we prove

(I) uniqueness of solutions to the original Cauchy problem,

(II) existence of bounded solutions on bounded (arbitrary) time intervals to the modified problem (3.3), we obtain also existence and non-negativity of the solution to the original problem. We will proceed along this way.

Theorem 3.2. (Uniqueness) *There is no more than one (bounded) solution to the Cauchy problem for equation (1.17) with the initial condition $f(v,0) = f_o(v)$.*

Proof. Let $\gamma = f_1 - f_2$, and

$$\phi_i(t) = \Phi \left[\int_0^{v_o} f_i(v,t) \, dv, \int_0^{v_o} f_i(v,t) v^{2/3} \, dv \right], \quad (i = 1, 2) \ ;$$

then, since

$$f_i(v,t) = f_o(v) + \int_0^t \phi_i(s) \left\{ L_c f_i(v,s) + L_b f_i(v,s) + L_s f_i(v,s) \right\} ds \ ,$$

we have

$$\left| \gamma(v,t) \right| \leq \int_0^t \left\{ \left| \phi_1(\theta) L_c f_1(v,\theta) - \phi_2(\theta) L_c f_2(v,\theta) \right| \right.$$

$$+ \left| \phi_1(\theta) L_b f_1(v,\theta) - \phi_2(\theta) L_b f_2(v,\theta) \right|$$

$$+ \left. \left| \phi_1(\theta) L_s f_1(v,\theta) - \phi_2(\theta) L_s f_2(v,\theta) \right| \right\} d\theta \ .$$

By adding and subtracting the appropriate terms and using the properties of the kernels α, β, τ_c and the boundedness of f_1, f_2, we easily arrive at the inequality

$$\left| \gamma(v,t) \right| \leq C \int_0^t \left[|\gamma(v,\theta)| + |\phi_1(\theta) - \phi_2(\theta)| \right] d\theta \ , \tag{3.8}$$

where C is a given positive constant.

Recalling the assumptions on ϕ, we have the estimate

$$|\phi_1(\theta) - \phi_2(\theta)| \leq C_1 \sup_{v \in (0,v_o)} |\gamma(v,t)| \ , \tag{3.9}$$

where C_1 is also known.

Therefore, if we set $\Gamma(t) = \sup_{v \in (0,v_o)} |\gamma(v,t)|$, from (3.8) and (3.9), we can derive the Gronwall inequality

$$0 \leq \Gamma(t) \leq C(1 + C_1) \int_0^t \Gamma(\theta) \, d\theta \ ,$$

implying $\Gamma(t) \equiv 0$ and hence uniqueness   □

Theorem 3.3. (Local Existence) *Under the assumptions (i)-(viii), the modified problem (3.3) has a solution ψ which is Lipschitz continuous in $[0, v_o] \times [0, \theta^*]$, for some positive θ^*.*

Proof. We take suitable time steps $0 = \theta_0 < \theta_1 < \ldots < \theta_n$ (to be defined later), and in each interval (θ_i, θ_{i+1}) we solve the problem

$$
\begin{cases}
\dfrac{\partial \psi^{(n)}}{\partial t}(v,t) = -\phi_i \psi^{(n)}(v,t) \left\{ \alpha(v) + \displaystyle\int_0^{v_o - v} \tau_{\mathrm{c}}(v,w) \left| \psi_i^{(n)}(w) \right| dw \right. \\[2ex]
\qquad + \left. \displaystyle\int_{v_o - v}^{v_o} \tau_{\mathrm{c}}(v,w) \left| \psi_i^{(n)}(w) \right| dw \right\} \\[2ex]
\qquad + \phi_i \displaystyle\int_0^{v/2} \tau_{\mathrm{c}}(w, v-w) \left(\psi_+^{(n)}\right)_i(w) \left(\psi_+^{(n)}\right)_i(v-w)\, dw \\[2ex]
\qquad + \phi_i \displaystyle\int_v^{v_o} \alpha(w)\beta(v,w) \left(\psi_+^{(n)}\right)_i(w)\, dw \\[2ex]
\qquad + \phi_i \displaystyle\int_{v_o - v}^{v_o} dw \displaystyle\int_{v+w-v_o}^{(v+w)/2} \left[\tau_{\mathrm{c}}(u, v+w-u)\beta(v+w,v) \left(\psi_+^{(n)}\right)_i(u) \right. \\[2ex]
\qquad \times \left. \left(\psi_+^{(n)}\right)_i(v+w-u) \right] du\,, \qquad \text{for } \theta_i < t < \theta_{i+1}\,, \\[2ex]
\psi^{(n)}(v,\theta_i) = \left(\psi_+^{(n)}\right)_i(v)\,.
\end{cases}
\tag{3.10}
$$

Here the subscript i means that the corresponding function is the one obtained at the previous time interval (θ_{i-1}, θ_i) and evaluated as $t \to \theta_i^-$. In particular, for $i = 0$ all the functions are evaluated for $t = 0$.

Equation(3.10) is just a linear o.d.e. that can be solved explicitly and for which we find the estimate

$$
\left| \psi^{(n)}(v,t) \right| \leq \left\| \left(\psi^{(n)}\right)_i \right\| \left[1 + c \left\| \left(\psi^{(n)}\right)_i \right\| (t - \theta_i) \right]\,,
\tag{3.11}
$$

where $\|\cdot\|$ denotes the sup-norm and c depends only on the kernels (note the use of assumptions (vii) on α and β; see also Remarks 3.1 and 3.2).

Thanks to (3.11) we can obtain by iteration

$$
\left\| \left(\psi^{(n)}\right)_{i+1} \right\| \leq \|f_0\| (1+\varepsilon)^{i+1}\,,
\tag{3.12}
$$

provided that

$$
\theta_{i+1} - \theta_i = \frac{1}{c\|f_0\|} \frac{\varepsilon}{(1+\varepsilon)^i}\,.
\tag{3.13}
$$

Thus, by taking $\varepsilon = 1/n$, we have uniform estimates for $\left| \psi^{(n)}(v,t) \right|$ and $\left| \dfrac{\partial \psi^{(n)}}{\partial t} \right|$. Also we note that θ_n is uniformly estimated from below by some $\theta^* > 0$ while it tends to a finite limit as $n \to \infty$. The argument for estimating $\dfrac{\partial \psi^{(n)}}{\partial v}$ is based on the differentiation of (3.10) with respect to v, which we perform assuming that

$\left(\psi^{(n)}\right)_i$ is differentiable, so that $\left(\psi_+^{(n)}\right)_i$ has an L^∞ derivative with respect to v. This procedure will turn out to be justified.

$$\frac{\partial}{\partial t}\frac{\partial \psi^{(n)}}{\partial v} = -\phi_i M_i \frac{\partial \psi^{(n)}}{\partial v} + \phi_i N_i$$

$$+ \phi_i \int_0^{v/2} \tau_c(w, v-w)\left(\psi_+^{(n)}\right)_i(w)\frac{\partial}{\partial v}\left(\psi_+^{(n)}\right)_i(v-w)\,dw$$

$$+ \phi_i \int_{v_o-v}^{v_o} dw \int_{v+w-v_o}^{v_o}\left[\tau_c(u, v+w-u)\beta(v+w,v)\left(\psi_+^{(n)}\right)_i(u)\right.$$

$$\left. \times \frac{\partial}{\partial v}\left(\psi_+^{(n)}\right)_i(v+w-u)\right]du \ ,$$

(3.14)

where M_i, N_i are easily estimated in terms of known quantities (here we have used assumption (vii)-(b) on β). We look at (3.14) as an o.d.e. for $\frac{\partial}{\partial v}\left(\psi^{(n)}\right)$ that can be immediately integrated.

Since $\left\|\frac{\partial}{\partial v}\left(\psi_+^{(n)}\right)_i\right\| \le \left\|\frac{\partial}{\partial v}\left(\psi^{(n)}\right)_i\right\|$, we obtain

$$\left|\frac{\partial}{\partial v}\psi^{(n)}\right| \le \left\|\frac{\partial}{\partial v}\left(\psi^{(n)}\right)_i\right\|(1+c_1\delta_i), \quad \delta_i = \theta_{i+1}-\theta_i \ ,$$

(3.15)

where $c_1 > 0$ is a known constant. Recalling (3.13) we have $1+c_1\delta_i \le 1+c_2\varepsilon$ and, by induction,

$$\left\|\frac{\partial}{\partial v}\psi^{(n)}\right\| \le \|f_o'\|(1+c_2\varepsilon)^n \ ,$$

(3.16)

that, with $\varepsilon = 1/n$, gives the desired uniform estimate, and at the same time proves that we were allowed to write equation (3.14).

By compactness in $C^0\left([0,v_o]\times[0,\theta^*]\right)$ we can now pass to the limit in the integrated form of (3.10) through some subsequence and get existence for the modified problem. □

We have already pointed out that the Theorem just proved implies local existence for the original problem too. We are now going to show that the solution of the original problem can be continued for all times.

Theorem 3.4. *(Global existence) Under the hypotheses of theorem 3.3, the solution $f(v,t)$ can be extended in $[0,v_o]\times[0,T]$ for any finite $T > 0$ with the same regularity properties.*

Proof. Recalling the proof of Proposition 2.1 it is rather easy to check that $\mathfrak{N}(t)$ obeys the following equation:

$$\frac{d\mathfrak{N}}{dt} = \phi(t)\left\{\int_0^{v_o}\alpha(v)f(v,t)\,dv - \frac{1}{2}\iint_E \tau_c(v,w)f(v,t)f(w,t)\,dwdv\right\} \ .$$

Let us recall that $\widetilde{\Phi}$ and $\widetilde{\alpha}$ are both finite; then, in the time interval $[0, T)$, where $f(v, t)$ exists and is positive, we get that

$$\frac{d\mathfrak{N}}{dt} \leq \widetilde{\phi}\widetilde{\alpha}\mathfrak{N} ,$$

that is

$$\mathfrak{N}(t) \leq \mathfrak{N}(0) \exp\left(\widetilde{\phi}\widetilde{\alpha}T\right) . \tag{3.17}$$

On the other hand, directly from equation (1.1) and the hypotheses we made on kernels, we can easily see that (by eliminating the negative terms and recalling that $f(v, t) \leq \max\limits_{[0,v_o]\times[0,t]} |f| = \|f\|_t$) the following estimate holds

$$f(v, t) \leq \|f\|_t \leq \sup_{[0,v_o]} f_o + C_1 \int_0^t \|f\|_\theta \left(\mathfrak{N}(\theta) + 1\right) d\theta , \tag{3.18}$$

where C_1 is a suitable positive constant. Because of (3.17), inequality (3.18) and Gronwall's Lemma immediately imply that

$$0 \leq \|f\|_t \leq C_2(T) < \infty .$$

On the other hand, we have seen that $\left|\dfrac{\partial f}{\partial v}\right|$ can be estimated in terms of $\sup f$. Therefore we can still use $f(v, T)$ as a new initial data for equation (1.1) and extend $f(v, t)$ after $t = T$. Well-known results on o.d.e.'s allow us to conclude that the solution exists over an arbitrary time interval.

□

We finally recall that uniqueness just required boundedness of f, so that Theorem 3.1 extends too.

4   Concluding Remarks

We have studied a model for the evolution of a liquid-liquid dispersion, emphasizing the necessity of introducing the so-called *volume scattering* effect in order to make the model consistent with the assumption that no droplet of volume larger than some given v_o can exist. The latter assumption is strongly suggested by experimental observations (see Figs. 4, 4). The model includes also an efficiency factor depending on the total number of droplets and on the total interfacial area of the dispersion. The volume scattering enters the evolution equation in a form similar to the scattering term in the Boltzmann's equation (hence its name). Global existence and uniqueness are proved under mild assumptions on the kernels.

Many questions are left open that deserve further research. Among these, we quote multiple breakage (and scattering), thermal effects. Numerical simulations will be presented in a forthcoming paper.

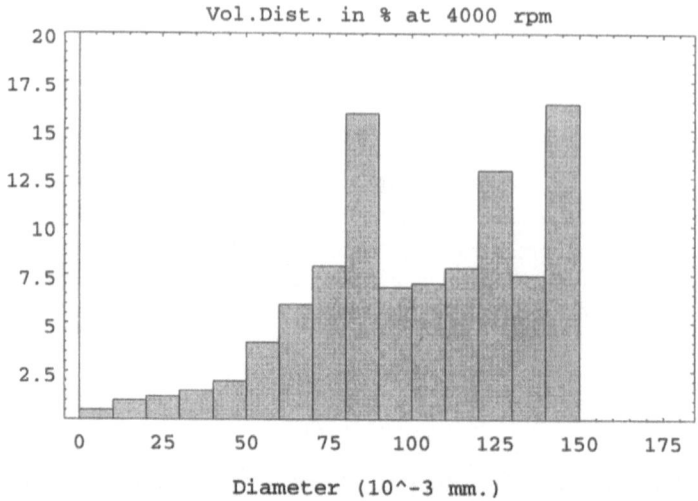

Fig. 1. Typical narrow shape after blade impeller mixing [courtesy of K. Panous-sopoulos]

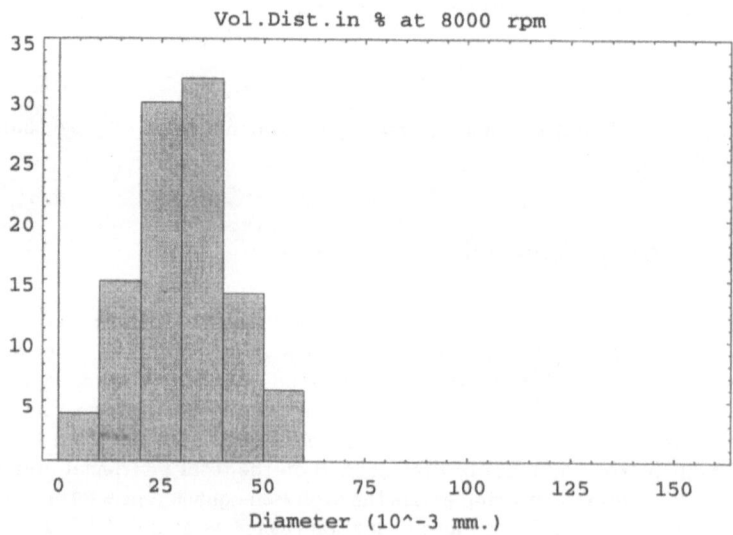

Fig. 2. Typical narrow shape after Ultra Turrax [courtesy of K. Panoussopoulos]

References

1. Bruno, O., Friedman, A., Reitich, F.: Asymptotic Behavior for a Coalescence Problem. Trans. Amer. Math. Soc. **338** (1993), 133–158
2. Carr, J.: Asymptotic Behaviour of Solutions to the Coagulation-Fragmentation Equations I Proc. Royal Soc. Edinburgh, Sect. A **121** (1992), 231–244
3. Chandrasekhar, S.: Stochastic Processes in Physics and Astronomy. Rev. Modern Phys. **15** (1943), 1–16
4. Dubovskii, P., B.: Mathematical Theory of Coagulation. Lecture Notes Series Number **23** (1994) Research Institute of Mathematics, Global Analysis Research Center, Seoul National University, Seoul, Korea
5. Dubovskii, P., B., Stewart, I., W.: Existence, Uniqueness, and Mass Conservation for the Coagulation-Fragmentation Equation. Math. Methods Appl. Sciences **19** (1996), 571–591
6. Fasano, A, Rosso, F.: Analysis of the Dynamics of Liquid-Liquid Dispersions. Progress in Industrial Mathematics at ECMI 98, Arkeryd, L., Bergh, J., Brenner, P., Pettersson, R., eds., Teubner, Stuttgart, 1999, 214–221
7. Friedman, A., Reitich, F.: Asymptotic Behavior of Solutions of Coagulation-Fragmentation Models. IMA Preprint Series **1479** (1997), 1–27
8. Herrero, M. A., Velàzquez, J. J. L., Wrzosek, D.: Sol-gel Transition in a Coagulation-Diffusion Model. Preprint, 1999
9. Laurençot,P., Wrzosek, D.: Fragmentation-Diffusion Model. Existence of Solutions and their Asymptotic Behaviour. Proc. Roy. Soc. Edimburgh Sec. A **128** (1998), 759–774
10. Mc Leod, J. B.: On the Scalar Transport Equation., Proc. London Math. Soc. **14** (1964), 445–458
11. Melzak, Z., A.: A Scalar Transport Equation. Trans. Amer. Math. Soc. **85** (1957), 547
12. Panoussopoulos, K.: Separation of Crude Oil-Water Emulsions: Experimental Techniques and Models. Ph.D. Thesis, Swiss Federal Institute of Technology Zurich, 1998
13. Shinnar, R.: On the Behaviour of Liquid Dispersions in Mixing Vessel. J. Fluid Mech. **10** (1961), 259–268
14. Spouge, J., L.: Analytic Solutions to Smoluchowski's Coagulation Equation: a Combinatorial Interpretation. J. Phys. A **18** (1985), 3063–3069.
15. Spouge, J. L.: An Existence Theorem for the Discrete Coagulation - Fragmentation Equations. II. Inclusion of Source and Efflux Terms. Math. Proc. Cambridge Phil. Soc. **98** (1985), 183–185
16. Spouge, J., L.: An Existence Theorem for the Discrete Coagulation - Fragmentation Equations. Math. Proc. Camb. Phil. Soc. **96** (1984), 351–357
17. Spouge, J., L.: A Branching-Process Solution of the Polydisperse Coagulation Equation. Adv. Appl. Prob. **16** (1984), 56–69
18. Valentas, K., J., Amundson, N., R.: Breakage and Coalescence in Dispersed Phase Systems. I& E C Fundamentals **5** (1966), 533–542
19. Valentas, K., J., Bilous, O., Amundson, N., R.: Analysis of Breakage in Dispersed Phase Systems. I& E C Fundamentals **5** (1966), 271–279
20. von Smoluchowski, M.: Versuch einer Mathematischen Theorie der Kogulationskinetic Kolloid Lösungen. Z. Phys. Chem. **92** (1917), 129–135

Optimal Control Problems for the Navier-Stokes Equations*

A. Fursikov[1], M. Gunzburger[2], L. S. Hou[3], and S. Manservisi[4]

[1] Department of Mechanics and Mathematics, Moscow State University, Moscow 119899, Russia
[2] Department of Mathematics, Iowa State University, Ames IA 50011-2064, USA
[3] Department of Mathematics, Iowa State University, Ames IA 50011-2064, USA and Department of Mathematics and Statistics, York University, Toronto, Ontario M3J 1P3, Canada
[4] DIENCA, Universita' degli studi di Bologna, Via dei Colli 16, 40136 Bologna, Italy. Current address: ITWM, Kaiserslautern University, Erwin-Schrödinger-Straße, Kaiserslautern, D-67663, Germany

Dedicated to Professor Karl-Heinz Hoffmann
on the occasion of his 60th birthday

Abstract. Optimal boundary control problems and related inhomogeneous boundary value problems for the Navier-Stokes equations are considered. The control is the data in the Dirichlet boundary condition. The objective functional is the drag on a body immersed in the fluid. The size of the control is limited through the application of explicit bounds or through penalization of the drag functional. A necessary step in the analysis of both the control problems and the related boundary value problems is the characterization of traces of solenoidal vector fields. Such characterization results are given in two and three dimensions as are existence results about solutions of the boundary value problems. Results about the existence of solutions of the optimal control problem are given in the two-dimensional case, as are results concerning the numerical approximation of optimal solutions.

1 Introduction

We are motivated by the study of optimal control problems for the time-dependent Navier-Stokes equations of incompressible, viscous flows. Our goal is to analyze and compute solutions of such problems. Here, we focus on one particular optimal control problem, namely the minimization of the drag on

* M. Gunzburger was supported by the National Science Foundation under grant number 9806358 and the Air Force Office of Scientific Research under grant F49620-95-1-040. L. S. Hou was supported by the Natural Science and Engineering Research Council of Canada under grant OGP-0137436. S. Manservisi was supported by the European Community under grant XCT-97-0117.

a body immersed in a fluid through adjustments on the velocity of the fluid at the boundary. Thus, the mathematical statement of the problem is to minimize, with respect to the velocity on the boundary, a functional that measures the drag on the body subject to the flow variables satisfying the Navier-Stokes system.

In order to study such optimal control problems, we must study two related problems. The first is the characterization of traces of solenoidal vector fields and the second is inhomogeneous boundary value problems for the Navier-Stokes equations. In Sect. 2, we review results for these two related problems in two and three dimensions. We also review results about the optimal control problem in the two-dimensional setting. Detailed presentations of these results are given in [2] and [3]. In §3, we briefly study the numerical approximation of solutions of the optimal control problem, based on results found in [6].

Of course, optimal control problems for the Navier-Stokes equations have been the subject of many other investigations in different contexts; the literature is too vast to mention here.

2   Analysis of Optimal Control Problems for the Navier-Stokes System

2.1   Model Problems in Drag Minimization

We use the following notation:

$$\Omega = \text{domain exterior to a bounded body } \partial\Omega$$
$$(0,T) = \text{specified time interval}$$
$$\mathbf{v}, p = \text{unknown velocity and pressure fields}$$
$$\mathbf{g} = \text{unknown velocity control on the boundary}$$
$$\mathbf{v}_0 = \text{given initial velocity field}$$
$$\mathbf{v}_\infty = \text{given constant velocity at infinity}$$
$$\rho, \mu = \text{constant density and viscosity coefficient}$$
$$\mathbf{w} = \mathbf{v} - \mathbf{v}_\infty = \text{difference flow}$$
$$\mathcal{D}(\mathbf{v}) = \tfrac{1}{2}\left(\nabla\mathbf{v} + (\nabla\mathbf{v})^T\right) = \text{rate of deformation tensor}$$
$$\mathcal{T} = -pI + 2\mu\mathcal{D}(\mathbf{v}) = \text{stress tensor}.$$

Then, the Navier-Stokes system describing incompressible, viscous flows is given by

$$\rho\partial_t\mathbf{v} - 2\mu\mathcal{D}(\mathbf{v}) + \rho\mathbf{v}\cdot\nabla\mathbf{v} + \nabla p = \mathbf{0} \quad \text{in } (0,T)\times\Omega \qquad (2.1)$$

$$\nabla\cdot\mathbf{v} = 0 \qquad \text{in } (0,T)\times\Omega \qquad (2.2)$$

$$\mathbf{v}|_{t=0} = \mathbf{v}_0 \qquad \text{for } \mathbf{x}\in\Omega \qquad (2.3)$$

$$\mathbf{v} \to \mathbf{v}_\infty \qquad \text{as } |\mathbf{x}| \to \infty \tag{2.4}$$

$$\mathbf{v}|_{\partial\Omega} = \mathbf{g} \qquad \text{in } (0,T) \times \partial\Omega. \tag{2.5}$$

This system will act as a constraint within the optimization problems we will consider. Note that the control function $\mathbf{g}$ acts on the whole boundary $\partial\Omega$; this is done for simplicity of exposition. The results hold as well for the case of control applied on only a portion of the boundary if one adds appropriate compatibility conditions on the control function.

The functional to be minimized is deduced from the following energy identity (see, e.g., [2] for a detailed derivation):

initial kinetic energy of the difference flow	$\dfrac{\rho}{2} \displaystyle\int_\Omega |\mathbf{w}(0,\mathbf{x})|^2 \, d\mathbf{x}$
$+$	$+$
work due to drag	$\displaystyle\int_0^T dt \int_{\partial\Omega} \mathbf{w} \cdot \mathcal{T}\mathbf{n} \, ds$
$=$	$=$
final kinetic energy of the difference flow	$\dfrac{\rho}{2} \displaystyle\int_\Omega |\mathbf{w}(T,\mathbf{x})|^2 \, d\mathbf{x}$
$+$	$+$
energy dissipated due to friction	$2\mu \displaystyle\int_0^T dt \int_\Omega \mathcal{D}(\mathbf{v}) : \mathcal{D}(\mathbf{v}) \, d\mathbf{x}$
$+$	$+$
work done by the boundary control	$\dfrac{\rho}{2} \displaystyle\int_0^T dt \int_{\partial\Omega} |\mathbf{w}|^2 \mathbf{v} \cdot \mathbf{n} \, ds.$

Then, the *drag functional* we will study is given by

$$\mathcal{J}(\mathbf{w}) = \frac{\rho}{4} \int_\Omega |\mathbf{w}(T,\mathbf{x})|^2 \, d\mathbf{x}$$
$$+ \mu \int_0^T dt \int_\Omega \mathcal{D}(\mathbf{v}) : \mathcal{D}(\mathbf{v}) \, d\mathbf{x} + \frac{\rho}{4} \int_0^T dt \int_{\partial\Omega} |\mathbf{w}|^2 \mathbf{v} \cdot \mathbf{n} \, ds. \tag{2.6}$$

Note that this functional is related to the work due to drag through the subtraction of the constant initial kinetic energy and multiplication by a factor of $1/2$. Thus, minimizing the functional in (2.6) is equivalent to minimizing the work due to drag. Also note that in many instances in the literature, the minimization of the energy dissipated due to friction is mistakenly assumed to be equivalent to drag minimization. The above energy identity shows that this is indeed not the case and that one should include the two additional terms in the functional in (2.6). We note that many other functionals can also be treated by the techniques described in this paper.

The problem we want to consider is to minimize the drag functional (2.6) over "suitable" boundary velocities **g**, subject to the Navier-Stokes system (2.1)–(2.5). However, in general, this problem is not well posed since there is no guarantee that the problem can be solved with bounded controls. Thus, one must limit the size of candidate control functions. We consider two means for limiting the size of the control. In the first approach, we *penalize the cost functional* by a measure of the size of the control. In two dimensions, we use the penalized functional

$$\mathcal{J}_N(\mathbf{v}) = \mathcal{J}(\mathbf{w}) + \rho N \left(\int_0^T \int_{\partial\Omega} |\mathbf{v}|^k \, ds \, dt + \int_0^T \int_{\partial\Omega} |\partial_t \mathbf{v}|^2 \, ds \, dt \right), \qquad (2.7)$$

where $k \geq 3$ and $N > 0$ with $N > \frac{1}{4}$ if $k = 3$. In the second approach, we *impose an explicit bound on the control*, e.g., in two dimensions we require that candidate controls satisfy

$$\int_0^T \int_{\partial\Omega} |\mathbf{v}|^k \, ds \, dt + \int_0^T \int_{\partial\Omega} |\partial_t \mathbf{v}|^2 \, ds \, dt \leq M, \qquad (2.8)$$

where $k \geq 3$ and for $M > 0$. The restrictions on k and N are governed by the mathematical analyses, specifically, the need to "dominate" the last term in the functional (2.6). The norms used in (2.7) and (2.8) to limit the size of the control are the weakest possible. Note that we have written the functional in terms of **v** and not **g**. Of course, on $\partial\Omega$, $\mathbf{v} = \mathbf{g}$; we are merely taking the view that since **g** is not known, the boundary conditions (2.5) may be deleted from the statement of the problem.

There is an interesting observation that has significant practical implications about what the analyses say is needed in order to limit the size of the control. Note that the terms added to the functional in (2.7) and used in the bound in (2.8) involve the time derivative of the control but do not involve spatial derivatives. Thus, to limit oscillations in time, one must limit the size of the time derivative of the control, but to limit spatial oscillations, it suffices to limit the size of an $L^k(\partial\Omega)$ norm with k sufficiently large. The particular choice of norm used to limit the size of the control has a significant effect on the optimality system from which optimal states and controls may be deduced. In particular, since no spatial derivatives are used, the necessary condition for optimality does not contain spatial derivatives, which effects a great simplification. For details, see [2].

The optimal control problems we consider are then given (in two dimensions) as follows.

Problem 1 Given Ω, T, μ, ρ, $\mathbf{v}_\infty$, $\mathbf{v}_0$, and N, find **v** such that
  – the Navier-Stokes system (2.1)–(2.4) is satisfied and
  – the functional $\mathcal{J}_N(\mathbf{v})$ in (2.7) is minimized.

Problem 2: Given Ω, T, μ, ρ, $\mathbf{v}_\infty$, $\mathbf{v}_0$, and M, find $\mathbf{v}$ and $\mathbf{g}$ such that
- the Navier-Stokes system (2.1)–(2.4) is satisfied,
- the functional $\mathcal{J}(\mathbf{v})$ in (2.6) is minimized,
- and the explicit bound (2.8) on the control is satisfied.

2.2   Summary of Results in Two Dimensions

A complete analysis of both optimal control problems and related inhomogeneous boundary value problems for the Navier-Stokes system are given in [2]. Here, we will only give the briefest of summaries of those results.

We begin by introducing some notation, beginning with the solenoidal spaces

$$\mathbf{V}^1(\Omega) = \left\{ \mathbf{u} \in \mathbf{H}^1(\Omega) \;:\; \nabla \cdot \mathbf{u} = 0, \int_{\partial\Omega} \mathbf{u} \cdot \mathbf{n}\, ds = 0 \right\}$$

and

$$\mathbf{V}_0^1(\Omega) = \left\{ \text{closure of } \mathbf{C}_0^\infty(\Omega) \cap \mathbf{V}^0(\Omega) \text{ in the } \mathbf{H}^1(\Omega)\text{-norm} \right\}$$

and the dual space

$$\mathbf{V}^{-1}(\Omega) = [\mathbf{V}_0^1(\Omega)]^* \,.$$

Then, the solution space for Navier-Stokes equations is given by (see, e.g., [9])

$$\mathcal{V}^{(1)}(Q_T) = \left\{ \mathbf{u} \in L^2(\mathbb{R}; \mathbf{V}^1(\Omega)) \;:\; \partial_t \mathbf{u} \in L^2(\mathbb{R}; \mathbf{V}^{-1}(\Omega)) \right\},$$

where $Q_T = (0, T) \times \Omega$.

The first result of [2] is the identification of the trace space of the solenoidal solution space for the Navier-Stokes system. The interesting feature is that the normal and tangential components of the trace of the velocity do not belong to the same space. In fact, if we have that

$$\mathbf{v}|_{\partial\Omega} = b_n \mathbf{n} + b_\tau \boldsymbol{\tau} \in \mathcal{V}^{(1)}(Q_T),$$

where $\mathbf{n}$ and $\boldsymbol{\tau}$ denote the unit vectors normal and tangential to the boundary of Ω, we have that

$$b_n \in L^2(0, T; H^{1/2}(\partial\Omega)) \cap H^{3/4}(0, T; H^{-1}(\partial\Omega))$$

$$b_\tau \in L^2(0, T; H^{1/2}(\partial\Omega)) \cap H^{1/4}(0, T; L^2(\partial\Omega)) \,.$$

This situation, i.e, that the normal and tangential components have different smoothness properties, is a result of the fact that the solution space for the Navier-Stokes system consists of *solenoidal* vector-valued functions. For general parabolic systems, one doesn't have this phenomena.

The second set of results of [2] are the existence, uniqueness, and a priori estimates for solutions of the Navier-Stokes system with *inhomogeneous* boundary conditions. We do not give the precise results here, but instead refer the reader to [2].

The space of admissible solutions to the optimal control problems is a subspace of the solution space for the Navier-Stokes system; specifically, the space of admissible solutions is given by

$$Y = \Big\{ \mathbf{w} \in \mathcal{V}^{(1)}(Q_T) \ : \ (\partial_t \mathbf{w})|_{\partial\Omega} \in L^2\left(0, T; \mathbf{L}^2(\partial\Omega)\right),$$

$$\int_{\partial\Omega} \partial_t \mathbf{w} \cdot \mathbf{n}\, ds = 0, \mathbf{w}|_{\partial\Omega} \in \mathbf{L}^k\left((0, T) \times \partial\Omega\right) \Big\},$$

where $k \geq 3$. The reason for this is that, in order to properly define the optimality system, we need for the stress vector on the boundary to be well defined; this is true for velocities in the space of admissible solutions (this is shown in [2]) but not necessarily true for velocities in the solution space for the Navier-Stokes system.

The other results found in [2] are that each of the two optimal control problem has a solution with the optimal velocity fields belonging to Y and the derivation of optimality systems both in weak form and in terms of a system of partial differential equations. The optimality system features the usual backward-in-time adjoint (with respect to the linearized Navier-Stokes system) system and a two-point boundary value problem in time along the boundary $\partial\Omega$. See [2] for details.

2.3   Preliminary Results in Three Dimensions

In three dimensions, proofs of the existence of optimal solutions and the derivation of optimality systems are currently in progress. However, the necessary trace theorems for solenoidal vector fields and results about the solution of the Navier-Stokes system with inhomogeneous boundary conditions have been obtained. These are summarized in this section.

We first introduce some notation, beginning with the function spaces $\mathcal{H}^{(s)}(Q_T)$ defined in terms of the norm

$$\|u\|^2_{\mathcal{H}^{(s)}(\mathbf{R}^4)} = \int_{\mathbf{R}^4} \left[(1 + |\boldsymbol{\xi}|^2)^s + (1 + |\tau|^2)(1 + |\boldsymbol{\xi}|^2)^{s-2} \right] |\hat{u}(\tau, \boldsymbol{\xi})|^2 \, d\boldsymbol{\xi}\, d\tau,$$

where $\hat{u}(\tau, \boldsymbol{\xi})$ denotes the Fourier transform.

If $Q_T = (0, T) \times \Omega$ and $Q'_T = (0, T) \times (\mathbf{R}^3 \backslash \Omega)$, then we let

$$\mathcal{H}^{(s)}(Q_T) = \mathcal{H}^{(s)}(\mathbf{R}^4) / \mathcal{H}^{(s)}_{Q'_T}(\mathbf{R}^4).$$

Spaces of solenoidal vector fields are defined as

$$\mathcal{V}^{(s)}(Q_T) = \{ \mathbf{v} \in [\mathcal{H}^{(s)}(Q_T)]^3 \ : \ \mathrm{div}\, \mathbf{v} = 0 \}.$$

We will also need the *logarithmic* spaces with norms

$$\|u\|^2_{\mathcal{H}^{(s)}_{\ln}(\mathbb{R}^3)}$$

$$= \int_{\mathbb{R}^3} \frac{(1+|\xi'|^2)^s + (1+|\tau|^2)(1+|\xi'|^2)^{s-2}}{\ln[2+(1+|\tau|^2)/(1+|\xi'|^2)^2]} |\hat{u}(\tau,\xi')|^2 \, d\xi' \, d\tau$$

and

$$\|u\|^2_{H^s_{\ln}(\mathbb{R};H^r(\mathbb{R}^2))}$$

$$= \int_{\mathbb{R}^3} \frac{(1+|\xi'|^2)^r(1+|\tau|^2)^s}{\ln[2+(1+|\tau|^2)/(1+|\xi'|^2)^2]} |\hat{u}(\tau,\xi')|^2 \, d\xi' \, d\tau$$

and the boundary space

$$\tilde{H}^\alpha(\partial\Omega) = \{v \in H^\alpha(\partial\Omega) \ : \ \int_{\partial\Omega} v \, d\mathbf{x} = 0\}.$$

Let $\mathbf{v}(t,\mathbf{x})$ denote a solenoidal vector field defined on Q_T, $\gamma\mathbf{v} =$ the restriction of $\mathbf{v}$ to the boundary $\Sigma_T = (0,T) \times \partial\Omega$, and $\gamma\mathbf{v} = \gamma_\tau\mathbf{v} + (\gamma_n\mathbf{v})\mathbf{n}$ be a decomposition of $\gamma\mathbf{v}$ into vectors tangent and normal to the boundary. We then define the tangential space $G^s_\tau(\Sigma_T)$

$$G^s_\tau(\Sigma_T) = \begin{cases} [\mathcal{H}^{(s-1/2)}(\Sigma_T)]^2 & \text{for } s \geq 5/2 \\ [L^2(0,T;H^{s-1/2}(\partial\Omega)) \cap H^{(2s-1)/4}(0,T;L^2(\partial\Omega))]^2 \\ \qquad \text{for } 1/2 < s \leq 5/2, \end{cases}$$

the normal space $G^s_n(\Sigma_T)$

$$G^s_n(\Sigma_T) = \begin{cases} \mathcal{H}^{(s-1/2)}(\Sigma_T) \cap L^2(0,T;\tilde{H}^1(\partial\Omega)) \\ \qquad \text{for } s \geq 3/2 \\ L^2(0,T;\tilde{H}^{s-1/2}(\partial\Omega)) \cap H^{(2s+1)/4}(0,T;\tilde{H}^{-1}(\partial\Omega)) \\ \qquad \text{for } 1/2 < s \leq 3/2, \end{cases}$$

and also

$$G^s(\Sigma_T) = G^s_\tau(\Sigma_T) \times G^s_n(\Sigma_T).$$

In addition, with

$$\partial\Omega = \cup^J_{j=1}\Gamma_j \qquad \text{and} \qquad \Gamma_i \cup \Gamma_j = \emptyset \quad \forall i \neq j,$$

we define the normal space $\hat{G}^s_n(\Sigma_T)$

$$\hat{G}^s_n(\Sigma_T) = \Big\{ u_n(t,x) \in G^s_n(\Sigma_T) \ : $$

$$\int_{\Gamma_j} u_n(t,x') \, dx' = 0 \text{ a.e. } t \in [0,T], \, j = 1,\ldots,J \Big\}$$

and also

$$\hat{G}^s(\Sigma_T) = G^s_\tau(\Sigma_T) \times \hat{G}^s_n(\Sigma_T).$$

The Problem of Characterization of Traces. We consider the following problem:

describe the space of Dirichlet traces onto Σ_T for the space $\mathcal{V}^{(s)}(Q_T)$.

In response to this problem, we have the following two results; for details, see [3].

The restriction theorem – *The operator γ can be extended by continuity into the following continuous operator:*

$$\gamma : \mathcal{V}^{(s)}(Q_T) \to G^s(\Sigma_T) \quad \text{for } s > 1/2,\ s \neq 3/2,\ s \neq 5/2.$$

If $s = 5/2$, then the restriction operator

$$\gamma = (\gamma_\tau, \gamma_n) : \mathcal{V}^{(5/2)}(Q_T) \to [\mathcal{H}_{\text{ln}}^{(2)}(\Sigma_T)]^2 \times \mathcal{H}^{(2)}(\Sigma_T)$$

is continuous. For $s = 3/2$, the restriction operator

$$\gamma = (\gamma_\tau, \gamma_n) : \mathcal{V}^{(3/2)}(Q_T) \to [L^2(0,T; H^1(\partial\Omega)) \cap H^{1/2}(0,T; L^2(\partial\Omega))]^2$$
$$\times [L^2(0,T; \widetilde{H}^1(\partial\Omega)) \cap H_{\text{ln}}^1(0,T; \widetilde{H}^{-1}(\partial\Omega))]$$

is bounded.

The extension (or lifting) theorem – *For $s > 1/2$, there exists a continuous extension operator*

$$R : \widehat{G}^s(\Sigma_T) \to \mathcal{V}^{(s)}(Q_T),$$

i.e., the operator R is such that $\gamma \circ R = I$, where $I : \widehat{G}^s(\Sigma_T) \to \widehat{G}^s(\Sigma_T)$ is the identity operator.

Some observations are in order. The theorems imply that the restriction operator γ is surjective for $s > 1/2$, $s \neq 3/2$ and $s \neq 5/2$. If $s = 3/2$, then $G^{3/2}(\Sigma_T) \subset \Im(\gamma)$, where $\Im(\gamma)$ is the image of the operator γ. Analogously, if $s = 5/2$, then $G^{5/2}(\Sigma_T) \subset \Im(\gamma)$, where $\Im(\gamma)$ is the image of the operator γ. For $s > 1/2, s \neq 3/2, s \neq 5/2$, we described the space $G^s(\Sigma_T)$ which is the restriction of $\mathcal{V}^{(s)}(Q_T)$ to Σ_T and we constructed the extension operator from $G^s(\Sigma_T)$ to $\mathcal{V}^{(s)}(Q_T)$. For $s = 3/2$ and $s = 5/2$, the space $G^s(\Sigma_T)$ for which we construct extension operators is narrower than the space $G_{\text{ln}}^s(\Sigma_T)$, where the latter is the space in which we seek the restrictions to Σ_T of functions belonging to $\mathcal{V}^{(s)}(Q_T)$. These results are fundamental in the study of inhomogeneous boundary value problems for systems involving solenoidal vector fields, and in particular, for the Navier-Stokes system.

We also make some remarks on the proofs of the theorems. The methods which were used to characterize traces of solenoidal vector fields for $s = 1$ in the two-dimensional case (see [2]) cannot be generalized to the three-dimensional case for $s > 1/2$. Roughly speaking, the proofs in three dimensions consists of two parts. The first part is the proof of the trace theorem for scalar functions belonging to the space

$$\mathcal{H}^{(s)}(Q_T) = \{\phi \in L^2(0,T; H^s(\Omega)) \ : \partial_t \phi \in L^2(0,T; H^{s-2}(\Omega))\}.$$

With the help of the method of localization, this task is reduced to the derivation of some estimates for Fourier representations. Of course, this approach is well known. Moreover, the proof of these estimates is also well known for $s > 5/2$. However, the cases of most interest in applications are the cases $1/2 < s < 5/2$. The second part is to work out the localization and rectification method for solenoidal vector fields. To establish the restriction theorem in case the smoothness index s of $\mathcal{V}^{(s)}(Q_T)$ is small, we increase the smoothness by transition from a solenoidal vector field $\mathbf{u}$ to a field $\mathbf{v}$ such that $\mathbf{curl\,v} = \mathbf{u}$ and then we express the traces of $\mathbf{u}$ with the help of the traces of $\mathbf{v}$. To realize this approach, we develop further some results of [1] on the solvability of the system

$$\mathbf{curl\,v} = \mathbf{u}, \quad \mathrm{div\,v} = 0, \quad \text{for } \mathbf{x} \in \Omega$$

$$v_n\big|_{\partial\Omega} = 0,$$

where v_n is the projection of the vector field $\mathbf{v}$ on the unit outward-pointing normal $\mathbf{n}$. To obtain the extension result we need to solve the following problem posed on the manifold $\partial\Omega$:

$$(\mathbf{curl\,v})\big|_{\partial\Omega} = \mathbf{u}\big|_{\partial\Omega},$$

where $\mathbf{u}\big|_{\partial\Omega}$ is a given vector field defined on $\partial\Omega$. $(\mathbf{curl\,v})\big|_{\partial\Omega}$ can be expressed in terms of $\mathbf{v}\big|_{\partial\Omega}$ and $(\partial\mathbf{v}/\partial n)\big|_{\partial\Omega}$, so that we can understand this as an equation for the unknowns $\mathbf{v}\big|_{\partial\Omega}$ and $(\partial\mathbf{v}/\partial n)\big|_{\partial\Omega}$. With the help of this curl problem, it is very convenient for us to transform the trace problem for solenoidal vector fields to the analogous problem for exterior differential forms. To make this transformation in a simple way, we are compelled to use only a special kind of local coordinates, the "orthogonal local coordinates." To solve for $\mathbf{v}\big|_{\partial\Omega}$ and $(\partial\mathbf{v}/\partial n)\big|_{\partial\Omega}$, we may use well-known results on the solvability of the Laplace operator in the classes of differential forms defined on the manifold $\partial\Omega$.

Inhomogeneous Boundary Value Problems for the 3D Stokes and Navier-Stokes Systems. As an application of the results of the trace results, we first consider the *Stokes equations* in three dimensions with *inhomogeneous* boundary conditions:

$$\begin{cases} \partial_t\mathbf{v} - \Delta\mathbf{v} + \nabla p = \mathbf{f} \quad\text{ and }\quad \mathrm{div\,v} = 0 \quad\text{on } Q_T \\ \mathbf{v}\big|_{\Sigma_T} = \mathbf{g} \quad\text{ and }\quad \mathbf{v}\big|_{t=0} = \mathbf{v_0}\,. \end{cases} \tag{2.9}$$

We invoke the following assumptions:

$$\mathrm{div\,v_0} = 0$$

$$(\mathbf{v_0}\cdot\mathbf{n})\big|_{\partial\Omega} = (\mathbf{g}\cdot\mathbf{n})\big|_{t=0}$$

$$\mathbf{f} \in L^2(0,T;\mathbf{V}^{s-2}(\Omega)) \quad \forall s \geq 1$$

$$\mathbf{v}_0 \in \mathbf{V}^{s-1}(\Omega) \quad \forall s \geq 1$$

$$\mathbf{g} \in \widehat{G}^s(\Sigma_T) \quad \forall s \geq 1$$

$$\int_{\Gamma_i} \mathbf{u} \cdot \mathbf{n}\, dx = 0 \quad \text{a.e. } t \in [0,T] \text{ for any connected component } \Gamma_i \text{ of } \partial\Omega$$

and

$$\text{for } s > 3/2, \quad \mathbf{v}_0\big|_{\partial\Omega} = \mathbf{g}\big|_{t=0}.$$

We then have the following result; see [3] for details.

Theorem – *Let $s \in [1,2]$ and assume that assumptions hold. Then, there exists a unique solution $(\mathbf{v}, \nabla p) \in \mathcal{V}^{(s)}(Q_T) \cap L^2(0,T;\mathbf{H}^{s-2}(\Omega))$ for the inhomogeneous boundary value problem (2.9) for the Stokes equations.*

We also consider the *Navier-Stokes equations* in three dimensions with *inhomogeneous* boundary conditions:

$$\begin{cases} \partial_t\mathbf{v} - \nu\Delta\mathbf{v} + \mathbf{v}\cdot\nabla\mathbf{v} + \nabla p = \mathbf{f}, & \text{and} \quad \operatorname{div}\mathbf{v} = 0 \quad \text{on } Q_T \\ \mathbf{v}\big|_{\Sigma_T} = \mathbf{g} \quad \text{and} \quad \mathbf{v}\big|_{t=0} = \mathbf{v}_0 \end{cases} \tag{2.10}$$

for which we have the following result (again, see [3] for details).

Theorem – *Let $s \in [3/2, 2]$ and assume that the assumptions hold. Suppose also that*

$$\|\mathbf{f}\|_{L^2(0,T;H^{s-2}(\Omega))} + \|\mathbf{v}_0\|_{V^{s-1}(\Omega)} + \|\mathbf{g}\|_{\widehat{G}^s(\Sigma_T)}$$

is sufficiently small. Then, there exists a unique solution $(\mathbf{v}, \nabla p) \in \mathcal{V}^{(s)}(Q_T) \times L^2(0,T;\mathbf{H}^{s-2}(\Omega))$ for the inhomogeneous boundary value problem (2.10) for the Navier-Stokes equations.

3 The Numerical Analysis of Optimal Control Problems for the Navier-Stokes System

We have also considered the development and analysis of finite-element based algorithms for finding approximate solutions of optimal control problems for the Navier-Stokes system in two dimensions. We have considered semidiscrete in time (backward Euler) and fully-discrete (finite elements in space) approximations and shown that the semidiscrete and fully discrete optimal control problems have solutions. We have rigorously justified the use of Lagrange multiplier principles for the discretized problems and derived optimality systems for both the semidiscrete and fully discrete optimal control problems. We have also shown that solutions of the semidiscrete and fully discrete optimality systems converge to solutions of the continuous optimality systems and proven various other properties of the discrete solutions. See, e.g., [4], [5], [6], [7], and [8] for details.

We have also applied the algorithms to a number of optimal control problems involving heat transfer problems, shape optimization problems, flow matching problems, drag reduction problems, transition delay, and crystal growth problems.

3.1 Flow Matching Through Boundary Control

We consider one concrete example, namely flow matching through boundary control. For details, see [6]. Let $\mathbf{U}$ denote a given velocity field, α, β, β_1, and β_2 given constants, Γ_c the part of boundary over which control is applied, and $\Gamma_o = \partial\Omega \setminus \Gamma_c$ the rest of the boundary $\partial\Omega$. We then consider the *Navier-Stokes system*

$$\begin{cases} \mathbf{u}_t + (\mathbf{u} \cdot \nabla)\mathbf{u} - \nu\Delta\mathbf{u} + \nabla p = 0 & \text{in } (0,T) \times \Omega \\ \nabla \cdot \mathbf{u} = 0 & \text{in } (0,T) \times \Omega \\ \mathbf{u} = \mathbf{g} & \text{on } (0,T) \times \Gamma_c \\ \mathbf{u} = 0 & \text{on } (0,T) \times \Gamma_o \\ \mathbf{u}|_{T=0} = \mathbf{u}_0 & \text{in } \Omega, \end{cases} \qquad (3.1)$$

the *matching functional*

$$\mathcal{J}(\mathbf{u}, \mathbf{g}) = \frac{\alpha}{2} \int_0^T \int_\Omega |\mathbf{u} - \mathbf{U}|^2 \, dxdt$$
$$+ \frac{\beta}{2} \int_0^T \int_{\Gamma_c} (|\mathbf{g}|^2 + \beta_1|\mathbf{g}_x|^2 + \beta_2|\mathbf{g}_t|^2) \, dxdt, \qquad (3.2)$$

and the *optimal control problem*

> find $\mathbf{u}, \mathbf{g}$ that minimize the functional $\mathcal{J}(\mathbf{u}, \mathbf{g})$ in (3.2) and satisfy the Navier-Stokes system (3.1).

Some observations are in order. Note that the functional is penalized by time and space derivatives of the control function $\mathbf{g}$. Analytical results for this problem include the existence of optimal solutions and the derivation of an optimality system of partial differential equations from which optimal solutions may be obtained. Semidiscrete-in-time and fully discrete finite element approximations to the optimal control problem and the corresponding optimality systems have been developed and their convergence properties analyzed.

The *optimality system* for the optimal control problem is stated in terms of the *state variables*, i.e., the velocity $\mathbf{u}$ and pressure p, the *control variable*, i.e., the velocity $\mathbf{g}$ along Γ_c, and the *adjoint variables*, i.e., the adjoint velocity $\mathbf{w}$ and adjoint pressure σ. The optimality system is given by the *Navier-Stokes system* (3.1) coupled to the *adjoint system*

$$\begin{cases} -\mathbf{w}_t + \nu\nabla^2\mathbf{w} + (\nabla\mathbf{u})^T \cdot \mathbf{w} - \mathbf{u} \cdot \nabla\mathbf{w} + \nabla\sigma \\ \qquad\qquad\qquad = \alpha(\mathbf{u} - \mathbf{u}) \quad \text{in } (0,T) \times \Omega \\ \nabla \cdot \mathbf{w} = 0 \quad \text{in } (0,T) \times \Omega \\ \mathbf{w} = 0 \quad \text{on } (0,T) \times \partial\Omega \\ \mathbf{w}|_T = 0 \quad \text{in } \Omega \end{cases} \qquad (3.3)$$

and the *optimality condition*

$$\begin{cases} -\beta_1 \mathbf{g}_{tt} - \beta_2 \Delta_s \mathbf{g} + \mathbf{g} + k(t)\mathbf{n} = \dfrac{1}{\beta}(\dfrac{\partial\mathbf{w}}{\partial n} - \sigma\mathbf{n}) \quad \text{on } (0,T) \times \Gamma_c \\ \displaystyle\int_{\Gamma_c} \mathbf{g} \cdot \mathbf{n}\, d\mathbf{x} = 0 \quad \text{in } (0,T) \\ \mathbf{g}(0,\mathbf{x}) = \mathbf{u}_0 \quad \text{on } (0,T) \times \Gamma_c \\ \mathbf{g}_t(T,\mathbf{x}) = 0 \quad \text{on } (0,T) \times \Gamma_c \\ \mathbf{g} = 0 \quad \text{on } (0,T) \times \partial\Gamma_c, \end{cases} \qquad (3.4)$$

where Δ_s denotes the surface Laplacian operator and $k(t)$ is a Lagrange multiplier introduced to enforce the constraint $\int_{\Gamma_c} \mathbf{g} \cdot \mathbf{n}\, d\mathbf{x} = 0$ in $(0,T)$.

Note that the optimality condition is a boundary value problem in space-time for a partial differential equation along the control boundary Γ_c. Also, note that the Navier-Stokes system is well posed as a forward-in-time problem while the adjoint system is well posed as a backward-in-time problem. As a result, the discretized Navier-Stokes system steps forward in time while the discretized adjoint system steps backward in time. Since these two systems are coupled with each other and with the optimality condition, we see that the optimality system is coupled in space and time. Solving the discrete Navier-Stokes equations and the coupled discrete adjoint equations and discrete optimality condition simultaneously is a formidable task. For this reason, in practice, one needs to develop a *decoupling algorithm*.

One simple decoupling algorithm is based on a *gradient method* for the optimal control problem. We have defined such an algorithm and established its convergence properties; see [6] for details. In the method, the gradient of the functional in (3.2) is determined by *successive* solutions of the state, e.g., Navier-Stokes, and adjoint state systems. Specifically, the algorithm is given as follows.

A gradient algorithm

Initialization (iteration counter $k = 0$)
    i) given $\mathbf{g}_h^{(0)}$, a tolerance τ, and $\epsilon = 1$
    ii) solve the discretized Navier-Stokes system with $\mathbf{g}_h^{(0)}$ for $\mathbf{u}_h^{(0)}$ and
       $p_h^{(0)}$

iii) evaluate $\mathcal{J}_h^{(0)}$, a discrete approximation to the functional evaluated at $\mathbf{u}_h^{(0)}, \mathbf{g}_h^{(0)}$

Main loop (for $k = 1, 2, \ldots$)

iv) solve the discretized adjoint system with $\mathbf{u}_h^{(k-1)}$ for $\mathbf{w}_h^{(k)}$ and $\sigma_h^{(k)}$

v) solve the optimality condition with $\mathbf{w}_h^{(k)}$ and $\sigma_h^{(k)}$ for $\widetilde{\mathbf{g}}_h$

vi) set $\mathbf{g}_h^{(k)} = \mathbf{g}_h^{(k-1)} - \epsilon(\mathbf{g}_h^{(k-1)} - \widetilde{\mathbf{g}}_h)$

vii) solve the discretized Navier-Stokes system with $\mathbf{g}_h^{(k)}$ for $\mathbf{u}_h^{(k)}$ and $p_h^{(k)}$

viii) evaluate $\mathcal{J}_h^{(k)}$, a discrete approximation to the functional evaluated at $\mathbf{u}_h^{(k)}, \mathbf{g}_h^{(k)}$

ix) if $|\mathcal{J}_h^{(k)} - \mathcal{J}_h^{(k-1)}|/\mathcal{J}_h^{(k)} \leq \tau$, stop

x) if $\mathcal{J}_h^{(k)} \leq \mathcal{J}_h^{(k-1)}$, set $\epsilon = 1.5\epsilon$ and go to (iv)

xi) if $\mathcal{J}_h^{(k)} > \mathcal{J}_h^{(k-1)}$, set $\epsilon = 0.5\epsilon$ and go to (vi) .

The most costly task (by far) in this algorithm is the Navier-Stokes solves required in Step (vii) (and Step (ii)). The adjoint solves in Step (iv) are also costly, but at least they involve a *linear* system of partial differential equations. Due the high costs of the Navier-Stokes solves, there is now much interest in developing *reduced order models* for the Navier-Stokes equations.

References

1. Foias, C., Temam, R.: Remarques sur les Equations de Navier-Stokes et les Phénomènes Successifs de Bifurcation. Annali Scuola Norm. Sup. di Pisa Series IV **V** (1978), 29–63
2. Fursikov, A., Gunzburger, M., Hou, L.: Boundary Value Problems and Optimal Boundary Control for the Navier-Stokes System: the Two-Dimensional Case. SIAM J. Cont. Optim. **36** (1998), 852–894
3. Fursikov, A., Gunzburger, M., Hou, L.: Trace Theorems for Three-Dimensional, Time-Dependent Solenoidal Vector Fields and their Applications. To appear
4. Gunzburger, M., Manservisi, S.: Analysis and Approximation of the Velocity Tracking Problem for Navier-Stokes Flows with Distributed Control. To appear in SIAM J. Numer. Anal.
5. Gunzburger, M., Manservisi, S.: The Velocity Tracking Problem for Navier-Stokes Flows with Bounded Distributed Controls. To appear in SIAM J. Cont. Optim.
6. Gunzburger, M., Manservisi, S.: The Velocity Tracking Problem for Navier-Stokes Flows with Boundary Control. To appear
7. Hou, L., Yan, Y.: Dynamics for Controlled Navier-Stokes Systems with Distributed Controls. SIAM J. Cont. Optim. **35** (1997), 654–677
8. Hou, L., Yan, Y.: Dynamics and Approximations of a Velocity Tracking Problem for the Navier-Stokes Flows with Piecewise Distributed Controls. SIAM J. Cont. Optim. **35** (1997), 1847–1885
9. Temam, R.: Navier-Stokes Equations, North-Holland, Amsterdam, 1979

Structural Identification of Nonlinear Coefficient Functions in Transport Processes through Porous Media

P. Knabner[1] and B. Igler[2]

[1] Universität Erlangen-Nürnberg, Angewandte Mathematik I, Martensstraße 3,
D-91058 Erlangen, Germany
[2] Andersen Consulting, München, Germany

Dedicated to Professor Karl-Heinz Hoffmann
on the occasion of his 60th birthday

1 Introduction

Mathematical models provide the starting point for the simulation of complex processes, which arise in the natural and engineering sciences. Characteristic properties of the considered systems are represented by model parameters or coefficients. These have to be determined by experiments. If the coefficients are not measured directly, as direct measurements are not possible or do not lead to satisfying results, numerical identification procedures have to be applied.

We have to distinguish between the identification of distributed (i.e. space - or time - dependent) parameter functions in linear or non-linear models and the identification of nonlinearities (i.e. parameter functions dependent on the solution). The first problem constitutes a basic task e.g. in groundwater hydrology. For a comparison of various numerical methods and the exposition of a specific approach for the first case we refer to [11]. In this paper we will be concerned with the latter case where typically the necessary over-specification is constituted by an additional boundary observation. We will outline the numerical approach and its analysis for a specific example, but it will become clear that similar results hold true for a wide class of problems.

1.1 Example: Convective-Diffusive Transport Coupled with Sorption Processes

The example of diffusive-convective transport of a sorbing chemical through a porous medium will illustrate the mathematical and numerical studies.

The transport is modelled by a mass balance equation with diffusive-convective flux:

$$\partial_t(\Theta u) = \nabla j + S$$
$$j = \Theta D \nabla u - qu$$

We use the notation

u dissolved concentration of the chemical
Θ water content
q specific discharge
D diffusion-dispersion coefficient $* \Theta$
S source/sink term

The sorption enters the mass balance equation as source/sink term:

$$S = -\rho \partial_t v$$

with
v sorbed concentration of the chemical
ρ mass fraction of the sorption site
Assuming equilibrium sorption we obtain a functional dependence of the sorbed chemical on the dissolved chemical:

$$v = \varphi(u).$$

The function $\varphi : \mathbb{R}_+ \to \mathbb{R}_+$, which determines the sorption process, is called *sorption isotherm*. For physical reasons it is assumed to be non-zero and monotone increasing. (A more detailed discussion of this model and related subjects can be found in [12].)

In the following we will assume that all other physical properties (Θ, q, D, ρ) are known and that only the sorption isotherm has to be identified by a soil column breakthrough experiment.

The most basic experimental set-up consists of a homogeneous, vertically oriented soil column which is subject to a stationary water flux. The concentration of the chemical in the water that enters the soil at the top $x = 0$ is adjusted to a predetermined value $f(t)$. The concentration $g(t)$ of the chemical is measured at the outlet at the bottom $x = L$.

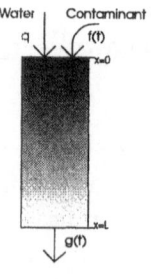

The experiment is motivated by the following idea: If the inflow concentration $f(t)$ is given and the outflow concentration $g(t)$ is measured, the isotherm $\varphi(u)$ can be identified. We will see in the following how this can be justified and how the identification is actually performed.

The physical properties of the experiment allow us to assume that it suffices to describe the transport and sorption processes in the soil column by a PDE in one spatial dimension, the so-called direct problem (DP):

$$\partial_t(\Theta u) + \rho \partial_t \varphi(u) - \partial_x(D\partial_x u - qu) = 0 \qquad x \in Q_T$$

$$u(x,0) = u_0(x) \; x \in \Omega$$

$$-(D\partial_x u - qu)(0,t) = qf(t) \; t \in (0,T]$$
$$\partial_x u(L,t) = 0 \qquad t \in (0,T]$$

$$\Omega := (0,L) \text{ and } Q_T := \Omega \times (0,T]$$

The measurement at the outflow is to be interpreted as a boundary condition:

$$u(L,t) = g(t) \qquad t \in (0,T].$$

For known φ the PDE (DP) yields a unique outflow measurement g ([12]). If φ is not known, g serves as an additional equation thus satisfying the rule of thumb, that the number of unknowns has to match the number of equations.

1.2 Inverse Problems are Ill-Posed

Given the coefficient φ we can solve the PDE (DP) and obtain the "measurement" g. This might be interpreted as a mapping

$$\varphi \mapsto g$$

which is composed of the *solution of the* direct problem and its observation. Direct problems are usually *well-posed* in the sense of Hadamard:

- Existence: A solution exists for given coefficients.
- Uniqueness: The solution is unique.
- Stability: The solution depends (in sensible norms) continuously on the coefficients.

In particular if stability is not guaranteed, small deviations in coefficients might lead to completely different simulation results. (Indeed, our example yields a well-posed direct problem ([12]).)

The direct problem has to be inverted in order to identify the coefficient function φ by measured values g:

$$g \mapsto \varphi.$$

This is denoted by the term *inverse problem*. It is usually **not** well-posed, i.e. at least one of the three above-mentioned properties is violated. The problem is then called *ill-posed*.

This means the following in the case of our example:

- It can be shown that monotone inflow concentrations f lead for all φ to monotone outflow concentrations g. However, even small measurement errors in the form of oscillations result in non-monotone measurements g^*. There cannot exist any φ for this g^* then.
- How many φ might lead to the same g? Are we really able to identify uniquely **one** φ?
- Small measurement errors in g might lead to large identification errors in φ.

As we cannot guarantee the existence of *the* solution to the identification problem, a generalized solution g_φ which is nearest to the measurement g^* can be found by the minimization of an error functional. In Sect. 2 the problem of uniqueness is examined further. Stabilization is primarily achieved by the discretization of the coefficient space which leads to a finite-dimensional problem which is well-posed. Further stabilization can be achieved by adding (positive definite, quadratic) terms to the error functional (Tikhonov regularization.) Error analysis for the discretized problem in Sect. 5 allows the explicit computation of indicators that illuminate the stability situation. (A general introduction to the theory of inverse and ill-posed problems can be found in the books of [1] and [15])

1.3   Numerical Identification by Output Least Squares

The actual identification problem is solved by an output least squares approach similar to [5], [6], [3], [14], [7]:

- The identification problem is transformed into the minimization of an error functional (g^* = measured g possibly disturbed by measurement errors):

$$J(\varphi) := \frac{1}{2}\|g_\varphi - g^*\|^2.$$

- The discretization of the coefficient function leads to a finite-dimensional optimization problem, where a real parameter vector $p = (p_1, \ldots, p_r)$ which minimizes

$$J(p) := \frac{1}{2}\|g_p - g^*\|^2$$

  has to be found.
- A multi-level ansatz for the discretization of φ accelerates and stabilizes the optimization procedure.
- Closed-form solutions for the direct problem do usually not exist. In this case the direct problem has to be discretized, too.

- Efficient optimization algorithms involve the computation of the (discrete) gradient $\mathcal{J}'$. The solution of an appropriate adjoint (discrete) problem provides fast algorithms to this end.

These steps are discussed in Sects. 3 to 5. A numerical example in Sect. 6 will show us how all this works out for our soil column breakthrough experiment.

2   Identifiability

The coefficient φ can be uniquely identified from exact, error-free measurements, i. e. the mapping

$$g \mapsto \varphi$$

is injective. In order to show this property of the inverse problem we apply the method of integral identities which originates in [4] and has been successfully applied to prove the uniqueness of solutions of various nonlinear identification problems [9,10,8].

Before we prove identifiability, we have to define precisely what we mean by identifiability. We have adopted the useful concept of distinguishability from [4]:

Definition. Two real, continuous functions $\alpha, \beta : [a, b] \to \mathbb{R}$ are called *distinguishable* on $[a, b]$, if there exists a finite *distinguishing partition* $a = z_0 < \cdots < z_n = b$, such that on every interval (z_{i-1}, z_i) either $\alpha < \beta$ or $\alpha > \beta$ or $\alpha = \beta$ is valid. One of the cases "<" and ">" has to be valid at least once.

We are now able to state the identifiability theorem:

Theorem. Let the concentration at the inlet be either given by a monotone increasing f with $f(0) = 0$ or by a step-like f with $f(0) = 0$, $f \geq 0$, f is monotone increasing on the interval $[0, \hat{T}]$ and monotone decreasing on the interval $(\hat{T}, T]$ for a $\hat{T} \in (0, T)$.

Let the initial condition be given by $u_0 = 0$, u and v denote the solutions of the direct problem (DP) for isotherms φ_1 and φ_2. If φ_1 and φ_2 are distinguishable on $[0, \max v]$, then $g_1 := u|_{x=L} \neq v|_{x=L} =: g_2$.

(For the sake of simplicity we will assume for the proof of the theorem that Θ, ρ, D and q are constant.)

The proof of the theorem uses an integral identity which is derived as follows. First, consider a weak formulation of the direct problem (DP) for appropriate test functions η (see [12]):

$$-\int_0^L \rho\varphi(u_0)\eta(.,0) - \int_0^T\int_0^L \rho\varphi(u)\partial_t\eta - \int_0^L \Theta(.,0)u_0\eta(.,0)$$

$$-\int_0^T\int_0^L \Theta u\partial_t\eta + \int_0^T\int_0^L (D\partial_x u - uq)\cdot\partial_x\eta$$

$$-\int_0^T f(t)(q\eta)(0,t) + \int_0^T g(t)\cdot(q\eta)(L,t) = 0$$

Then, we consider the difference of the weak formulation for φ_1, u, g_1 and φ_2, v, g_2 and rearrange the terms. If we choose η to be a test function that satisfies the adjoint problem (AP):

$$-\int_0^T\int_0^L (\rho p+\Theta)w\partial_t\eta + \int_0^T\int_0^L (D\partial_x w - wq)\partial_x\eta + \int_0^T (qw\eta)(L,t) = \int_0^T h(t)w(L,t)$$

with

$$p := \int_0^1 \varphi_1'(u + s(v-u))\,ds$$

for a given (but arbitrary) h and appropriate test functions w (e.g.: $w = u - v$) or – as a classical PDE:

$$(\rho p + \Theta)\partial_t\eta + \partial_x(D\partial_x\eta + q\eta) = 0 \quad \text{on } (0,L)\times(0,T)$$
$$\eta(x,T) = 0 \quad \text{in } (0,L)$$
$$\partial_x\eta(0,t) = 0 \quad \text{in } (0,T]$$
$$(D\partial_x\eta + q\eta)(L,t) = h(t) \quad \text{in } (0,T],$$

then the above-mentioned difference of the weak formulation of (DP) results in the desired integral identity:

$$\int_0^T \Delta g(t)\cdot h(t) = \int_0^T\int_0^L \rho\Delta\varphi(v)\partial_t\eta$$

with $\Delta\varphi(v) := \varphi_1(v) - \varphi_2(v)$ and $\Delta g(t) := g_1(t) - g_2(t)$.

We now return to the proof of the theorem:

Proof. As $\partial_t v \geq 0$ for $t \in [0,\hat{T}]$ and $v(t) \leq \max_{x\in(0,L)} v(\hat{T})$ for $t \in [0,T]$, it is sufficient to consider a monotone inflow concentration f.

Let i_1 be the smallest $i \in \{1, \ldots, n\}$, such that $\Delta\varphi \neq 0$, say $\Delta\varphi > 0$, on the interval (z_{i-1}, z_i) of the distinguishing partition and define t_1 to be the largest $t \in (0, T]$, such that $\max_x v(x, t) \leq z_{i_1}$. If $h := \begin{cases} 1 \text{ in } [0, t_1] \\ 0 \text{ in } (t_1, T] \end{cases}$, then $\partial_t\eta \begin{cases} < 0 \text{ in } [0, t_1) \\ = 0 \text{ in } (t_1, T] \end{cases}$. This proves

$$\int_0^T (u(L, t) - v(L, t))h(t) = \int_0^T \int_0^L \rho\Delta\varphi(v)\partial_t\eta < 0,$$

hence $u|_{x=L} \neq v|_{x=L}$. $\qquad\qquad\qquad\qquad\qquad\qquad\qquad\qquad\Box$

3   How to Compute the Gradient

Efficient optimization algorithms need at least the gradient of the error functional $\mathcal{J}(\varphi)$. We may write (DPW), the weak formulation of the direct problem (DP), as

$$F(u, \varphi) = 0$$

for an appropriate

$$F : U \times \Phi \to V^*$$

with Banach spaces U (solution space), Φ (coefficient space) and V (test function space) and interpret $u(\varphi)$ as implicit function of $F = 0$. The error functional is a mapping

$$\mathcal{J} : \Phi \to \mathbb{R}$$

which can be decomposed into

$$\mathcal{J}(\varphi) = \tilde{\mathcal{J}}(Tu(\varphi))$$

for an observation operator (our example: trace operator $u \mapsto g$)

$$T : U \to W$$

and a functional (our example: $g \mapsto \frac{1}{2}\int_0^T (g(t) - g^*(t))^2 \, dt$)

$$\tilde{\mathcal{J}} : W \to \mathbb{R}_+.$$

The derivative $\mathcal{J}'$ can be computed explicitly. First, we apply the chain rule:

$$\langle \mathcal{J}'[\varphi], \delta\varphi \rangle = \left\langle \tilde{\mathcal{J}}'[Tu], T\frac{du}{d\varphi}\delta\varphi \right\rangle = \left\langle T^*\tilde{\mathcal{J}}'[Tu], \frac{du}{d\varphi}\delta\varphi \right\rangle.$$

Then, if η solves the adjoint problem (AP')

$$\frac{\partial F}{\partial u}[u, \varphi]^*\eta = T^*\tilde{\mathcal{J}}'[Tu],$$

we obtain

$$\langle \mathcal{J}'[\varphi], \delta\varphi \rangle = \left\langle \frac{\partial F}{\partial u}[u, \varphi]^* \eta, \frac{du}{d\varphi}\delta\varphi \right\rangle = \left\langle \eta, \frac{\partial F}{\partial u}[u, \varphi]\frac{du}{d\varphi}\delta\varphi \right\rangle.$$

Total differentiation of $F(u(\varphi), \varphi) = 0$:

$$\frac{dF}{d\varphi}(u(\varphi), \varphi) = \frac{\partial F}{\partial u}[u, \varphi]\frac{du}{d\varphi} + \frac{\partial F}{\partial \varphi}[u, \varphi] = 0$$

finally yields

$$\langle \mathcal{J}'[\varphi], \delta\varphi \rangle = -\left\langle \eta, \frac{\partial F}{\partial \varphi}[u, \varphi]\delta\varphi \right\rangle = -\left\langle \frac{\partial F}{\partial \varphi}[u, \varphi]^* \eta, \delta\varphi \right\rangle.$$

Now consider again our example. Let us compute the terms in (AP'). F is given by the weak formulation of (DP) and $\tilde{\mathcal{J}}$ equals $\frac{1}{2}\int_0^T (g(t) - g^*(t))^2\, dt$. The two expressions in the adjoint problem are found to be:

$$\left\langle \eta, \frac{\partial F}{\partial u}[u, \varphi]\delta u \right\rangle = -\int_0^T \int_0^L \rho\frac{d\varphi}{du}(u)\delta u \partial_t \eta - \int_0^T \int_0^L \Theta \delta u \partial_t \eta$$

$$+ \int_0^T \int_0^L (D\partial_x \delta u - \delta u q) \cdot \partial_x \eta + \int_0^T q \cdot (T\delta u)(t) \cdot \eta(L, t),$$

$$\left\langle \tilde{\mathcal{J}}'[Tu], T\delta u \right\rangle = \int_0^T (g - g^*)T\delta u\, dt,$$

hence η has to solve

$$-\int_0^T \int_0^L \rho\frac{d\varphi}{du}(u)\delta u \partial_t \eta - \int_0^T \int_0^L \Theta \delta u \partial_t \eta + \int_0^T \int_0^L (D\partial_x \delta u - \delta u q) \cdot \partial_x \eta$$

$$+ \int_0^T q \cdot (T\delta u)(t) \cdot \eta(L, t) = \int_0^T (g - g^*)T\delta u\, dt.$$

This equation is equivalent to the adjoint problem (AP) of Sect. 2, if h is set to $g - g^*$, w is replaced by δu and $p = \int_0^1 \varphi_1'(u + s(v - u))\, ds$ by $\frac{d\varphi}{du}$.

Computing the φ-derivative of F

$$\left\langle \eta, \frac{\partial F}{\partial \varphi}[u, \varphi]\delta\varphi \right\rangle = -\int_0^T \int_0^L \rho\delta\varphi(u)\partial_t \eta$$

yields the gradient of the error functional for our example

$$\langle \mathcal{J}'[\varphi], \delta\varphi \rangle = \int\limits_0^T \int\limits_0^L \rho\delta\varphi(u)\partial_t\eta,$$

where η solves the adjoint problem and $u = u(\varphi)$.

The similarity between the adjoint problems that has been derived in this section and the one that has been obtained for the integral identity in Sect. 2 is striking. Indeed, the gradient $\mathcal{J}'[\varphi]$ also can be computed as limit of a sequence of difference ratios involving the integral identity.

Furthermore it is possible to use basically the same ideas as in the previous section to show that for exact measurements g^* the equality

$$\mathcal{J}'[\varphi] = 0$$

implies that $g_\varphi = g^*$ and hence, that a zero gradient is only obtained for the unique solution φ of the identification problem.

4   Discretization

The value of the error functional $\mathcal{J}(\varphi)$ and its gradient $\mathcal{J}'(\varphi)$ have to be computed in order to perform the numerical identification via minimization of $\mathcal{J}$. In both cases numerical computations have to be performed. We will sketch the discretization for our example and will indicate the general scheme for the computation of the discrete gradient.

4.1   Direct Problem

We have to solve the PDE (DP) in order to get the solution of the direct problem. We discretize the PDE according to [13] with linear Finite Elements. For the sake of simplicity we confine the following studies to constant coefficients Θ, D, q, implicit Euler steps for time discretization and equidistant time and space discretization ($0 = t_0 < \ldots < t_n = T$, $t^{k+1} - t^k = \Delta t$, $k = 0, \ldots, n-1$, $0 = x_0 < \ldots < x_m = L$, $x^{i+1} - x^i = \Delta x$, $i = 0, \ldots, m-1$). We use the notation

$$c_i^k := u(x_i, t^k)$$
$$f^k := f(t^k)$$

and define

$$(M)_{i,j} := \int_0^L \Theta \psi_i \psi_j \quad \text{(mass matrix)},$$

$$(M_\rho)_{i,j} := \int_0^L \rho \psi_i \psi_j \quad \text{(mass matrix)},$$

$$(K)_{ij} := \int_0^L D\partial_x \psi_j \partial_x \psi_i - q\partial_x \psi_i \psi_j$$

$$+(q\psi_j\psi_i)|_{x=L} \quad \text{(stiffness matrix)},$$

$$\Phi(c^k, \varphi) := \text{diag}(\varphi(c_0^k), \dots, \varphi(c_m^k))$$

$$b_0^k := q\Delta t f^k,$$

$$b_i^k := 0 \quad \text{for } i \neq 0$$

where ψ_i denote the basis vectors of the Finite Element space. The PDE (DP) is then approximated by a set of nonlinear equations for $k = 1, \dots, n-1$:

$$(M + \Delta t K)c^{k+1} + M_\rho \Phi(c^{k+1}, \varphi) = Mc^k + M_\rho \Phi(c^k, \varphi) + b^k$$

and a start value c_0 which is obtained by interpolation of v_0. The nonlinear equations are solved by a Newton algorithm. (Convergence analysis for the Finite Element discretization for transport processes of sorbing solutes with equilibrium sorption can be found in [2].)

Finally, the discrete error functional $\mathcal{J}(\varphi) = \tilde{\mathcal{J}}(g)$ can be computed with

$$\tilde{\mathcal{J}}(g) = \sum_{k=0}^n \alpha_k (g^k - g^{k,*})^2$$

for weighting coefficients α_k (possibly representing a specific quadrature rule), simulated outflow values $g^k := c_m^k$ and measurement values $g^{k,*} := g(t^k)$, $g^0 - g^{0,*} = 0$.

4.2   Gradient

The discrete gradient can be computed in three ways:

– Finite Difference
$$\frac{\mathcal{J}(\varphi + \delta\varphi) - \mathcal{J}(\varphi)}{\delta\varphi}$$

– Discretization of the adjoint problem in Sect. 3 and evaluation of $\langle \mathcal{J}'[\varphi], \delta\varphi \rangle =$
$$-\int_0^T \int_0^L \rho \delta\varphi(u)\partial_t \eta \quad \text{by quadrature rules.}$$

– Computing the derivative of the discretized direct problem with respect to φ.

We will discretize according to the third procedure for several reasons:

– less computational effort,
– easier to code,
– consistency of functional evaluation and gradient,
– easier to analyse.

In order to differentiate the discretized g with respect to φ the system of nonlinear equations which is used to discretize the direct problem is rewritten as:

$$F^{k+1}(c^{k+1}, c^k, \varphi) = 0$$

for $k = 0, \ldots, m - 1$ and nonlinear mappings

$$F^{k+1} : \mathbb{R}^{m+1} \times \mathbb{R}^{m+1} \times \Phi \to \mathbb{R}^{m+1}$$
$$(c^{k+1}, c^k, \varphi) \mapsto (M + \Delta t K)c^{k+1} + M_\rho \Phi(c^{k+1}, \varphi)$$
$$- Mc^k - M_\rho \Phi(c^k, \varphi) - b^k.$$

This is equivalent to the "weak formulation"

$$\eta^T F(c, \varphi) = 0$$

for all $\eta \in \mathbb{R}^{(m+1) \cdot n}$ and

$$c := (c^1, \ldots, c^n)^T,$$
$$\eta := (\eta^1, \ldots, \eta^n)^T,$$
$$F(c, \varphi) := (F^1(c, \varphi), \ldots, F^n(c, \varphi))^T.$$

Interpreting $c = c(\varphi)$ as implicit function of $F = 0$, we are able to apply the ideas described in Sect. 3. First, we have to find an η which satisfies the adjoint problem:

$$\eta^T \frac{\partial F}{\partial c}[c, \varphi]\delta c = \tilde{J}'[Tc]^T T \delta c$$

for all $\delta c \in \mathbb{R}^{(m+1) \cdot n}$ (i.e. $\delta c^0 = 0$) and

$$\left(\frac{\partial F}{\partial c}[c, \varphi]\right)_{i,k} = \frac{\partial F^i}{\partial c^k}[c, \varphi]$$

$$= \begin{cases} M + \Delta t K + M_\varrho \frac{\partial \Phi}{\partial c^k}[c^k, \varphi] & \text{for } i = k \\ -M - M_\varrho \frac{\partial \Phi}{\partial c^k}[c^k, \varphi] & \text{for } i = k + 1 \\ 0 & \text{otherwise} \end{cases}$$

$$\frac{\partial \Phi}{\partial c^k}[c^k, \varphi] = \text{diag}(\varphi'(c_0^k), \ldots, \varphi'(c_m^k))$$
$$(\tilde{J}'[Tc]^T T.)_k = (0, \ldots, 0, \alpha_k(g^k - g^{k,*})).$$

Such an η can be found as solution to the discrete adjoint problem:

$$(M + \Delta t K + M_\rho \frac{\partial \Phi}{\partial c^k}[c^k, \varphi])^T \eta^k = (M + M_\varrho \frac{\partial \Phi}{\partial c^k}[c^k, \varphi])^T \eta^{k+1}$$
$$+ (0, \dots, 0, \alpha_k(g^k - g^{k,*}))^T,$$
$$(M + \Delta t K + M_\varrho \frac{\partial \Phi}{\partial c^n}[c^n, \varphi])^T \eta^n = (0, \dots, 0, \alpha_n(g^n - g^{n,*}))^T.$$

The coefficient function φ has to be discretized in order to compute the φ-derivatives of Φ, i.e.:

$$\varphi = \varphi_p(c)$$

for a parameter vector $p = (p_1, \dots, p_r)$. (Our actual parametrization choice will be explained in the next section.) We thus obtain

$$\left(\frac{\partial F}{\partial p}[c, \varphi] \right)_k = M_\varrho \frac{\partial \Phi}{\partial p}[c^k, \varphi] - M_\varrho \frac{\partial \Phi}{\partial p}[c^{k-1}, \varphi]$$
$$\left(\frac{\partial \Phi}{\partial p}[c^k, \varphi] \right)_{i,j} = \frac{\partial \varphi}{\partial p_j}(c_i^k)$$

and hence

$$\frac{dJ}{dp}[\varphi] = -\frac{\partial F}{\partial p}[c, \varphi]^T \eta = -\sum_{k=1}^{n} \left(M_\varrho \frac{\partial \Phi}{\partial p}[c^k, \varphi] - M_\varrho \frac{\partial \Phi}{\partial p}[c^{k-1}, \varphi] \right)^T \eta^k.$$

Consequently, the computation of the gradient of the error functional involves essentially two steps:

- Solving the adjoint problem which uses matrices that already have been computed in the discretization of the direct problem. (The discrete adjoint problem can be solved with less computational effort than the discrete direct problem.)
- Multiplying the solution of the discrete adjoint problem with the p-derivatives of Φ and summing up the products.

Further examples and remarks on the adjoint state method for the computation of gradients in groundwater modeling inverse problems may be found in the book of [17].

5   Numerical Identification

We are now ready to transform the original identification problem into a discrete minimization problem, which can be actually solved on a computer. Let us put the last pieces together:

5.1   Unbiased Parametrization

We have to discretize the coefficient function φ by an appropriate parametrization $\varphi(c) = \varphi_p(c)$ for a parameter vector $p = (p_0, \ldots, p_r)$ of dimension $r + 1$ – also called the number of degrees of freedom. We are looking for an unbiased parametrization, which does not take any a-priori shape information (like *Freundlich:* $\varphi(c) = Ac^p$ or *Langmuir:* $\varphi(c) = \frac{akc}{a+kc}$ for parameters A, p, a, k) for granted. Of course, as long as parametrizations are involved, something has to be assumed about the possible shape of the unknown coefficient function. Piecewise polynomial functions provide an acceptable compromise. We choose piecewise linear functions in order to keep the computations as simple as possible: The parameter vector p provides the function values $\varphi(c_i) = p_i$ for a given c-discretization $c_0 < \ldots < c_r$. (In practice, polynomials of higher order are only indispensable, if more smoothness is needed for the coefficient function or the error functional.)

5.2   Constrained Optimization with Gradients

The resulting minimization problem reads as:

- Find the minimum of $\mathcal{J}(p)$ for $p \in P$ (= parameter space)
- under constraints $G(p) \geq 0$.

Several terms might be added to the error functional in order to obtain more stable results (Tikhonov regularization), e.g. the square of the norm of the coefficient function or of its second derivative. Constraints further stabilize the minimization procedure: e.g. monotonicity or concavity of the coefficient function. Piecewise linear parametrizations result in transformations of these constraints into simple inequalities, e.g. $\varphi' \geq 0 \Leftrightarrow p_{i+1} \geq p_i$.

Optimization algorithms which need the gradient of the error functional are usually faster than algorithms which only need the value of the error functional. As the gradient of $\mathcal{J}$ can be computed efficiently (Sect. 4.2), quasi-Newton algorithms (e.g. the SQP method described in [16]) outperform optimization without gradients.

5.3   Multi-Level

Major problems ensue the minimization of highly nonlinear functionals:

- high sensitivity on the start value (local minima),
- slow convergence.

These problems become especially relevant for a high number of degrees of freedom. A multi-level ansatz (see also [6], [7]) accelerates and stabilizes the optimization procedure:

1. Start the optimization with the least possible number of degrees of freedom (if possible: $r = 1$).

2. Interpolate the optimization result for a parametrization with one or more degrees of freedom added and use this as next start value.
3. Go to (2) as long as the optimization result is not heavily disturbed by oscillations.

We note the similarity between this ansatz and the application of hierarchical bases and wavelets for the identification of distributed coefficient functions (see e.g. [14]).

5.4   Error Analysis

An error analysis provides a criterion for the quantification of the rather fuzzy term "heavily disturbed by oscillations". We perform a linear error analysis similar to [7]: A first order expansion of the error functional at its minimum is given by

$$0 = \delta \mathcal{J} = \|\frac{\partial g}{\partial p} \delta p - \delta g^*\|^2,$$

where δg^* denotes the measurement deviation and δp denotes the according coefficient deviation. These deviations are coupled by the sensitivity matrix $M := \frac{\partial g}{\partial p}$, i.e.:

$$\frac{\partial g}{\partial p} \delta p \approx \delta g^*,$$

A singular value decomposition of the sensitivity matrix (see e.g. [15] for more detailed remarks on singular values and generalized solutions) yields a singular system $\{v_n, u_n; \sigma_n\}$ with the following properties:

- $\{\sigma_j\}$ are the square roots of the eigenvalues of the matrix $M^T M$,
- $\{u_j\}$ form an orthonormal basis of range(A),
- $\{v_j\}$ form an orthonormal basis of range(A^T),
- The identification-measurement-error relation reads:

$$\delta p \approx \sum_j \sigma_j^{-1} \langle \delta g^*, u_j \rangle v_j.$$

We thus obtain two characteristic numbers

- maximum error amplification: $\sigma := \max_j \sigma_j^{-1}$
- (generalized) condition number: $\kappa := \max_{j,k} \frac{\sigma_j}{\sigma_k}$

The numbers σ and κ can serve as boundaries for confidence intervals or as stopping criterion for a multi-level process.

An interpretation in the sense of [7] is possible by observing that

$$\text{Hess}\mathcal{J} = M^T M + (g - g^*)S$$

for a matrix S, which contains the second derivatives, and concluding that for $\mathcal{J} \approx 0$:

$$\text{Hess}\mathcal{J} \approx M^T M.$$

In this case the singular value decomposition describes the properties of the Hessian of $\mathcal{J}$ and the singular values are the eigenvalues of the Hessian of $\mathcal{J}$.

6   Example

The algorithms and methods described in Sects. 3-5 have been used to write a C++ software package that provides a numerical solver for the direct problem (simulation of the breakthrough experiment), the computation of the gradient via the adjoint problem and the numerical identification procedure. The optimization code was taken from an already existing Fortran program ([16]).

The identification code was tested for several artificial outflow measurements, which were generated by simulating the direct problem numerically and adding white noise to the synthetic breakthrough curve. Good identification results were obtained.

After that the identification procedure was applied to data obtained in several soil column breakthrough experiments. We illustrate the results for one example:

The chemical anthracene was filtrated through an initially clean ($c|_{t=0} = 0$) soil column of length $7.26cm$ packed homogeneously with spodic material. The experimental conditions were determined by a specific discharge of $5.48\frac{cm}{s}$, a dispersion coefficient of $1.46cm$ and a water content of 0.23. (This results in a Peclet number of 5.) The inflow concentration was set constant to $f(t) = c_0 = 27.4 \cdot 10^{-9}\frac{g}{cm^3}$.

Figs. 1-5 depict the successive steps in the multi-level process for a growing number of degrees of freedom. New points on the c-axis were inserted hierarchically. The corresponding singular values (Fig. 6) confirm, that higher degrees of freedom lead to higher error amplification σ and worse condition numbers κ, i.e. the degree of ill-posedness increases.

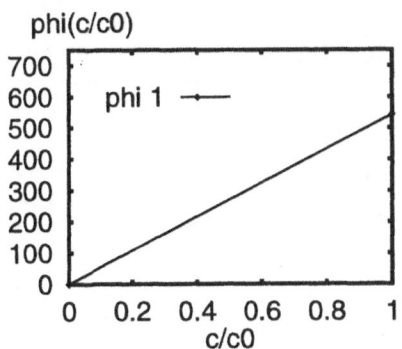 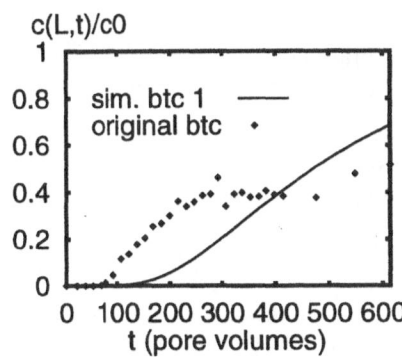

Fig. 1. Identification for 1 degree of freedom at $c = 1$

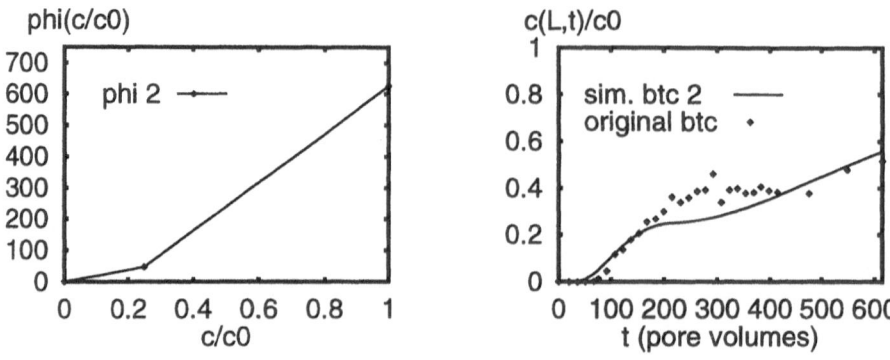

Fig. 2. Identification for 2 degrees of freedom at $c = 0.25, 1$

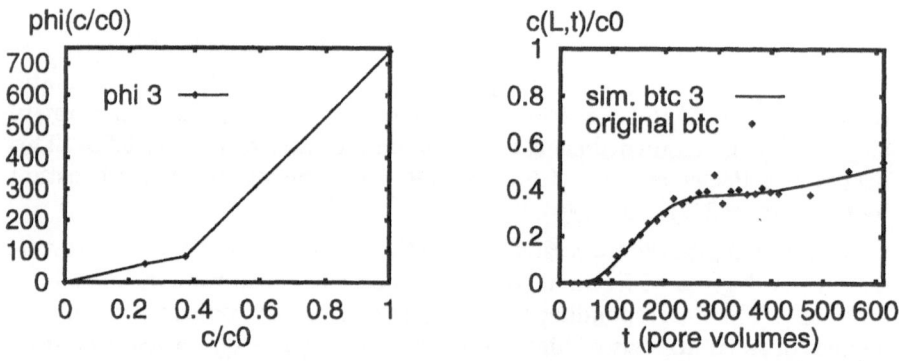

Fig. 3. Identification for 3 degrees of freedom at $c = 0.25, 0.375, 1$

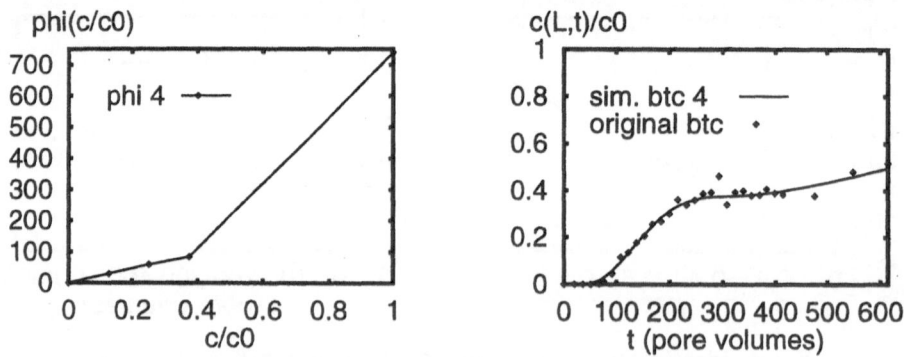

Fig. 4. Identification for 4 degrees of freedom at $c = 0.125, 0.25, 0.375, 1$

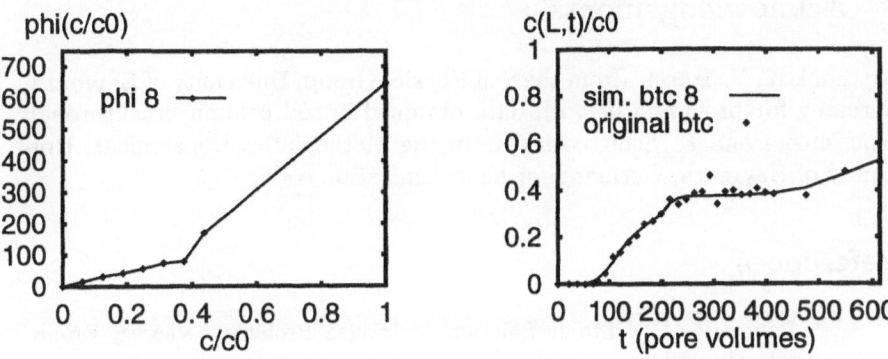

Fig. 5. Identification for 8 degrees of freedom at
$c = 0.06125, 0.125, 0.1875, 0.25, 0.3125, 0.375, 0.4375, 1$

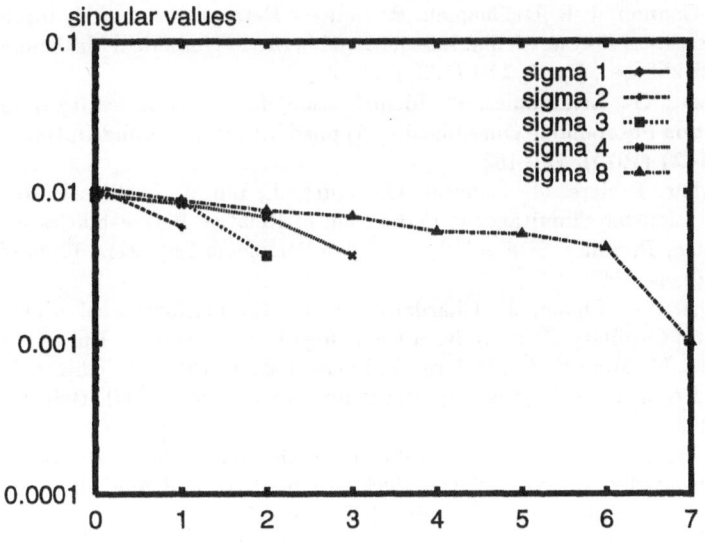

Fig. 6. Singular values for $1, 2, 3, 4, 8$ degrees of freedom

7   Acknowledgements

We thank K. U. Totsche from the Soil Physics Group, University of Bayreuth, Germany for providing us with data obtained in soil column breakthrough experiments and K. Schittkowski from the Mathematics Department, University of Bayreuth, Germany for his optimization code.

References

1. Baumeister, J.: Stable Solution of Inverse Problems. Vieweg, Braun-schweig, 1987
2. Barrett, J. W., Knabner, P.: Finite Element Approximation of Transport of Reactive Solutes in Porous Media. Part 2: Error Estimates for Equilibrium Adsorption Processes. SINUM, **34(2)** (1997), 455–479
3. Chardaire, C., Chavent, G., J. Jaffré, J., Liu, J.: Relative Permeabilities and Capillary Pressure Estimation through Least Square Fitting. In P. R. King, editor, The Mathematics of Oil Recovery,Clarendon Press Oxford, 1992
4. J. R. Cannon, J. R.,DuChateau, P.: Indirect Determination of Hydraulic Properties of Porous Media. International Series of Numerical Mathematics, Birkhäuser, Basel, **114** (1993), 37–50
5. Chavent, G., Lemmonier, P.: Identification de la Non-Linearité d'une Equation Parabolique Quasilineaire. Applied Mathematics and Optimization **1(2)** (1974), 121–162
6. Chardaire-Riviere, C., Chavent, G., Jaffre, J., Liu, J.: Multiscale Representation for Simultaneous Estimation of Relative Permeabilities and Capillary Pressure. SPE 20501, Society of Petroleum Engineers, Richardson, Texas, 1990
7. Chavent, G., Zhang, J., Chardaire-Riviere, C.: Estimation of Mobilities and Capillary Pressure from Centrifuge Experiments. In Bui. H. D., Tanaka, M., Bonnet, M., Maigre, H., Luzzato, E., Reynier, M., editors, Inverse Problems in Engineering Mechanics, Rotterdam, (1994), Balkema, 265–272
8. DuChateau, P.: An Inverse Problem for the Hydraulic Properties of Porous Media. In Proc. of the 1994 Groundwater and Modeling Conference, Colorado State University, (1994) 95–103
9. DuChateau, P.: An Introduction to Inverse Problems in Partial Differential Equations for Engineers, Physicists and Mathematicians, a Tutorial. In Gottlieb, J., DuChateau, P., editors, Proceedings of the Workshop on Parameter Identification and Inverse Problems in Hydrology, Geology and Ecology, Kluwer Aca. Publ., (1995), 3–50
10. DuChateau, P.: Monotonicity and Invertibility of Coefficient-To-Data Mappings for Parabolic Inverse Problems. SIAM Journal on Mathematical Analysis, **36(6)** (November 1995), 1473–1487
11. Hoffmann, K.-H., Knabner, P., Seifert, W.: Adaptive Methods for Identification Problems in Groundwater. Adv. in Water Resources, **14** (1991), 220–239

12. Knabner, P.: *Mathematische Modelle für Transport und Sorption gelöster Stoffe in porösen Medien.* P. Lang, Frankfurt, 1991
13. Knabner, P.: Finite-Element-Approximation of Solute Transport in Porous Media with General Adsorption Processes. In S.-T. Xiao, editor, Flow and Transport in Porous Media, World Scientific Publishing, Singapore (1992), 223–292
14. J. Liu. A multiresolution method for distributed parameter estimation. SIAM Journal on Scientific Computing, **14(2)** (March 1993),389–405
15. Louis, A. K.: Inverse und schlecht gestellte Probleme. Teubner, Stuttgart, 1989
16. Schittkowski, K.: Solving Constrained Nonlinear Least Squares Problems by a General Purpose SQP-Method. In Hoffmann, K.-H., Hiriart-Urruty, J.-B., Lemarechal, C., and Zowe, J., editors, Trends in Mathematical Optimization, volume **83**, Birkhäuser, 1988
17. Sun, N.-Z.: Inverse Problems in Groundwater Modeling. Kluwer Academics, Dordrecht, 1994

Part II

Contributions from the SFB 438

Development of Innovative Osteosynthesis Techniques by Numerical and In Vitro Simulation of the Masticatory System

A. Neff[1], A. Kuhn[1], H. Schieferstein[1,3], A. M. Hinz[2], E. Wilczok[2],
G. Mühlberger[1], H.-F. Zeilhofer[1], R. Sader[1], H. Deppe[1], and H.-H. Horch[1]

[1] Klinik und Poliklinik für Mund-Kiefer-Gesichtschirurgie der Technischen
Universität München, Ismaninger Straße 22, D-81675 München, Germany
[2] Forschungseinheit für Mathematische Modellbildung, Zentrum Mathematik,
Technische Universität München, Germany, Arcisstraße 21, D-80290 München,
Germany
[3] Klinik für Orthopädie und Sportorthopädie der Technischen Universität
München, Abteilung Biomechanik, Connollystraße 32, D-80809 München,
Germany

Dedicated to Professor Karl-Heinz Hoffmann
on the occasion of his 60th birthday

Abstract. Maxillofacial surgery still faces a considerable number of biomechanical
questions. An attempt to cope with the complexity of the masticatory system is
introduced: synchronizing numerical simulation and in vitro experiments for mutual validation in order to clarify some events out of the vast number of unknown
variables in the chewing process, such as muscle forces, bone stress and strain. This
interdisciplinary project can be considered a promising approach based on the lat-
est technology in medicine, mathematics and mechanical engineering. In addition,
the application of adaptive materials will help to optimize maxillofacial surgery.

1 Introduction

A broad range of fundamental questions regarding the biomechanics of the
mandible is still unanswered. As the events which accompany functional loading even under physiological conditions are far from being fully understood,
the development and application of new osteosynthesis materials traditionally
had – and still has – to use highly invasive recording techniques. Nevertheless,
observations made for example in hominids have to be based on extrapolation
due to differences in morphology and function. Another disadvantage of measurements taken either in living subjects or alternatively in physical models
is related to their trial and error character. In addition, physical models are
frequently based on oversimplifications due to the measuring method or the
dissection technique employed in biomechanical test stands, creating more

or less artificial fractures. Taking into account the complex elastomechanical properties and events in the human mandible and the enormous variety of individual conditions given in a living subject, we may ask to what extent current biomechanical simulations are really consistent with the underlying physiological events. Computer modelling offers a promising alternative approach in this regard. Recently, several mathematical models, predominantly referring to the mandible, have been developed [31,36,38,39,51,52] in order to predict regional stresses and strains even in locations otherwise inaccessible in physical models. However, it was clearly demonstrated by these studies, that accurate solutions for the strain distribution in the entire mandible cannot be established using two-dimensional models, three-dimensional bars or dissected mandible halves. In addition, the arbitrary selection of different boundary conditions makes comparisons between the studies presented so far very difficult. Thus, as a matter of fact, in the field of oral and maxillofacial surgery (OMFS) the majority of current medical implants for osteosynthesis or reconstructive surgery are still manufactured without employing finite element based experimental models both for development and design optimization.

2   Principles of Internal Fracture Fixation in OMFS

2.1   Compression Plate Osteosynthesis

Starting with internal wire fixation as late as in the fifties, several types of internal fracture fixation devices have been used in OMFS. Screws and plate-osteosynthesis techniques superseded internal wire fixations in the mid-seventies, using the principles of compression osteosynthesis established by the AO-working group[1] in general surgery and orthopedics. Although mechanical properties and stability of the AO-plates (2.7 mm DCP/dynamic compression plate or 2.4 mm LC-DCP/limited contact dynamic compression plate), modified plates like the LUHR-system or the 1.5 mm Leibinger®-system are far superior [43] to internal wirings or external splints, avoiding a long term immobilization of the patients' masticatory system and making contact healing[2] possible, some major disadvantages present themselves in

- difficult handling, especially concerning plate-adaptation and plate-bending, due to the great dimensions of the compression plates and the materials employed like stainless steel, Vitallium® or titanium alloys (cf. Table 1),

---

[1] "Arbeitsgruppe für Osteosynthese", abbr. AO

[2] i.e. primary healing of the bone's Haversian systems. Thus, healing takes place without forming a so called "fixation callus", which requires at least six or more weeks to transform the hematoma within the fracture gap into cancellous bone, before remodelling into cortical bone in the run of several months, cf. Fig. 2.

— the plate's position near the buccal, lower rim of the mandible, making an extraoral approach obligatory.

Besides visible scaring, this procedure, however, implies a risk of damaging the facial nerve (resulting in a palsy of the muscles of the ipsilateral lower lip, causing a severe functional and aesthetic impairment). Damages to the inferior alveolar nerve (resulting in a numbness of the ipsilateral gums and lower lip) or the roots of the inferior teeth may be caused by the screws fixating the plate, which are anchored both in the buccal and the lingual cortical bone. As a consequence of the plate's eccentric position, another major problem is posed by the fact, that

— a compression plate positioned at the lower border of the mandible cannot compensate the strains within the occlusal area caused by the masticatory muscles (see Fig. 1), or gaps in the lingual cortical bone resulting from compression on the side near the plate (see Fig. 2).

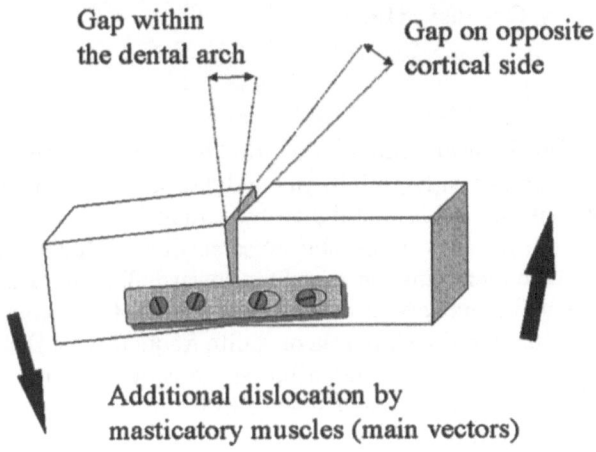

Fig. 1. Distraction of the fractured mandible by compression-type plate osteosynthesis and masticatory muscles (redrawn after [17])

To avoid these system-inherent disadvantages, modifications regarding position and orientation of the spheric gliding slots were introduced in order to counterbalance tensile stresses within the alveolar area instead of applying additional external splints within the dental arch. Recently, by introducing 2.0 mm AO-Mini-DCP-plates and 1.5 mm thick compression plates by Leibinger®, a tendency towards miniaturization of the plates is visible [43].

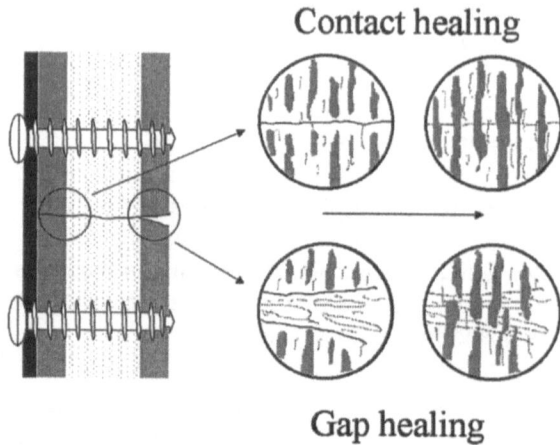

Fig. 2. Distraction of fragments by compression-type plates with zones of contact and gap healing (redrawn after [66])

2.2   Lag Screw Osteosynthesis

Lag screw osteosynthesis also follows the principle of compressive strength between the bone fragments. Especially in condylar process fractures, this type of osteosynthesesis can achieve stability by a functionally stable medullary nailing. Forces are applied inside the cross section of the fractured bone. Thus, there is no tendency in the mandible towards lingual or alveolar gaps as described above in the case of compression plates. Other locations of application are fractures in the chin area or, quite frequently, of the mandibular angle, as presented e. g. in orthognathic surgery. As a common principle, lag screws are inserted with a cortical or anchor thread into the fragment to be fixed through a gliding channel in the fragment on the side of the screw head. However, the head of the lag screw and its bed in the underlying bone is one of the most critical issues in lag screw design [43,93]. The small head has to transmit the tensile forces, which can exceed 1000 N in the symphysis [22,93] to a relatively small supporting area of the cortical bone plate of the mandible, cancellous bone being an inadequate support for the screw. As countersinking weakens the cortical bone and cannot be performed in all locations, screw heads or washers with a concave supporting area were designed as well as self adapting washers [93] designed for oblique insertions. Incongruencies in countersinks or washers should be avoided, since resulting strain concentrations can lead to fatigue fracture as load is transmitted over a very small area.

2.3 Miniaturized Plates

Osteosynthesis with miniaturized plates follows a completely different principle. They were introduced first by Champy et al. [22] in 1975, who, on the basis of simple mathematical models and photoelastic analysis that used polarized light on araldite bars representing the mandible concluded that tension zones occurred at the upper border, compression areas at the lower border of the mandible. The incisor and canine regions present a zone of torsional movements. In contrast to compression osteosynthesis miniaturized plates are inserted in the basal area of the alveolar process, i. e. the area of maximum tensile stress [84]. From a biomechanical point of view, this localization is more favorable than the lower border, as additional leverages do not have to be considered. From a surgical point of view, miniaturized plating systems are much more convenient, as most fractures can be stabilized over an intraoral approach with monocortical screws. Due to their small dimensions, handling of the plates is easier, the risk of lesions to dental or nerval structures is greatly reduced in comparison to compression osteosynthesis. A major advantage of miniaturized plates lies in their applicability for midfacial fractures or fractures of the frontal sinus or skull. An even further miniaturization led to the introduction of microplate systems, allowing osteosynthesis of small fragments, especially in midfacial structures (e. g. naso-ethmoidal complex) or fractures of the alveolar process [43].

The indications for the use of different types of plates however must be assessed critically. Specific indications for mono- or bicortical osteosynthesis have been described. For instance, comminuted and infected fractures contraindicate miniplates. Sometimes, e. g. in case of primary instable fractures, miniplates cannot provide satisfying torsional and vertical stability. The application of lag screws or compression plates, on the other hand, is largely restricted by given anatomical factors.

3 Biological Interface

Biomaterials are indispensable components of all medical implants and devices. Over the last decade biomaterials research as a specific discipline enabled important scientific breakthroughs in understanding interactions of cells and materials leading to fundamental improvements in the way we treat and repair trauma, congenital malformations or aging related diseases. Especially in the field of dental implants the understanding of tissue reactions induced a continuous implementation of advanced surface coating techniques based e. g. on plasma-ion-immersion-implantation (PIII). Nevertheless, medical device designers are still limited to a relatively small number of established materials that were not originally designed to be used in medical implants [50]. A survey of the materials commonly used in the manufacture of medical implants and devices is given in Table 1.

Table 3.1. Materials commonly used in the manufacture of medical implants and devices [50]

Type of Material	Specific Examples
Biostable polymers and resins	Polyurethanes, silicone rubber, Teflon®, Dacron®, nylon, polymethylmethacrylate (PMMA)
Biodegradable polymers	Polylactic acid, polyglycolic acid, polydioxanone
Natural and semisynthetic products	Treated porcine or bovine grafts, processed cellulose, processed collagen, Algipore®, Bone Source®
Metals	316 and 316L stainless steel, Vitallium®, titanium alloys, Co-Cr alloy, Co-Cr-Mo alloy
Ceramics	Aluminium oxides, calcium aluminates, titanium oxides, pyrolytic carbon, Bioglass®, hydroxyapatite
Composites	Apatite composites, carbon coated metals, carbon reinforced polymers

The majority of these industrial plastics and metals do not provide a biologically functional interface with the surrounding tissue. As a consequence, virtually all currently available implants provoke a mild foreign body response at the implant site and are considered to be more or less out-dated [50]. As far as osteosynthesis materials are concerned, biodegradable synthetic polymers have been introduced successfully in orthognathic surgery and, only recently, in the area of trauma and reconstructive procedures including pediatric mid-face and craniofacial trauma and reconstruction, tumor reconstruction and craniotomy flap fixation, bone-graft procedures in the mandible and mid-face or craniofacial skeleton. Unlike titanium plating, which remains in the body unless removed in a second surgery, biodegradable polymers are hydrolyzed by the patient's natural metabolism and evacuate from the surrounding tissue by forming e. g. glycolic or lactic acid and, finally, carbon dioxide and water. Nevertheless, so far, all synthetic polymers lead to some inflammatory response either by surgical trauma or by interactions of the tissue with the implant. The inflammation response resembles the well-known foreign body reaction. In this sense cellular and molecular components of the specific and nonspecific immune responses are involved. For this effect, three major factors are thought to be responsible [40]:

– Implantation is followed by adsorption of different proteins on the implant's surface. Due to adsorption or cellular enzymes these proteins

are denaturized or show changes to the conformation, thus invoking immune responses.
- The implant's surface itself, if not covered by proteins, leads to direct humoral or cellular responses of the organism.
- Products resulting from the polymer's degradation caused by enzymes or aggressive metabolites may induce immune responses. According to their conformation and dimensions these structural subunits may represent themselves as antigenes or haptenes, inducing specific and unspecific response mechanisms.

As long term results fibrotic reactions, autoimmunological responses and even malignant transformations in the surrounding tissues are discussed. Thus, with safety and efficacy in addition being obligatory demands, it will be necessary to redesign a large fraction of the currently available medical implants by synthetic or semisynthetic materials with a better biological interface. The innovative field of tissue engineering offers promising advances in reconstructive surgery leading to replacements for both bone and soft tissue defects. Osteosynthesis and fixation materials applied in reconstructive surgery will have to cope with these demands.

4   Smart Materials in Osteosynthesis

4.1   Experiences with NiTiNOL

Material selection is seldom based on a single property, but instead a combination of several properties. Stainless steel, titanium and other metals are quite stiff as compared to biological materials, yielding little if at all in response to pressures from surrounding tissue. The extraordinary compliance of NiTiNOL[3] clearly makes it the metal most similar to biological materials mechanically (see Figs. 3 and 4).

Whereas NiTiNOL is the exception with respect to the world of metallurgy, stainless steel is the misfit in the world of biology [29]. Due to this improved physiological similarity, ingrowth of bone and proper healing by sharing loads with the surrounding tissues are facilitated. NiTiNOL has been applied so far successfully as hip implants, bone spacers, bone staples, skull plates and the like. In OMFS internal wiring, bone staples and even compression osteosynthesis were performed or under clinical evaluation using NiTiNOL [19] as early as in the seventies. These NiTiNOL-plates, however, were based on the principle of AO-compressive plating. They had to be inserted over an extraoral approach and tried to achieve compressive strains using NiTiNOL's specific shape recovery effect. Due to unfavourable dimensions and the difficulties in handling inherent in AO-type compressive plating, osteosynthesis was rather complicated than improved by application of shape memory alloys. Tensile stresses were diffult to calculate, with gaps resulting

---

[3] Nickel-Titanium-Naval-Ordonance-Laboratory, abbr. NiTi

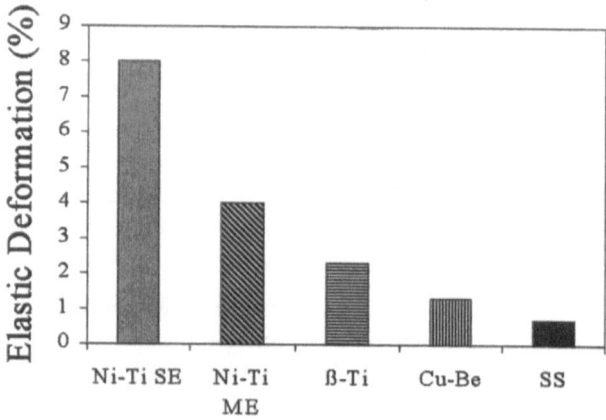

Fig. 3. Elastic deformation comparing NiTiNOL to different metals (redrawn after [29])

in the occlusal area and on the opposite cortical bone. Even strain induced necrosis of the bone adjoining to the fracture gap was observed. As a result, the development of shape memory alloy plates in OMFS was discontinued after phase II experiments on animals. In other medical fields, however, NiTiNOL was promoted, leading e. g. in orthopedics to spinal vertebrate spacers. This later application is particularly interesting by utilizing porous NiTiNOL, allowing better bone ingrowth. Combustion synthesis, or using the heat of fusion to "ignite" the formation of NiTi from Nickel and Titanium, has been shown to be an effective way to produce a porous macrostructure, with densities from 40-90% [29]. This sponge maintains superelastic as well as shape memory properties, and shows a reduced modulus of elasticity. Thus, bone ingrowth and adhesion to the surrounding tissue is improved.

4.2   Biocompatibility of NiTiNOL

As far as biocompatibility is concerned, it is important to emphasize that from among the large group of so-called "Smart Materials" only NiTi based alloys appear to be chemically and biologically compatible with the human body. NiTiNOL alloys for medical applications contain approximately 50 atomic percent of each nickel and titanium as binary alloys. With approximately 49 atomic percent of nickel the alloy shows athermal shape memory effect. Between 50.6 and 51 atomic percent of nickel the alloy shows superelastic properties, being integral to the design of a variety of medical products, e. g. orthodontic self-strengthening wires, intravascular and non-vascular stents, self expanding filters and baskets, smart guide wires and punction needles and a lot of other medical devices using superelasticity for interventional

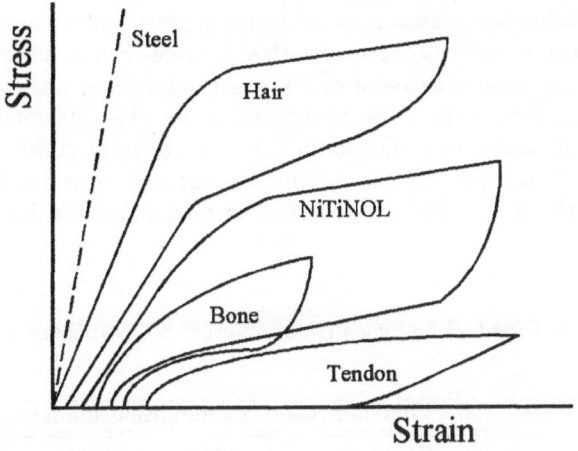

Fig. 4. Stress/strain curves comparing NiTiNOL to different biomaterials (redrawn after [29])

and endoscopic procedures. In dentistry, superelasticity is sucessfully applied in NiTi-endodontic files for the preparation of curved root channels. Unfortunately, the high percentage of nickel as compared to conventional 316L stainless steel instruments or implants often leads to a critical attitude in potential applicants, who recognize that nickel itself is considered allergic and toxic [29]. From a metallurgical point of view, however, this rather obvious statement has little meaning. Depending upon the treatment, NiTiNOL oxidizes with a small TiO_2 surface layer with no nickel or - alternatively - with small islands of pure nickel present at the surface. Polarization testing in Hank's solution repeatedly proved NiTiNOL to be chemically more stable and less corrosive than stainless steel. As compared to pure titanium, however, it is less stable. Biocompatibility thus being clearly superior at least to stainless steel, some NiTiNOL medical implants were approved as permanent implants for use in the USA by the FDA[4] [29]. Surface treatment, however, as the most important factor besides the definition of starting chemistries and thermomechanical treatments must be carefully considered in future assessments of NiTiNOL's biocompatibility before it can be firmly determined, how biocompatible the material really is. Although it clearly outperformances stainless steel, it has to compete with titanium and – quite new as a material for osteosynthesis – with biodegradable polymers (cf. Table 2). Laser and plasma treatments have already been shown to improve the cytocompatibility of NiTi samples to a level similar to Ti alloys [29]. These treatments decreased the corrosion rate by one to two orders of magnitude in simulated physiological Hank's solution. It seems that both solution treatments create

---

[4] Food and Drug Administration

a protective oxide layer able to reduce the quantity of released ions in the medium and the quantity of ions absorbed by cells. Smoother surface conditions achieved by laser treatment or polishing to a mirror surface is beneficial for the cell attachment and cell proliferation. Besides, plasma treated NiTi samples corresponding to a plasma polymerized tetrafluoroethylene (PTFE) coating allowed to separate the metallic substrate from the cells by the hydrophobe PTFE film. This was considered to explain the improvement of cytotoxity [96].

Table 4.1. Pros (+) and cons (−) of different osteosynthesis materials

Properties	Stainless Steel	Titanium	Biodegradables	NiTiNOL
Corrosion	−	++	++	+
Stiffness/Elasticity	−	+	+/?	++
Cytotoxity	−	+	−	+/−
Allergic potential	−	++	+	+
Handling	−	+	−	?
2nd Operation (removal)	−	−	++	−
Osseointegration	−	+	− −	+
Radio-opacity	+	+	− −	+
MRI-compatibility	− −	−	+	+/−

4.3   Compatibility with Diagnostic and Invasive Procedures

For applications in the medical field, compatibility of osteosynthesis materials with current diagnostics can be considered almost as essential as biocompatibiliy itself. Conventional X-ray imaging is increasingly replaced by computer-aided diagnostic procedures like computed tomography (CT) [67], ultrasound imaging (US) and magnetic resonance imaging (MRI).

With ionizing radiation not involved and scanning time relatively unlimited due to missing side effects, MR imaging becomes a more and more indispensable diagnostic tool in all fields of OMFS, especially in traumatology in conjunction with the temporomandibular joint [77]. Osteosynthesis material inserted in the image plane, however, reduces the signal intensity. Magnetic field inhomogeneities around the osteosynthesis material as caused by their paramagnetic effects decrease the relaxation time of the protons rendering their signal undetectable within the echo time. The size of the artifacts depends on the material, the applied sequences, and the angle of the sample within the magnetic field. As a consequence, material properties such as ferromagnetism, paramagnetism, diamagnetism and magnetic susceptibility are decisive. Thus, in the field of MR imaging, NiTi is considered to be superior to stainless steel and even to pure titanium [29,67].

5   Mandibular Morphology

Over the last years, finite element based simulations have been introduced in the development of new osteosynthesis materials and the optimization of current systems [30,31,36,38,39,51,52,84]. Mostly however, these biomechanical models represent gross oversimplifications due to the computing facilities available and, most of all, due to mandibular morphology itself. The human mandible is a very complex three-dimensional object and a biomechanical simulation has to cope with several major problems:

- non-homogeneity of the mandible, composed of cortical bone, cancellous bone, the teeth with their periodontal ligaments and the fibrocartilage of the temporomandibular joint,
- high grade anisotropy of the mandibular bone,
- elasticity of the mandible changing from elastic to viscoelastic properties when deformed more than 0.3%,
- elasticity of the mandible influenced by the velocity of loading forces,
- mandibular morphology showing an enormous range of interindividual variations, due to race, sex, age, skeletal growth patterns (e. g. mandibular prognathism or retrognathism), number of missing teeth and progress of resultant atrophic processes, fully dentulous vs. edentulous mandible etc. [57,94] (cf. Fig. 5),
- fresh human bone specimens for comparing biomechanical test rows hardly available, with stiffness in dry bone specimens up to 20% higher,
- availability and reliability of suitable data for the variables involved (homogeneous and inhomogeneous boundary conditions) [5,6,47,76,83,95].

In order to give an authentic description of the mechanical events, a mathematical simulation should be able to describe not only the individual geometric characteristics but also the inhomogeneous properties of the mandible, its elastic and viscoelastic properties and, as an essential factor, its anisotropy. If, e. g., bone tissue is modelled as being only elastically isotropic, an error up to 45 degree in the prediction of principal stress axes could happen [25]. In addition, the definition of static and dynamic actions which the bone may be submitted to, should be considered for its relevance in a numerical approach such as the finite element method.

Bone shows some biomechanical peculiarities which distinguish it from materials used in the engineering field, such as the possibility of varying structure and mechanical characteristics according to functional stresses and strains when submitted to action. Available biomechanical data as presented in the literature are often contradictory due to the anatomical localization and size of the specimens, pre-treatment preparation, temperature and level

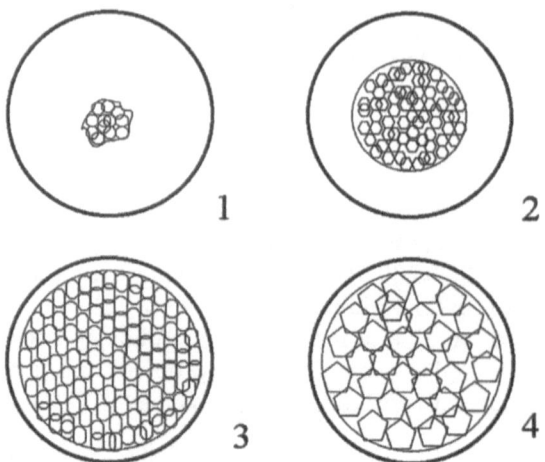

Fig. 5. Categories of bone quality (1-4) regarding the proportion and quality of cortical and cancellous bone (redrawn after [57])

of moisture during testing. In macroscopic terms, bone consists of compact or cortical bone with porosity from 3 to 5% and, on the other hand, of cancellous or trabecular bone, with porosity up to 90%. Among the mechanical properties considered of primary significance [83] are:

- elastic constants (relating stress to strain),
- viscoelastic parameters (relating stress, strain, and time),
- plastic parameters (describing permanent deformation),
- strength or ultimate properties (stating conditions of fracture or failure).

Ultrasonographic wave tests [6] from a suitably prepared specimen can give the required values for defining the characterizing coefficients (nine in case of definition as orthotropic material) in a certain localization, whereas the determination of these parameters using mechanical tests is problematic because only one Young modulus and two Poisson coefficients can be evaluated on a single specimen at the same time. Given the anisotropy and non-homogeneity of bone, three mechanical tests must therefore be carried out on three different specimens, which may have different characteristics [76] as there is an irregular distribution within a section and along the side of a specimen. The determination of the mechanical properties of trabecular bone is even more complex because it is composed of bone material variously arranged in columns or plates making up a fabric of variable density and porosity [76,95]. Though there are similar properties in cancellous and compact bone, the varying porosity causes a profoundly different overall mechanical behaviour (cf. Fig. 6). Elastic modulus values strongly depend on density (average value 0.3 g/cm^3), while the strain rate only exerts a small influence. As it seems quite unrealistic to determine all the anisotropic features of the material, different models for numerical simulations of trabecular bones are proposed:

- assumption as a network of cortical bone,
- assumption as a cellular model with cells formed of a cubic or hexagonal network of columnar elements at low densities, and closed by plate-like elements at higher densities.

To sum up, experimental data can give a necessary basis for biomechanical analysis, although overall knowledge cannot be considered exhaustive [76]. Besides, regarding callus behaviour, hardly any knowledge is available in the literature. Numerical approaches must pay close attention to the fact that the adoption of uncertain features in bone data has to be accepted as inevitable for the present. The data published for the human mandible indicate that it is elastically homogeneous but anisotropic. Its exhibited elastic properties are similar to, if slightly less stiff, human femora. Elastically, it may be compared to a long bone bent into the shape of a horseshoe [6].

Nevertheless, for the purpose of validation of finite element based simulations, biomechanical in vitro test stands are still indispensable. Due to the complexity of the stomatognathic system, direct measurements are needed to substantiate the mathematical models by comparing results and consecutive optimization.

6 Physiological and Biomechanical Principles of the Masticatory System

Biomechanical examinations of the jaw and facial skeleton are obviously not feasible in vivo. Nevertheless, it is possible to analyze mastication as a complex dynamical process in infinitely many snapshots and, based on this knowledge, to simulate it in a variety of load situations. The analysis and proper in vitro simulation of such load situations requires an exact knowledge of the physiology of the masticatory system and thus of its anatomical components, the dental arches, the temporomandibular joints and the neuromuscular system.

The masticatory system functions as a feedback loop with individual structures permanently influencing each other. Any change in the biostatics (geometry, position, force) and the biodynamics (motion, velocity) is inevitably linked to a change of the efficiency and thus the energy consumption of the system. The most important factor in this process are receptors within the periodontal ligament, muscle spindles and receptors in ligaments, tendons and joint capsules [7,21,68,102].

Border movements, functional movements and parafunctional movements are distinguished in examinations of jaw movements. Border movements of the mandible are unphysiological and thus rare, with the exception of the cranial end position due to the occlusal contacts. All excursion movements are functionally limited primarily by the musculature directly depending on the muscle tone. Every change of the muscle tone modifies the movement constraints whereby the jaw movements are safeguarded through tendon receptors and muscle spindles [26,33,100]. The border movements limited by the ligaments are proprioceptively safeguarded and represent the last protection prior to a destruction of the joints. Receptors within the tendons of the musculature, the ligaments and in the bilaminary zone are activated and initiate an intensified muscle tone and other neuromuscular reflex patterns [23,63]. Occlusal contacts limit the adduction of the mandible to its cranial end position. Periodontal ligament receptors, tendon receptors and muscle spindles have a monitoring function.

The chewing cycles based on learned and conditioned reflexes are prominent within the functional motion cycles of the mandible. The patterns of the motion cycle are decisively influenced and engrammed with the eruption of the first and the permanent dentition. These motion patterns are constantly modified throughout one's lifetime by perturbances and pathological changes of individual structures of the masticatory system [56,69,75]. Parafunctional motion patterns and their pathogenesis will not be considered here.

The temporomandibular joints play a dominant role in proprioceptively monitoring the motion cycle. Highly sensitive receptors in the bilaminary zone are most important in this context. False positions and loads are signalled to the central nervous system via afferent nerves resulting in changes of muscle tone and muscular activity. In all functional movements of the temporomandibular joint, the articulary disk is kept in a stretched condition by the upper belly of the lateral pterygoid muscle and bundles of elastic fibres. The position of the condyle at the maximal intercuspidation is fixed neuromuscularily based on individually conditioned reflexes. This structural position of the condyle is located slightly anterior of the center of the mandibular fossa on the eminentia articularis. From an anatomical perspective the thickest cartilage layer on both the condyle and the eminentia is at exactly this position [24,70,82].

The neuromusculary system is the propulsion of all mandibular movements. Given structural and functional harmony in the masticatory system, the musculature works very efficiently with the lowest action potentials. Perturbances in single structures like the temporomandibular joints or the occlusal relation of the dental arches cause a neuromuscular reengramming via the receptor systems. Most perturbances are thus compensated by a primary muscular hypertonus. If the primary muscular hypertonus is superposed by a secondary hypertonus over a longer time period due to additional perturbances or exogenous influences, a dysfunctional disease of the masticatory system may be the result [35,37,54,92].

In Biomechanics, the mandible is defined as a general lever, i. e. a body which can be rotated around different axes and on which forces can be applied at different points. Basically three ways to apply forces to the mandible are distinguished: first the passive forces of the condyles on the mandibular fossae, second the passive force of the nutrition bolus on the dental arches and third the active force consisting of all muscular forces. The vectorial sum of all active and passive forces applied to the mandibula is zero at any time [73]. This theory emphasizes the advantages of a complete mandibula as a study basis in contrast to simplified models. The question of the size of lever does not arise then. The emphasis is instead put on the amount and direction of the applied forces. The passive pressure load of the temporomandibular joint is being discussed controversely in the literature [18,32,88,99]. Histological examinations of the joint permit the hypothesis of pressure load in the joints during the chewing cycles. Chondroid cells agglomerate in the middle of the discus articularis thus indicating a pressure load. The functional architecture of the condylar process supports this hypothesis. Photoelastic experiments on models, measurements of the bone density and electromyographical examinations all led to the conclusion that pressure can be exerted through the respective pivot perpendicular onto the joint surface under physiological conditions [20,59,71,101,86]. Amount and direction of this compressive force depend on the pivot of the nutrition bolus, which should be understood as hypomochlion. The joint is thus not loaded in any case, but depending on the jaw movement [62]. The unilateral occlusion loads the joint of the balance side (contralateral) more heavily than the joint of the working side (ipsilateral) [11,12,81]. The parallel development of a biomechanical test stand and a three-dimensional mathematical model of the masticatory system led to first qualitative results regarding the condylary load on the working and balancing side [39]. Applying a muscular force of 100% results in an occlusal load at the molars in the range of 45 to 65% whereas the condyles on the working side were loaded with 10 to 15% and on the balancing side with approximately 25% of the applied force [30]. The assumption that excessive forces on the temporomandibular joints lead to dysfunctional perturbances and pathological changes in the joint structures is the driving factor for the ongoing interest in examining these forces. Quantitative results regarding amount and direction of the condylary forces based on biomechanical experiments and mathematical models are nevertheless still missing.

Besides the joints the dental arches are also subjected to a passive force which is directly dependent on the resistance of the nutrition bolus. These physiological forces for making nutrition smaller are governed and limited by the neuromusculary system via receptors in the periodontal ligament. The loading capacity of the teeth increases from mesial to distal. In an analysis of the maximally possible bite forces the limiting factor was found to be the pain threshold of the periodontal ligament at 100 N/cm^2 [10]. First in vivo examinations of the bite forces were already done in the fifties using force sensors in tooth fillings and replacements followed by a variety of similar experiments. The measured values vary interindividually and depending on the type of nutrition and the exact measurement position between 5 N and 120 N [2,44]. The discrepancy between actually applied bite forces and possible maximal forces is rather prominent. Bakke et al. [9] describe the dependency of the maximal forces in the region of the first molar between 300 N and 570 N on age, sex and occlusal factors.

The only active forces of the masticatory system are just the forces of the musculature. The maximal forces of the individual chewing muscles were determined numerically from their physiological muscle cross section. These computations however do not allow any statement regarding their participation in the chewing process [86]. For this purpose electromyographical examinations were carried out which allowed to correlate different levels of activity of the individual muscles to typical situations of the chewing process. Some authors used photoelastic examinations, model experiments and theoretical considerations to assign typical load situations to physiological muscular forces. A quantitative analysis of these examinations, however, is extremely difficult [8,48,55].

7   In Vitro Experimental Setup – Technical Equipment and Proceeding

Biomechanical investigations of the human mandible were based on photoelastic methods or strain gauges up to now [81,89,91]. Mechanical loads simulating muscular forces or biting loads were realized by weights or springs [53]. These methods allowed merely statical tests ignoring physiological kinematics. Some simulators extended this setup to dynamical tests using electric drives, thus applying movements instead of forces to the sample. These studies were primarily concerned with surgical or osteosynthetic implants. Chewing movements of a single tooth as well as groups of teeth were realized using robotics with a limitation of the maximal force [97], but all these experiments concentrated on teeth or implants, not the whole mandible and the temporomandibular joints. The *Mandibulator* of the Dept. of Oral and Maxillofacial Surgery of the University of Technology Munich is a new and variable test stand with the objective to adapt the simulation step by step to the in vivo situation:

- Step I: Dynamical simulation of *biting loads*. Application of dental forces, i. e. reactive forces of the bolus onto the teeth; no chewing movement simulated.
- Step II: Dynamical simulation of the *chewing motion*; no dental forces applied.
- Step III: Combination of Steps I and II. Movements will be applied at the mandible; forces at the teeth according to physiological observations. The kinematics of the artificial joints will be adjusted to physiological or pathological movements. Considering diagnostical data from axiography [79] and literature, the artificial joints will be modified.

All steps include full computer control, i. e. arbitrary force, motion, velocity and acceleration profiles are transmitted by the computer to the respective control units. The same computer simultaneously controls a set of measurement devices and acquires experimental data. The novel idea of the Mandibulator is to use a three-dimensional measurement of the joint forces at the simulated jaw joint. Furthermore, in contrast to similar setups, the movements and deformations of the jaw are measured by a three-dimensional motion capture system.

7.1   Step I

A freely adjustable rig holds the artificial joints, actuators and sensors. The arrangement is intended to realize maximal view at and access to the object during test observation and preparation. The specimen is fixed at the processus coronoidei. Its condyles rest in the artificial joint cups. Five hydraulic cylinders apply the dental forces ("active forces") via wire ropes according to in vivo measurements. Based on classical cases from diagnostics or literature, single teeth as well as groups of teeth can be loaded, with either symmetrical or asymmetrical loading configuration. Despite dynamic loading, the mandible does not show rigid body movement. The artificial joints and wire ropes which act as mm. masseteres give balance to the specimen. Actuator forces are considered active forces, masseter and joint forces are passive. Piezoelectric sensors measure the three-dimensional joint forces, simple strain gauged sensors the "active forces" as feedback to the control units. As a "simple" case, both active and passive forces have vertical orientation, upper and lower jaw articulate. Depending on material isotropy or anisotropy, the applied load causes varying stress flows in the specimen. For the determination of local stresses, spatial dislocations and movements of selected points are acquired by video analysis.

The control computer and a set of interface cards are placed in an industrial 19" rack. Hard- and software are carefully chosen to meet the high requirements in measurement and automation. Prior to the start of the experiment, the frequency, the number of test cycles, the load profiles including their maxima and minima and the abort criteria (excess of motion or force limits) for the respective control units are entered. The load profiles can be individually chosen for each unit and may have arbitrary shape (e. g., a sine profile or profiles adjusted to data measured in vivo). The program calculates the corresponding voltage signals and outputs them via D/A converters to the control units. During the experiment, desired and actual values are displayed online on the screen and saved on disc for offline analysis. Realistic conditions of geometry, lubrication and friction given, the two three-dimensional force sensors at the condyles allow the measurement of the joint reaction forces thus reducing the number of unknown forces involved at the mandible. Using piezoelectric sensors avoids the disadvantage of overlapping signals of strain gauges when measuring three-dimensional forces [39].

Hydraulic drives are used in simulator engineering in both industrial and research, especially biomechanical laboratories, because of their sophisticated controls and the advantages of hydrodynamic lubrication. Moving parts do not touch as shaft and bearing in electric drives. Thus minimizing wear, endurance tests as those for total hip arthroplasties at higher frequencies become possible. Biomechanical tests designed for the understanding of physiological circumstances rather than for the proof of component fatigue resistance allow non-hydraulic drives.

For the measurement of deformations of the surface, strain gauges are usually fixed to specific points of simple or simplified test specimens. The human mandible is a highly complex component: its mechanical properties vary depending on the location, its elasticity varies depending on direction. A video based motion capture system is a contact free method which is already in use for medical (speech and gait analysis) and sport biomechanics (optimization of training and movements), because these systems employ simple passive markers. Due to the lack of wires and telemetry, they can be considered non-retroactive. Currently a spatial resolution of 0.2 mm out of a cube with edges of 1 m view can be realized. Given that the mandible fits in a cube of length 25 cm, the manufacturer estimates a resolution better than 10 μm, which is sufficiently accurate.

7.2   Step II

In situ, muscles are attached to the bone over a certain area and not at a point. Even simple movements can hardly be realized in detail by the given tools. The masticatory model of Osborn et al. [78] consisted of 26 muscles (13 on each side). Considering this, our temporary model has to be reduced to the most important muscle groups simulated using electric drives for the mm. temporalis and masseter. Data derived from jaw motion analysis and electromyography are used as controlling parameters for distances, velocities, accelerations and their timing correlations at selected points.

Angle, angular velocity and acceleration of modern electric drives can be controlled at high precision by supplied voltage, while the maximum moment can be reduced by limiting the maximum current in order to protect specimen and drive against destruction. Because the smart control units are programmable, this kind of drives represent a good alternative to hydraulic path controlled drives for movement simulating tasks.

7.3   Step III

Due to the lack of quantitative knowledge of various physiological parameters in the mandibular region like muscular and joint forces [91], each of which can differ significantly from the values measured at the teeth, the simulation of motion cycles and the resulting loads on bones and implants is rather difficult. Using the Mandibulator test stand, data measured in vivo (motion patterns, electromyographical results and forces exerted by the teeth) and experimental data are combined.

The muscles mentioned above are simulated by electric drives applying movements, whereas hydraulic cylinders apply dental forces. In contrast to robots which merely make a single tooth or groups of teeth chew, our setup considers the whole mandible including teeth and temporomandibular joints. Forces are applied to the teeth and movements to the jaw, while the forces needed for the movements are measured.

When realizing chewing, a bunch of unknown quantities has to be assumed temporarily. Without the possibility of measuring in vivo, Osborn et al. [78] suggested an additional equation for the solution of the indetermined equation system. The mathematical model consisted of the jaw lever and the attacking forces. The additional equation demanded the minimization of muscle forces for a biting force given. Similarly, the force of the most active muscle was minimized by Koolstra et al. [51]. Alternatively, single teeth can be equipped with force sensors. An exceeded force limit can be defined as "pain": avoiding pain then represents another control algorithm. Analogously, the limitation of joint dislocations or requested power can be used to define a controlling concept.

7.4 Outlook

The mechanical demands on the mandible are more complex than on humerus or femur, e. g., which are treated by implants substituting bone and joints routinely today. Current reconstructive maxillofacial implants still show many complications [98]. The described setup, simulating both movements and forces of chewing, is supposed to discover loads which cannot be measured in vivo. The observed and measured knowledge on mandibular bone and joints formulates new demands on implants for oral surgery. Identifying characteristics of the masticatory system and its operational parameters by experimental structure analysis together with numerical simulation allows to optimize implant design, material and planning.

8 Computer Simulation of the Human Mandible with Implants

8.1 Goals

There are a number of good reasons to design a mathematical model and to develop a computer simulation device for the mechanical behaviour of the human mandible. From the clinical point of view it is a rather non-invasive technique to examine a patient's status and a promising tool for planning the treatment. For biomechanics it provides an alternative way to analyze (thermo-)elastic properties of the jaw bone together with implant materials. To achieve these goals, the model has to be able to deal with highly individual personal data and has to be capable of incorporating different (adaptive) materials. In the long run, the simulation will come with an adequate interface and should be fast enough to be used by the surgeon. In the first phase, however, it will be employed for mutual validation in experiments with the biomechanical test stand.

A lot of problems are associated with these objectives. They range from data acquisition (for both geometry and material), through visualization and segmentation (bone from the rest, mandible from the skull) to the choice of a suitable mathematical model (depending, e. g., on the internal character of bone material, on the implant material envisaged, and on the load situation) and of an efficient way to realize it in a simulation, together with a calibration for comparison with biomechanical tests. The medical and experimental aspects of these difficulties have been mentioned in earlier sections. The mathematical points will be discussed in the course of the outline of our method which follows.

8.2   Outline of the Method

Though other diagnostic techniques like ultrasound and magnetic resonance imaging are available, X-ray computerized tomography (CT) is the preferred way to obtain and record patient data for the kind of problems envisaged here. CT provides a material density distribution in 3D, which then has to be calibrated (cf. [27]) in order to identify bone (or other) material for the purpose of separating (cf. [74]) the mandible and of visualizing (cf. [45] and [80]). A disadvantage of CT is the different resolution obtained in the successive image planes on one hand and in the perpendicular direction on the other hand (cf. [61]), which has consequences for the size of the data set and for the choice of a 3D grid to be generated to represent the geometry in the mathematical model. The other necessary individual input, namely material data like density and elasticity coefficients, are even more crucial. While the density can be obtained to some extent from the CT, mechanical properties of the bone material have to be extrapolated from independent sources (see, e. g., [76,83,4,5]). As mentioned earlier, these values vary considerably among individuals, so that this stage of acquiring data has to be carried out with great care and is a main possible source of error.

For segmentation we use software developed at our Research Unit, namely the SIPFAS code (for Simulated Interactive Plastic FAcial Surgery); see [90] for a description. It allows to define threshold values for the density coming from the CT data to specify bone material and to cut away by hand those parts of the skull which will not be considered in the mathematical model. At this stage it is also possible to eliminate artifacts arising from existing implants and the like.

In lack of existing software that allows to transfer the geometry given by a 3D mass density distribution directly into a suitable tetrahedral grid, SIPFAS generates a surface triangulation using the *marching cubes* algorithm (cf. [60]), which can then be employed for both, visualization and data storage for conversion into a 3D grid. It should be noted that all the steps mentioned so far, CT data creation and transfer, segmentation, visualization and 3D grid generation, are currently developing all over the world, but different data structures and formats make it very difficult to use existing software.

The 3D grid is the geometric starting point for both the biomechnical experiment and the mathematical model and simulation. In the test stand we can use models made from a plastic material as close to bone material as possible produced by a *rapid prototyping* technique, for instance. The simulation of the mandible requires three ingredients: the geometry, given by points, tetrahedra, and surface triangles, the mathematical material model, based on elasticity theory, and the numerical *finite element* (FE) method.

The FE approach (see [87,14,1]) has proved appropriate for the femur in connection with hip surgery (cf., e. g., [58,46,65,85] and literature cited there) and also for the mandible (cf., e. g., [49,52,38]). Due to the higher geometrical complexity of the latter, however, the results so far can only be taken as qualitative and show that considerable computer power is needed to cope, for instance, with the fine structure of the dental area. We use a package developed at our Research Unit called FeliCs (for **F**inite **el**ements **i**n **C** plus plus), realized in C++ and which is able to support FE grids and systems of differential equations in weak formulation with different boundary conditions defined by the user and to assemble the systems of equations. It is equipped with an interface for access to modern algorithms of numerical linear algebra and will be further developed to support parallelization and adaptive techniques (cf. [15]).

This approach gives us great flexibility in incorporating different mathematical models, namely systems of partial differential equations together with appropriate boundary conditions, from elasticity theory (see, e. g., [34] and [3]) and for various adaptive implants. These models yield the displacements of the knots in the grid or, alternatively, stress or strain, as response to outer forces and other influences like temperature. Contrary to bone material, which for our purpose is considered elastic (note, however, the remarks about plasticity made earlier), materials like those with shape memory exhibit *hysteresis* effects (see [41] and literature cited there). More elaborate theoretical models are currently being developed at our Research Unit and will be cast into numerical algorithms in the future.

Another feature of our simulation code will be the possibility to define the boundary conditions by hand or semi-automatic. This is necessary if we want to calibrate the simulation with the experiments on the test stand and for a realistic inclusion of the influence of the rather complicated system of masticatory muscles (cf. [13]) and, for instance, chewing loads. Furthermore, different regions of materials with internal structure ranging from homogeneous to anisotropic have to be accounted for. Finally, defining boundary conditions will also be decisive in connection with the inclusion and optimization of implants.

8.3 Conclusion

Our collaboration of surgeons, engineers and mathematicians has the big advantage to unite clinical expertise, in vitro experiment and numerical modelling in a close group. This allows for an interactive comparison of the results of simulation and test stand and for an optimal adaptation to clinical exigencies. For example, all of the abovementioned models in literature have been compared with experimental results from other places and consequently there were problems with calibration and verification. We are, of course, aware of difficulties ensuing, for instance, from the use of different data formats and in data transfer. Also the intrinsically complicated nature of the object under investigation will lead to many discussions about acceptable simplifications and modifications. And finally we have to rely on further progress in computer power to allow for a stable and rapid operation of the code. But on progress we always have to rely!

References

1. Ames, W. F.: Numerical Methods for Partial Differential Equations, 3rd ed., Academic Press, Boston MA, 1992
2. Anderson, D. J.: A Method of Recording Masticatory Loads. J. Dent. Res. **32** (1953), 785
3. Antman, S. S.: Nonlinear Problems of Elasticity, Springer, New York, 1995
4. Arendts, F. J., Sigolotto, C.: Standardabmessungen, Elastizitätskennwerte und Festigkeitsverhalten des Human-Unterkiefers, ein Beitrag zur Darstellung der Biomechanik der Unterkiefer - Teil I. Biomed. Technik **34** (1989), 248–255
5. Arendts, F. J., Sigolotto, C.: Mechanische Kennwerte des Human-Unterkiefers und Untersuchung zum ≫in-vivo≪-Verhalten des kompakten Knochengewebes, ein Beitrag zur Darstellung der Biomechanik des Unterkiefers - Teil II. Biomed. Technik **35** (1990), 123–130
6. Ashman, R. B., van Buskirk, W. C.: The Elastic Properties of a Human Mandible, Adv. Dent. Res. 1(1) (1987), 64–67
7. Auriol, M., Coutand, A., Crintez, V., Chomette, G., Doumit, A., Lucht, M.: Proprioceptive Sensibility and Orofacial Functions. Rev. Stomatol. Chir. Maxillo. Fac. **86** (1985), 137–146
8. Bakke, M.: Mandibular Elevator Muscles: Physiology, Action and Effect of Dental Occlusion. Scan. J. Dent. Res. **101** (1993), 314–331
9. Bakke, M., Holm, B., Hensen, B. L., Michler, L., Moller, E.: Unilateral, Isometric Bite Force in 8-68-Year-Old Women and Men Related to Occlusal Factors. Scand. J. Dent. Res. **98** (1990), 149–158
10. Balabanoff, M.: Funktionell mechanische Reize beim Kauakt. Zahnärztliche Praxis, München **9** (1958), 69–70
11. Barbenel, J.: The Biomechanics of the Human Temporomandibular Joint: A Theoretical Study. J. Biomechanics **5** (1972), 251–265
12. Barbenel, J.: The Mechanics of the Temporomandibular Joint - a Theoretical and Electromyographical Study. J. Oral Rehab. **1** (1974), 19–27
13. Baron, P., Debussy, T.: A Biomechanical Functional Analysis of the Masticatory Muscles in Man. Archs. Oral Biol. **24** (1979), 547–553

14. Bathe, K.-J.: Finite Element Procedures, Prentice-Hall, Englewood Cliffs NJ, 1996
15. Bauer, H.-J.: Objektorientierter Entwurf und Implementierung multilevel-adaptiver Finite-Elemente-Methoden, Thesis, Munich, 1999
16. Baumgart, F., Bensmann, G., Hartwig, J.: Mechanische Probleme bei der Nutzung des Memory-Effektes für Osteosyntheseplatten. Tech. Mitt. Krupp, Forsch.-Ber. **35**(3) (1977), 157–171
17. Becker, R., Austermann, K.-H.: Frakturen des Gesichtsschädels. In: Schwenzer, N., Grimm, G. (eds): Zahn-Mund-Kiefer-Heilkunde, Vol. 2, Spezielle Chirurgie, Georg Thieme Verlag, Stuttgart, New York (1981), 506
18. Benninghoff, A.: Die Architektur der Kiefer und ihrer Weichteilbedeckung. Paradentium **6** (1934), 51–68
19. Bensmann, G., Baumgart, F., Hartwig, J., Haasters, J.: Untersuchungen der Memory-Legierung Nickel-Titan und Überlegungen zu ihrer Anwendung im Bereich der Medizin. Tech. Mitt. Krupp, Forsch.-Ber. **37**(1) (1979), 21–33
20. Boucher, L. J.: Anatomy of the Temporomandibular Joint as it Pertains to Centric Relation. J. Prosth. Dent. **12** (1962), 464–472
21. Caffesse, R. G., Carraro, J. J., Albano, E. A.: Influence of Temporomandibular Joint Receptors on Tactile Occlusal Perception. J. Periodont. Res. **8**(6) (1973), 400–403
22. Champy, M., Lodde, J. P.: Syntheses Mandibulaires. Location des Syntheses en Fonction des Contraintes Mandibulaires, Rev. Stomatol. **77** (1976), 971–979
23. Clark, R. K. F., Wyke, B. D.: Contributions of Temporo Mandibular Articular Mechanoreceptors to the Control of Mandibular Posture: an Experimental Study. J. Dent. **2**(3) (1974), 121–129
24. Couly, G.: The Temporomandibular Articulation and Functional Masticatory Interrelations. Actual Odontostomatol **114** (1976), 233–252
25. Cowin, S. C., Hart, R. T.: Errors in the Orientation of the Principal Stress Axes if Bone Tissue is Modeled as Isotropic. J. Biomechanics **23**(4) (1990), 349–352
26. Cueni, H., Graber, G.: Limited Movements of the Mandible under normal Conditions and under Experimental Stress. A Comparative Study Using Stereographic Recording. Schweiz. Monatsschr. Zahnmed. **96** (1986), 861–878
27. Daegling, D. J.: Biomechanics of Cross-Sectional Size and Shape in the Hominoid Mandibular Corpus. Am. J. Phys. Anthropology **80** (1989), 91–106
28. De Boever, J. A., McCall, W. D., Holden, S., Ash, M. M.: Functional occlusal forces: An Investigation by Telemetry. J. Prost. Dent. **40**(3) (1978), 326–333
29. Duerig, T. W., Pelton, A. R., Stöckel, D.: The Use of Superelasticity in Medicine, Metall **9** (1996), 569–574
30. Faulkner, M. G., Hatcher, D. C., Hay, A.: A Three-Dimensional Investigation of Temporomandibular Joint Loading. J. Biomechanics **20** (1987), 997–1002
31. Ferré, J. C., Legoux, R., Helary, J. L., Albugues, F., Le Floc'h, C., Bouteyre, J., Lumineau, J. P., Chevalier, C., Le Cloarec, A. Y., Orio, E., Marquet, F., Barbin, J. Y.: Study of the Deformations of the Isolated Mandible under Static Constraints by Simulation on a Physicomathematical Model, Anat. Clin. **7** (1985), 183–192
32. Findlay, J. A.: Mandibular Joint Pressures. Res. **43** (1964), 140–148
33. Gernet, W.: Die kinesiographische Aufzeichnung der Unterkieferbewegung. Dtsch. zahnärztl. Z. **37** (1982), 327–331

34. Gould, P. L.: Introduction to Linear Elasticity, Springer, New York, 1983

35. Graber, G.: Was leistet die funktionelle Therapie und wo findet sie ihre Grenzen? Dtsch. zahnärztl. Z. **40** (1985), 165–169

36. Gupta, K. K., Knoell, A. C., Grenoble, D. E.: Mathematical Modeling and Structural Analysis of the Mandible, Biomat., Med. Dev., Art. Org. **1**(3) (1973), 469–479

37. Hannam, A. G., Inkster, W. C., Scott, J. D.: Peak Electromyographic Activity and Jaw-Closing Force in Man. J. Dent. Res. **54** (1975), 694–703

38. Hart, R. T., Hennebel, V. V., Thongpreda, N., Van Buskirk, W. C., Anderson, R. C.: Modeling the Biomechanics of the Mandible: a Three-Dimensional Finite Element Study, J. Biomechanics **25** (1992), 261–286

39. Hatcher, D. C., Faulkner, M. G., Hay, A.: Development of Mechanical and Mathematic Models to Study Temporomandibular Joint Loading. J. Prost. Dent. **55**(3) (1986), 377–384

40. Henze, U., Zwadlo-Klarwasser, G., Klosterhalfen, B., Höcker, H., Richter, H., Mittermayer, C.: Kunsttsoffe für den medizinischen Einsatz als Implantatmaterialien, Dt. Ärztebl. **96**(15 A) (1999), 979–986

41. Hoffmann, K.-H., Haller, H., Hörmann, A.: Adaptive Materialien und Strukturen — Mathematische Modellierung und Simulation. In: [42], 141–150

42. Hoffmann, K.-H., Jäger, W., Lohmann, T., Schunck, H. (Hrsg.): Mathematik Schlüsseltechnologie für die Zukunft, Springer, Berlin, 1997

43. Horch, H.-H., Herzog, M.: Traumatologie des Gesichtsschädels. In: Horch, H.-H. (ed): Mund-Kiefer-Gesichtschirurgie I, Praxis der Zahnheilkunde, Vol 10/I, Urban und Schwarzenberg, München-Wien-Baltimore, 3rd edition, 1997, 55–163

44. Jaeger, K., Graber, G., Schrutt, L.: Aufbau und Test einer Me3kette zur Kaukraftmessung. Schweiz. Monatsschr. Zahnmed. **99**(6) (1989), 670–675

45. Jäger, W., Quien, N., Simon, J., Wirth, J.: 3D-Visualisierung von Tomogrammdaten zur Operationsplanung, Operationssimulation und optimalen Positionierung dentaler Implantate. In: [42], 409–420

46. Kaddick, C., Stur, S., Hipp, E.: Mechanical Simulation of Composite Hip Stems. Med. Eng. Phys. **19** (1997), 431–439

47. Katz, J. L., Meunier, A., The Elastic Anisotropy of Bone. J. Biomechanics **20**(11,12) (1987), 1063–1070

48. Kessler, W.: Das spannungsoptische Oberflächenschichtverfahren zur mechanischen Spannungsmessung am menschlichen Unterkiefer unter physiologischer Belastung. Zahnmed. Diss. München, 1980

49. Knoell, A. C.: A Mathematical Model of an *in vitro* Human Mandible. J. Biomechanics **10** (1977), 159–166

50. Kohn, J.: Biomaterials Science at a Crossroads: Are Current Product Liability Laws in the United States Hampering Innovation and the Development of Safer Medical Implants? Pharmaceutical Research **13**(6), (1996), 815–819

51. Koolstra, J. H., Van Eijden, T. M. G. J., Weijs, W. A., Naeije, M.: A Three-Dimensional Mathematical Model of the Human Masticatory System Predicting Maximum Possible Bite Forces. J. Biomechanics **21**(7) (1988), 563–576

52. Korioth, T. W. P., Romilly, D. P., Hannam, A. G.: Three-Dimensional Finite Element Stress Analysis of the Dentate Human Mandible. Am. J. Physical Anthropology **88** (1992), 69–96

53. Kroon, F. H. M., Mathisson, M., Cordey, J. R., Rahn, B. A.: The Use of Miniplates in Mandibular Fractures: An in Vitro Study. J. Cranio-Max.-Fac. Surg. **19** (1991), 199–204

54. Kubota, K., Masegi, T.: Muscle Spindle Supply to the Human Jaw Muscle. J. Dent. Res. **56** (1977), 901–909

55. Küppers, K: Analyse der funktionellen Struktur des menschlichen Unterkiefers. Ergeb. Anat. Entwicklungsgesch. **44**(6) (1971), 3–90

56. Lee, R. L.: Frontzahnführung. Hanser-Verlag, München (1985), 95–112

57. Lekholm, U., Zarb, G. A.: Patientenselektion und Aufklärung der Patienten. In: Brånemark, P.-I., Zarb, G. A., Albrektsson, T. (eds.): Gewebeintegrierter Zahnersatz, Quintessenz Verlags-GmbH, Berlin (1985), 197–198

58. Lengsfeld, M., Kaminsky, J., Merz, B., Franke, R. P.: Automatisierte Generierung von 3-D Finite Elemente Codes des menschlichen Femurs. Biomed. Technik **39** (1994), 117–122

59. Loos, S.: Die Mechanik der Kiefergelenke. Urban & Schwarzenberg, Wien, 1946

60. Lorensen, W. E., Cline, H. E.: Marching Cubes: High Resolution 3D Surface Construction Algorithm. Computer Graphics **21**(4) (1987), 163–169

61. Louis, A. K., Dietz, R.: Algorithmen für die 3D-Computer-Tomographie bei zerstörungsfreien Prüfverfahren. In: [42], 397–407

62. Ludwig, P.: Untersuchungen zur Frage der Unterkieferverlagerung bei isometrischer Kontraktion der Kaumuskulatur. Dtsch. zahnärztl. Z. **28** (1973), 901

63. Lundeen, H. C., Gibbs, C. H.: Kieferbewegungen und ihre klinische Bedeutung. Phillip Journal **2** (1987), 87–97

64. Mao, J., Osborn, J. W.: Direction of a Bite Force Determines the Pattern of Activity in Jaw-Closing Muscles. J. Dent. Res. **73**(5) (1994), 1112–1120

65. Martin, H., Zacharias, T., Fethke, K., Holzmüller-Laue, S., Gerhardt, H., Schmitz, K.-P.: Entwicklung und Validierung eines Finite-Elemente-Modells für Femurknochen. VDI Berichte **1463** (1999), 73–78

66. Meeder, P. J., Weller, S.: Traumatologie des Schädels, des Haltungs- und Bewegungsapparates. Frakturen und Luxationen. In: Reifferscheid, M., Weller, S. (eds): Chirurgie, 7th edition. Georg Thieme, Stuttgart (1986), 670

67. Melzer, A., Schmidt, A., Kipfmüller, K., Grönemeyer, D., Seibel, R.: Technology and Principles of Tomographic Image-Guided Interventions and Surgery. Surg. Endosc. **11** (1997), 946–956

68. Mercurio, A. R.: Nervous Control of Occlusion. Dent. Clin. N. Am. **25** (1981), 381–394

69. Meyenberg, K., Kubiks, S., Palla, S.: Relationships of the Muscles of Mastication to the Articular Disc of the Temporomandibular Joint. Schweiz. Monatschr. Zahnmed. **96** (1986), 815–834

70. Moffett, B.: Eine biologische Betrachtung der zentrischen Relation aufgrund der skelettalen und bindegewebigen Reaktion. In: Celenza, F. V., Nasedkin, J.N.: Okklusion-Der Stand der Wissenschaft. Quintessenz, Berlin (1979), 13–20

71. Molitor, J.: Untersuchungen über die Beanspruchung des Kiefergelenks. Z. Anat. Entwickl. Gesch. **128** (1969), 109

72. Møller, E., Bakke, M., Rasmussen, O. C.: Bildfunktionslaere. Odontologisk Boghandels Forlag, Kopenhagen (1985), 1–84

73. Motsch, A.: Spannungsoptische Experimente zur funktionellen Anatomie des Unterkiefers. Med. habil. Univ. Freiburg i. Br., 1965

74. Müller, R., Hildebrand, T., Rüegsegger, P.: Non-Invasive Bone Biopsy: a New Method to Analyse and Display the Three-Dimensional Structure of Trabecular Bone. Phys. Med. Biol. **39** (1994), 145–164

75. Mushimoto, E., Mitani, H.: Bilateral Coordination Pattern of Masticatory Muscle Activities During Chewing in Normal Subjects. J. Prosth. Dent. **48** (1982), 191–197

76. Natali, A. N., Meroi, E. A.: A Review of the Biomechanical Properties of Bone as a Material. J. Biomed. Eng. **11** (1989), 266–276

77. Neff, A., Kolk, A., Deppe, H., Horch, H.-H.: Neue Aspekte zur Indikation der operativen Versorgung intraartikulärer und hoher Kiefergelenkluxationsfrakturen. Mund Kiefer GesichtsChir. **3** (1999), 24–29

78. Osborn, J. W., Baragar, F. A.: Predicted Pattern of Human Muscle Activity during Clenching Derived from a Computer Assisted Model: Symmetric Vertical Bite Forces. J. Biomechanics **18**(8) (1985), 599–612

79. Piehslinger, E., Celar, A., Celar, A. R., Slaviek, R.: Orthopeadic Jaw Movement. Part V: Transversal Condylar Shift in Protrusive and Retrusive Movement. J. Craniomand. Prac. **12**(4) (1994), 247–251

80. Pöppe, C., et al.: Medizinische Bildverarbeitung. Spektrum der Wissenschaft 6/1997, 102–124

81. Ralph, J., Caputo, A.: Analysis of Stress Patterns in the Human Mandible. J. Dent. Res. **54** (1975), 814–821

82. Ramfjord, S. P.: Die Ziele einer idealen Okklusion und Unterkieferstellung. In: Solberg, W. K., Clark, G. T.: Kieferfunktion — Diagnostik und Therapie. Quintessenz, Berlin (1985), 81–100

83. Reilly, D. T., Burstein, A. H.: The Mechanical Properties of Cortical Bone. J. Bone and Joint Surgery **56-A**(5) (1974), 1001–1022

84. Rozema, F. R., Otten, E., Bos, R. R. M., Boering, G., van Willigen, J. D.: Computer-Aided Optimization of Choice and Positioning of Bone Plates and Screws Used for Internal Fixation of Mandibular Fractures. Int. J. Oral Maxillofac. Surg. **21** (1992), 373–377

85. Schmitt, J., Lengsfeld, M., Alter, P., Leppek, R.: Die Anwendung voxelorientierter Femurmodelle zur Spannungsanalyse. Biomed. Technik **40** (1995), 175–181

86. Schuhmacher, G. H.: Funktionelle Morphologie der Kaumuskulatur. G. Fischer, Jena, 1961

87. Schwarz, H. R.: Methode der finiten Elemente. Teubner, Stuttgart, 1980

88. Scott, J. H.: Muscle Growth and Function in Relation to Skeletal Morphology. Amer. J. Phys. Anthropol. **15** (1957), 197–234

89. Sonnenburg, M., Haertel, J.: Biomechanische Untersuchungen verschiedener Osteosyntheseverfahren am spannungsoptischen Kiefermodell. Stomatol. DDR. **28** (1978), 83–91

90. Spiegl, A.: Mathematische Modellierung von menschlichem Gewebe zur präoperativen Planung in der Gesichtschirurgie. Thesis, Munich, 1998

91. Standlee, J. P., Caputo, A. A., Ralph, J. P.: The Condyle as a Stress-Distributing Component of the Temporomandibular Joint. J. Oral Reh. **8**(5) (1981), 391–400

92. Stohler, C. S.: A Comparative Electromyographic and Kinesiographic Study of Deliberate and Habitual Mastication in Man. Archs. oral. Biol. **31**(19) (1986), 669–678

93. Terheyden, H., Mühlendyck, C., Feldmann, H., Ludwig, K., Härle, F.: The Self-Adapting Washer for Lag Screw Fixation of Mandibular Fractures: Finite Element Analysis and Preclinical Evaluation. J Craniomaxillofac Surg. **27** (1999), 58–67

94. Tsukasa, I., Gibbs, C. H., Marguelles-Bonnet, R., Lupkiewicz, S. M., Young, H. M., Lundeen, H. C., Mahan, P. E.: Loading on the Temporomandibular Joints with Five Occlusal Conditions. J. Prosthet. Dent. **56**(4) (1986), 478–484

95. Turner, C. H., Cowin, S. C., Rho, J. Y., Ashman, R. B., Rice, J. C.: The Fabric Dependence of the Orthotropic Elastic Constants of Cancellous Bone. J. Biomechanics **23**(6) (1990), 549–561

96. Villermaux, F., Tabrizian, M., Rhalmi, S., Rivard, C., Meunier, M., Czeremutskin, G., Piron, D. L., Yahia, L'H.: Cytocompatibility of NiTi Shape Memory Alloy Biomaterials. In: Pelton, A., Hodgson, D., Russel, S., Duerig, T. (eds): SMST-97: Proceedings of the Second International Conference on Shape Memory and Superelastic Technologies, SMA Inc., Santa Clara (1997), 417–422

97. Wang, L., Sadler, J. P., Breeding, L. C.: A Robotic System for Testing Dental Implants. Mechanism and Machine Theory **33**(5) (1998), 593–597

98. Watson, C. J., Ogden, A. R., Tinsley, D., Russell, J. L., Davison, E.M.: A 3- to 6-Year Study of Overdentures Supported by Hydroxyapatite-Coated Endosseous Dental Implants. Int. J. Prosthodont. **11**(6) (1998), 610–619

99. Weber-Thedy, K. W.: Kritischer Beitrag zur Artikulationslehre. Dtsch. zahnärztl. Z. **17** (1962), 1266–1279

100. Williamson, E. H., Marshall, D. E. Jr.: Myomonitor Rest Position in the Presence and Absence of Stress. Facial Orthop. Temporomandibular Arthrol. **3**(2) (1986), 14–17

101. Zenker, A., Zenker, W.: Die Tätigkeit der Kiefermuskeln und ihre elektromyographische Analyse. Z. Anat. Entwickl. Gesch. **119** (1955), 174–185

102. Zimny, M.: Mechanoreceptors in Articulator Tissues. Am. J. Anat. **182** (1988), 16–32

Rate Independent Hysteresis

M. Brokate

Zentrum Mathematik, Technische Universität München, D-80290 München

Dedicated to Professor Karl-Heinz Hoffmann
on the occasion of his 60th birthday

Abstract. We outline some basic features of the mathematical theory of hysteresis operators and indicate connections to applications, in particular magnetics and mechanics.

1 Introduction

The mathematical theory of rate independent evolutions has been developed during the last 50 years within different areas of mathematics and other sciences, sometimes in parallel. We describe here, from a mathematical perspective, some models which have proved to be fundamental in the field, and offer some comments on their basic features and on the present state of knowledge. We do not try to give an in-depth presentation; instead, we give references to the current literature.

2 Hysteresis Loops and Rate Independence. Scalar Hysteresis Operators

Let us consider a system which generates a time-dependent output function $w = w(t)$ from a given time-dependent input function $v = v(t)$. If both functions are scalar-valued, we can look at the curve c defined by $c(t) = (v(t), w(t))$ in $\mathbb{R}^2$ to visualize the input-output evolution. Let now v be an input function which takes the value $v_1 = v(t_1)$ at some time t_1, increases monotonically to $v(t_2) = v_2 > v_1$ for some $t_2 > t_1$ and decreases back to $v(t_3) = v_1$ for some $t_3 > t_2$. We speak of a *hysteresis loop*, if the curve c forms a loop in the input-output-plane like the one depicted in Fig. 1. A hysteresis loop becomes a particularly prominent feature of the input-output behaviour of the system if it is invariant with respect to velocity changes in the input. For the example above this means that all input functions increasing from the value v_1 to v_2 and then back to v_1 generate the same loop in the input-output plane. A derivative-free statement of this invariance property, called the property of *rate independence*, runs as follows: If w is the output

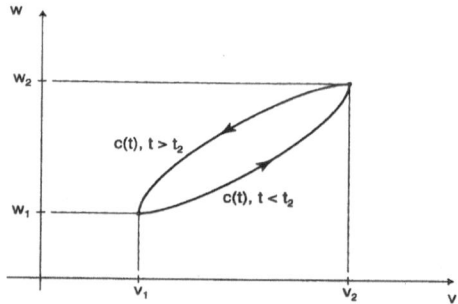

Fig. 1. A closed hysteresis loop

generated by v, and if φ is a nondecreasing transformation of time, then the input $v \circ \varphi$ generates the output $w \circ \varphi$. Using the operator notation

$$w = \mathcal{W}[v]\,, \tag{2.1}$$

the rate independence property becomes

$$\mathcal{W}[v \circ \varphi] = \mathcal{W}[v] \circ \varphi\,, \tag{2.2}$$

which has to hold for all inputs and all time transformations within a suitable class of functions.

Let $v : [0,T] \to \mathbb{R}$ be a piecewise monotone function. Intuition tells us that, if $\mathcal{W}$ is rate independent, the output $\mathcal{W}[v]$ is essentially determined by the values of the local maxima and minima of v. This observation leads to a canonical construction of rate independent operators in discrete time. Let us define the set S_A of *alternating strings* as the set S of all finite strings s of the form

$$s = (v_0, \dots, v_N)\,, \quad N \geq 0, \, v_i \in \mathbb{R}\,, \tag{2.3}$$

satisfying the condition

$$(v_{i+1} - v_i)(v_i - v_{i-1}) < 0\,, \quad \text{if } 0 < i < N\,. \tag{2.4}$$

If $\tilde{\mathcal{H}} : S_A \to \mathbb{R}$ is an arbitrary map, we can define an operator $\tilde{\mathcal{W}} : S_A \to S$ by

$$\tilde{\mathcal{W}}(s) = (\tilde{\mathcal{H}}(v_0), \tilde{\mathcal{H}}(v_0, v_1), \dots, \tilde{\mathcal{H}}(s))\,, \tag{2.5}$$

if $s = (v_0, \dots, v_N) \in S_A$. Moreover, for any piecewise monotone function $v : [0,T] \to \mathbb{R}$ we obtain a function $w = \mathcal{W}[v]$ on $[0,T]$ if we set

$$w(t) = \tilde{\mathcal{H}}(s_t)\,, \tag{2.6}$$

where s_t is the finite string of local maxima and minima of v on the time interval $[0, t]$. It turns out that *every rate independent and causal operator* $\mathcal{W}$ defined on the space $C_{pm}[0, T]$ of continuous and piecewise monotone real-valued functions on the interval $[0, T]$ can be obtained in this manner, see [3] for the proof. Here, causality means as usual that the value $\mathcal{W}[v](t)$ at time t is completely determined by the values $v(s)$, $s \leq t$.

Definition 2.1. A rate independent and causal operator $\mathcal{W} : C_{pm}[0, T] \to M[0, T]$, the set of all real-valued mappings on $[0, T]$, is called a *scalar hysteresis operator*.

Note that the discrete-time construction of a scalar hysteresis operator outlined above does not involve any limit process; indeed one can study such models extensively without having to use (or even to mention) the concept of a function space, see e.g. [5]. This is no longer possible, of course, if one wants to develop a mathematical theory of dynamical systems, where hysteretic behaviour appears in differential or integral equations, because such a theory is based on general principles of analysis which rely on contraction or compactness properties and thus requires the setting of complete spaces. One therefore is led to the question for which function spaces X, Y a given hysteresis operator $\mathcal{W}$ possesses a unique extension $\mathcal{W} : X \to Y$, and what are the regularity properties of such an extension. Indeed, a large proportion of the pioneering work of the group around Krasnosel'skiĭ has been devoted to such problems, see [1].

3   The Preisach Model

For various reasons, the Preisach model represents a focus for studies of scalar hysteresis ([1–8]). It features a hierarchy of nested hysteresis loops with several (in fact: an arbitrary number of) levels and thus accounts for a rather complex hysteretic behaviour; this is important in some areas of application, in particular in magnetics. It can be constructed from a linear superposition of hysteretic relays $\mathcal{R}_{x,y}$ with thresholds x for downward switching to the value -1, and y for upward switching to the value 1, namely

$$\mathcal{W}[v](t) = \int\!\!\int_{x<y} \omega(x, y) \mathcal{R}_{x,y}[v](t)\, dx\, dy \,. \tag{3.1}$$

It can also be represented as a nonlinear superposition

$$\mathcal{W}[v](t) = \int_0^\infty q(r, \mathcal{P}_r[v](t))\, dr \,, \quad q(r, s) = 2 \int_0^s \omega(r, \sigma)\, d\sigma \,, \tag{3.2}$$

where $\mathcal{P}_r$ denotes another basic hysteresis operator, called the *play operator*, which is the solution operator to a certain variational inequality to be discussed below. Furthermore, the *memory* or *internal state* $\psi(t)$ of the Preisach operator at time t, which represents all the information from the input v prior to t needed to determine the future evolution from the input values $v(s)$, $s \geq t$, is given by the collection $(\mathcal{P}_r[v](t))_{r \geq 0}$. For every t, we thus have $\psi(t) \in \Psi$ for some suitable space Ψ of real-valued functions defined on $\mathbb{R}_+ = [0, \infty)$. The graph of $\psi(t)$ has an immediate interpretation as the curve dividing the relays which are on $+1$ from the relays which are on -1, see Fig. 2 for a typical $\varphi = \psi(t) \in \Psi$. Moreover, the time evolution of this

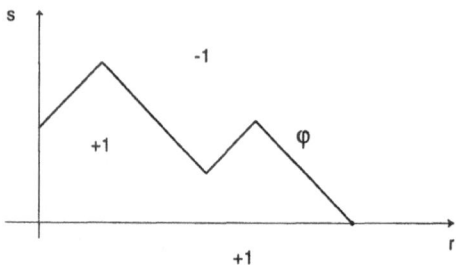

Fig. 2. The Preisach memory state

curve admits a simple update rule.

The basic regularity and continuity properties of the Preisach model have been investigated thoroughly [1–4]. A lot of variants and extensions of the Preisach model have been developed, mostly within the context of magnetics research [5–8]. Following [5], there are some results which connect the analytical definition of such a model to the nested structure and the shape of the hysteresis loops. Of a general structure theory of scalar hysteresis operators only the beginnings are visible, however, see [1,3,4]; in [3], for example, the class of operators of *Preisach type* is introduced as the class of all operators of the form

$$W[v](t) = Q(\psi(t)), \tag{3.3}$$

where $Q : \Psi \to \mathbb{R}$ is an arbitrary mapping and $\psi(t)$ is the internal state of the Preisach operator. The latter is recovered if we set

$$Q(\varphi) = \int_0^\infty q(r, \varphi(r)) \, dr, \tag{3.4}$$

compare (3.2).

Concerning numerical approximation and identification, only few contributions of mathematicians exist, see e.g. ([19–21]). It seems that during the last years, there has been a lot of activity concerning modeling and simulation among researchers in magnetics for which there is almost no counterpart in the mathematics community.

4   Vector Hysteresis Operators

A vector hysteresis operator is a causal and rate independent operator which maps input functions $v : [0, T] \to V$ to output functions $w : [0, T] \to W$, where V and W are vector spaces. One observes immediately that the property of rate independence in the non-scalar case is a much less stringent restriction than in the scalar case. Rate independence means that two input functions traveling along the same (directed) curve in input space yield – modulo time transformation – identical outputs. In contrast to the non-scalar case, only two directions of movement (increasing and decreasing) are possible for scalar inputs. In the non-scalar case, there is much more freedom, and the scalar notion of a closed hysteresis loop as well as the role of the space of piecewise monotone functions as the natural domain of definition of the hysteresis operator do not have an adequate counterpart. One might use the space of piecewise affine linear functions instead; in this manner, the close relation to discrete-time sequences of input values is maintained, but a limit process is necessary to define the operator on simple input curves like circles. Also, one might use other means of interpolation instead the linear one. One might also define a hysteresis operator on inputs whose intrinsic time scale coincide with arc length, that is $|\dot{v}(t)| = 1$, and then extend the operator to arbitrary inputs by the requirement of rate independence. In any case, we cannot easily separate structural properties of the hysteresis memory from the continuity and regularity properties of the specific hysteresis model under consideration; in contrast to that, structural properties of the hysteresis model and smoothness properties of the operator can be separated to a large extent in the scalar case.

In the vector case, too, it appears natural to start with some basic models whose significance in theory or application is obvious, and then construct more elaborate ones.

5   The Vector Play and Stop

Among models for rate independent hysteresis with vector inputs, these two are the ones which have attracted the greatest amount of attention. They both arise as solution operators for the following evolution variational inequality problem. Let Z be a convex and closed subset of some separable Hilbert space H. Given a function $v : [0, T] \to H$ and an element $z^0 \in Z$, we want to determine functions $w, z : [0, T] \to H$ which satisfy, for (almost) all $t \in [0, T]$,

$$v(t) = w(t) + z(t), \quad z(t) \in Z, \tag{5.1}$$

$$\langle \dot{w}(t), z(t) - \zeta \rangle \geq 0, \quad \forall \zeta \in Z, \quad z(0) = z^0. \tag{5.2}$$

The corresponding solution operators

$$w = \mathcal{P}[v], \quad z = \mathcal{S}[v], \tag{5.3}$$

are called the *play* and the *stop* respectively. One can visualize the behaviour of the functions w and z as follows. If at some time t we have $z(t) \in \text{int}(Z)$, then (5.2) implies that $\dot{w}(t) = 0$ and consequently $\dot{z}(t) = \dot{v}(t)$. If, on the other hand, $z(t)$ belongs to the boundary ∂Z of Z, and if $\dot{v}(t)$, considered as a vector with base point $z(t)$, points in an outward direction, then $\dot{w}(t)$ and $\dot{z}(t)$ are the projection of $\dot{v}(t)$ on the normal and tangential cone to Z at $z(t)$ respectively, see Fig. 3.

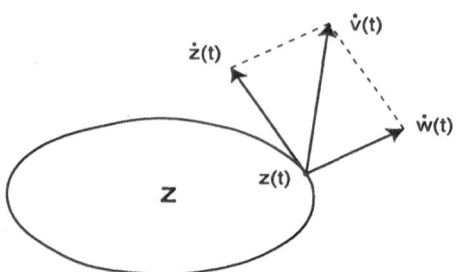

Fig. 3. Vector play and stop

In the scalar case $H = \mathbb{R}$, if we set $Z = [-r, r]$ for $r > 0$, we obtain the scalar play $\mathcal{P}_r$ which appeared already in the representation (3.2) of the Preisach model above.

Problem (5.1),(5.2) constitutes a special case (namely, the case of *normal reflection*) of the *Skorokhod problem* [23], which plays an important role in the study of the boundary behaviour of stochastic evolutions; in this context, the operator S is called the *Skorokhod map*; let us refer to [24] for a recent detailed exposition. Moreover, (5.1),(5.2) arises as a special case of the *sweeping process* whose investigation was initiated by Moreau [26,27]; see also [14,28]. Indeed, contact and friction problems as well as the formulation of elastoplastic constitutive laws and the corresponding boundary value problems lead to systems of ordinary or partial differential equations coupled to evolutions of the form (5.1),(5.2), see [2,15–17,29].

Thus, a lot of properties of the vector play and stop have been discovered (and rediscovered) by now. In [22], one finds an up to date exposition concerning the question, on which function spaces the vector play and stop are well defined, and concerning their continuity properties; let us just offer some remarks here. Not surprisingly, those properties often depend on the geometry of the convex constraint Z. The two main cases are where Z is a polyhedron and where Z is strictly convex and has a smooth boundary. Some of the more refined properties of $\mathcal{P}$ and S hold in both cases, but not in general; moreover, separate proofs are required for those two cases.

6   Vector Models with Complex Memory

For the vector play and stop, the knowledge of either $w(t_0)$ or $z(t_0)$ suffices to determine the values $w(t)$ and $z(t)$ for $t \geq t_0$ for any given input function $v : [t_0, T] \rightarrow H$; thus, the memory at any given time t is represented by a single element of H. This is adequate, for example, for the basic constitutive laws of (rate independent) elastoplasticity. Here, H is a subspace of the space of 3×3 tensors, thus of finite dimension. The Prandtl-Reuss model has the form

$$\sigma = \mathcal{S}[a : \varepsilon] \,, \tag{6.1}$$

where $\sigma, \varepsilon : [0, T] \rightarrow H$ are the stress and the strain respectively, and a is the fourth order tensor which represents the linear elastic law. The Melan-Prager model for linear kinematic hardening (with hardening constant C_0) can be written as

$$\varepsilon = a : \sigma + \frac{1}{C_0} \mathcal{P}[\sigma_d] \,, \tag{6.2}$$

where σ_d denotes the deviatoric part of σ. In both models, the boundary ∂Z of the convex set Z in (5.1) coincides with the *yield surface*. There is one tensorial internal variable, usually to be chosen as the plastic strain

$$\varepsilon^p = \varepsilon - \varepsilon^e = \varepsilon - a : \sigma \,. \tag{6.3}$$

Multisurface models have a more elaborate structure. There are various ways to construct such models. One can, for instance, begin with a one-parameter family, say $(\mathcal{P}_r)_{r \geq 0}$, of vector plays and define

$$\mathcal{W}[v](t) = \int_0^\infty \mathcal{P}_r[v](t) \, d\nu(r) \,, \tag{6.4}$$

where ν is a measure on $\mathbb{R}_+$. A natural way to obtain $\mathcal{P}_r$ would be to parametrize the convex set as $Z_r = rZ_1$ for some fixed closed convex set Z_1. Such models are called *Prandtl-Ishlinskiĭ* models, they have been investigated in [2], see also [22]. The model (6.4) is the vector analogue of the representation (3.2) of the Preisach model. If ν is a finite linear combination of Dirac functionals, one gets a superposition of a finite number of plays and thus a finite-dimensional memory; in general, the memory is infinite-dimensional. However, so far no description of the internal state is known which is comparable in simplicity to the Preisach state $\varphi = \psi(t)$ from Fig. 2. A model which keeps the piecewise linear structure of the internal state as well as the simplicity of the update rule is given by the one-parameter *Mróz model* [30]; however, its regularity properties appear to be weaker than of the Prandtl-Ishlinskiĭ model [31]. In the latter, a family Z_r, $r \geq 0$, of surfaces appears explicitly as part of the definition of the play $\mathcal{P}_r$. A different method of constructing multisurface models is used in the context of *nonlinear kinematic hardening* models like the model of Chaboche. Here, the *backstress* is decomposed into a sum whose individual summands satisfy a nonlinear differential equation of Armstrong-Frederick type and serve as additional internal variables. This time, the auxiliary surfaces are not part of the definition of the model, but they appear implicitly as bounds for the backstress components, due to the Armstrong-Frederick equation. The definition of the corresponding hysteresis operator $w = \mathcal{W}[v]$ involves the solution operator of the Cauchy problem

$$\dot{u} = \dot{v} + \mathcal{M}[u] \cdot |\dot{\xi}|, \quad \xi = \mathcal{P}[v], \tag{6.5}$$

where $\mathcal{P}$ is the vector play and $\mathcal{M}$ is another hysteresis operator, see [29,35] for more details.

7   The Duhem Model

For the basic scalar version of this model, the hysteresis operator $\mathcal{W}$ is defined to be the solution operator of the Cauchy problem

$$\dot{w} = g_+(v, w)(\dot{v})_+ - g_-(v, w)(\dot{v})_-, \quad w(0) = w^0. \tag{7.1}$$

Here $g_+, g_- : \mathbb{R}^2 \to \mathbb{R}$ are given functions and

$$(\dot{v})_+ = \max\{0, \dot{v}\}, \quad (\dot{v})_- = \max\{0, -\dot{v}\}, \tag{7.2}$$

are the positive and negative parts of $\dot{v}$ respectively. Thus, the evolution (which is rate independent) is governed by a differential equation with two right hand sides, triggered by the sign of the input velocity. A natural generalization of (7.1) to the vector case is

$$\dot{w} = g\left(v, w, \frac{\dot{v}}{|\dot{v}|}\right), \quad w(0) = w^0. \tag{7.3}$$

The wellposedness of the Cauchy problem and the corresponding regularity and continuity properties of the resulting hysteresis operator $w = \mathcal{W}[v]$ have been studied in [1,2]. Here, we only want to point out an interesting variant of it, namely the rate independent control system

$$\dot{x} = |\dot{v}| \cdot Ax + B\dot{v}, \quad x(0) = 0, \tag{7.4}$$
$$w = Cx. \tag{7.5}$$

Here, $A \in \mathbb{R}^{n,n}$, $B \in \mathbb{R}^{n,m}$ and $C \in \mathbb{R}^{k,n}$ are given matrices; the internal state $x : [0, T] \to \mathbb{R}^n$ and the output $w : [0, T] \to \mathbb{R}^k$ have to be determined from the input $v : [0, T] \to \mathbb{R}^m$. Motivated by problems in dry friction, Bliman and Sorine have investigated system (7.4),(7.5) to some extent [32–34], further studies should certainly prove to be of interest.

References

1. Krasnosel'skiĭ, M. A., Pokrovskiĭ, A. V.: Systems with Hysteresis. Springer-Verlag, Heidelberg, 1989; Russian edition: Nauka, Moscow, 1983
2. Visintin, A.: Differential Models of Hysteresis. Springer-Verlag, Berlin, 1994
3. Brokate, M., Sprekels, J.: Hysteresis and Phase Transitions. Springer-Verlag, New York, 1996
4. Krejčí, P.: Hysteresis, Convexity and Dissipation in Hyperbolic Equations. Gakkōtosho, Tokyo, 1996
5. Mayergoyz, I. D.: Mathematical Models of Hysteresis. Springer-Verlag, New York, 1991
6. Bertotti, G.: Hysteresis in Magnetism for Physicists, Materials Scientists, and Engineers. Academic Press, San Diego, 1998
7. Della Torre, E.: Magnetic Hysteresis. IEEE Press, 1999
8. Iványi, A: Hysteresis Models in Electromagnetic Computation. Akadémiai Kiadó, Budapest, 1997
9. Göcke, M.: Starke Hysteresis im Außenhandel. Heidelberg, 1993
10. Moreau, J. J., Panagiotopoulos, P. D., Strang, G. (eds.): Topics in Nonsmooth Mechanics. Birkhäuser, Basel, 1988
11. Visintin, A. (ed.): Models of Hysteresis. Pitman Research Notes in Mathematics **286**, Longman, Harlow, 1993
12. Visintin, A. (ed.): Phase Transitions and Hysteresis. LN Mathematics **1584**. Springer-Verlag, Berlin, 1994
13. Drábek, P., Krejčí, P., Takáč: Nonlinear Differential Equations. CRC Research Notes in Mathematics **404**, CRC Press, London, 1999
14. Monteiro Marques, M. D. P.: Differential Inclusions in Nonsmooth Mechanical Problems. Shocks and Dry Friction, Birkhäuser, Basel, 1993
15. Duvaut, G., Lions, J.-L.: Inequalities in Mechanics and Physics. Springer-Verlag, Berlin, 1976; French edition: Dunod, Paris, 1972
16. Han, W., Reddy, B. D.: Plasticity. Mathematical Theory and Numerical Analysis. Springer-Verlag, New York, 1999
17. Alber, H.-D.: Materials with Memory. Initial-Boundary Value Problems for Constitutive Equations with Internal Variables. LN Mathematics **1682**, Springer-Verlag, Berlin, 1998
18. Brokate, M., Siddiqi, A. H. (eds.): Functional Analysis with Current Applications in Science, Technology and Industry. Pitman Research Notes in Mathematics **377**, Longman, Harlow, 1998
19. Verdi, C., Visintin, A.: Numerical Approximation of the Preisach Model for Hysteresis. Math. Model. Numer. Anal. **23** (1989), 335–356
20. Hütter, T: Ein Verfahren zur näherungsweisen Berechnung des Preisachoperators mit Anwendungen bei gewöhnlichen Differentialgleichungen. Doctoral Dissertation, Technische Universität Berlin, Berlin, 1991
21. Hoffmann, K.-H., Meyer, G. H.: A Least Squares Method for Finding the Preisach Hysteresis Operator from Measurements. Numer. Math. **55** (1989), 695–710
22. Krejčí, P.: Evolution Variational Inequalities and Multidimensional Hysteresis Operators. In [13]
23. Skorokhod, A. V.: Stochastic Equations for Diffusion Processes in a Bounded Region. Theor. Probab. Appl. **6** (1961), 264–274

24. Dupuis, P., Ramanan, K.: Convex Duality and the Skorokhod Problem – I. Prob. Theor. Rel. Fields 1999, to appear; Technical Report LCDS 96-5, Brown University, Providence, 1996

25. Dupuis, P., Ramanan, K.: Convex Duality and the Skorokhod Problem – II. Prob. Theor. Rel. Fields 1999, to appear; Technical Report LCDS 96-5, Brown University, Providence, 1996

26. Moreau, J.-J.: Problème d'Evolution Associé à un Convexe Mobile d'un Espace Hilbertien. C.R. Acad. Sci. Paris Sér. A-B **273** (1973), A791–A794

27. Moreau, J.-J.: Evolution Problem Associated with a Moving Convex Set in a Hilbert Space. J. Diff. Eq. **26** (1977), 347–374

28. Castaing, C., Duc Ha, T. X., Valadier, M.: Evolution Equations Governed by the Sweeping Process. Set-Valued Analysis **1** (1993), 109–139

29. Brokate, M.: Elastoplastic Constitutive Laws of Nonlinear Kinematic Hardening Type. In [18], 238–272

30. Chu, C. C.: A Three-Dimensional Model of Anisotropic Hardening in Metals and its Application to the Analysis of Sheet Metal Formability. J. Mech. Phys. Solids **32** (1984), 197–212

31. Brokate, M., Dreßler, K., Krejčí, P.: On the Mróz model. Eur. J. Appl. Math. **7** (1996), 473–497

32. Bliman, P.-A., Sorine, M.: A System-Theoretic Approach of Systems with Hysteresis. In Proc. 2nd Eur. Cont. Conf., Groningen (1993), 1844–1849

33. Bliman, P.-A., Sorine, M.: Easy-to-Use Realistic Dry Friction Models for Automatic Control. In Proc. 3nd Eur. Cont. Conf., Roma (1995), 3788–3794

34. Bliman, P.-A., Krasnosel'skiĭ, A. M., Sorine, M.: Dither in Systems with Hysteresis. INRIA Rapport de recherche **RR-2690**, Paris, 1995

35. Brokate, M., Krejčí, P.: On the Wellposedness of the Chaboche Model. In Control and Estimation of Distributed Parameter Systems, Desch, W., Kappel, F., Kunisch, K., eds., Birkhäuser, Basel, 1998, 67–79

The Oxidation Process of Silicon

W. Merz

Zentrum Mathematik, Technische Universität München, D-80290 München,
Germany

Dedicated to Professor Karl-Heinz Hoffmann
on the occasion of his 60th birthday

1 Introduction

In electronic devices the electrical insulation of different areas is often achieved
by layers of silicon-dioxide. These are usually created by exposing the silicon
wafers at process temperatures between 700–1200⁰C to a gas flow containing
oxygen or to a stream. Using different oxygen isotrops tracer experiments
have shown that the new oxide is created at the interface between silicon and
silicon-dioxide.

In the sixties Deal and Grove [3] developed an oxidation model in one
space dimension. This model provides an excellent description of the linear-
parabolic growth rate of the oxide thickness with time. For very short instants
of time the assumed linear reaction rate at the interface limits the oxide
growth, whereas for large times the oxygen diffusion through the oxide-layer
limits the growth rate. This rate depends mainly on the considered temperature and the composition of the gas flow.

In the two or three dimensional case, the situation is much more complicated due to the masks that are used to define the position and the lateral
extension of the oxide-layers. These masks are impervious to oxygen, so the
lateral distribution of oxygen as well as the consumption of silicon becomes
non-homogeneous. Moreover, the newly created silicon-dioxide requires more
than twice of the volume of the consumed silicon. This volume expansion
inside the material causes a mechanical deformation of the whole layer structure.

A mathematical model of the oxidation process has to describe the diffusion-
convection effects and the reaction of oxygen, the volume expansion in the
chemical reaction, the mechanical behaviour of all substances and the changes
in the layer system due to the growth or consumption of layer materials.

The area of simulation, where these equations have to be solved, is in
general extremely complicated in shape. Very thin layers with an extension
of only a few nanometers are mixed with structures of several hundred mi-
crometers in size. Also the oxide thickness increases during the oxidation
process from an initial expansion of about 1.5nm–50nm to a final widening
from several hundred nanometers to around one micron. This leads to a very

complex and complicated mathematical system consisting of nonlinear free and moving boundary value problems.

In order to enable the numerical simulation of the full problem several simplifications are necessary, where some of them are justified by experimental results.

In [1], [9] and [12] different models describing the mechanical behaviour of the oxide have been proposed and in different temperature regimes a purely elastic, viscoelastic or viscous material law has been assumed for the oxide.

In general the stresses influence the oxidation process. This complicates the simulation of the oxygen diffusion and reaction additionally. For this see [2] and [10].

In [11] the authors replaced the sharp reaction front between silicon and silicon-dioxide by an extended interface zone. This region is filled with both substances. It can be regarded as a smooth transition zone from one material to the other. The basic idea of this approach was to overcome some "standard" difficulties in the numerical treatment of free boundaries, such as the necessity of remeshing the oxide region in every time step.

In [5]–[7] a new oxidation model has been introduced and the idea of an extended reaction zone has been further developed, simply as a mathematical tool to regularize the free boundary in order to formulate an existence result concerning the mathematical model.

In the present contribution we start with a model containing a sharp reaction front and turn over to a "smoothed" model where the free reaction front is eliminated from the model. For this we regard the silicon and the oxide as components of a mixture in which oxygen is dissolved.

Next we state a local existence result concerning the smoothed model, see [7] for instance. Then we use the techniques of formal asymptotic analysis to the smoothed model in order to derive again a model with a sharp reaction front, see [8], between the two phases and compare it with the model stated at the beginning of the present work.

2   Mathematical Model

The simulation of the thermal oxidation process requires the computation of the change of layer structures with time. At a process temperature which typically lies between 700–1200^0C, the silicon wafers are exposed for 30–500 minutes to a gas atmosphere containing oxygen (O_2). Parts of the structure are covered with a nitride mask (Si_3N_4) impervious to oxygen. Oxygen enters from outside the silicon-dioxide-layer (SiO_2), diffuses towards the the silicon-layer (Si) and reacts there producing new silicon-dioxide according to

$$Si + O_2 \rightarrow SiO_2 \, .$$

As silicon is consumed, the boundary between silicon and silicon-dioxide moves further into the silicon-layer. The freshly created oxide requires more than twice the volume of the consumed silicon, for what reason the oxide-layer grows. The reaction rate and the volume expansion are usually non-uniform along the reaction front and this results in non-homogeneous deformations of the whole layer system (see Fig. 1).

gas

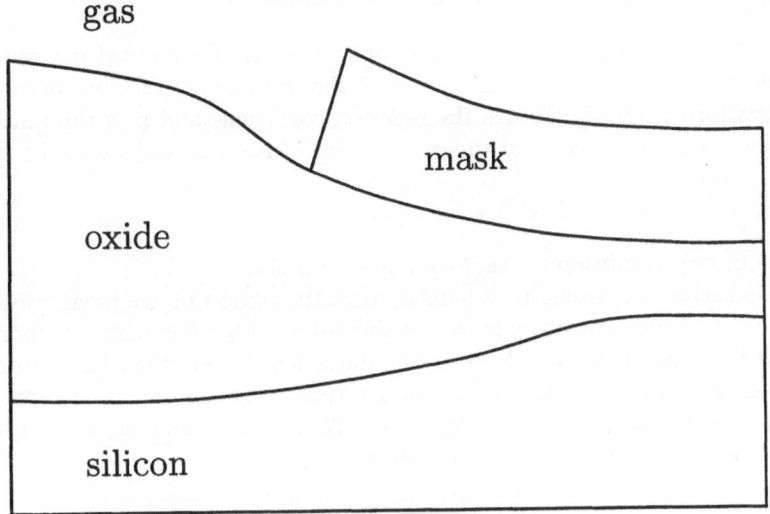

Fig. 1. Bird's beak structure

2.1   A Model with a Sharp Reaction Front

The transport of dissolved oxygen can be described by a diffusion-convection equation in the oxide layer with certain boundary conditions. One may neglect the oxygen transport in the silicon and the other layers, but has to balance the fluxes at the interfaces to these materials. The transition of oxygen into the oxide-layer is modelled by boundary conditions of mixed type or by Dirichlet-boundary-data. The oxygen consumption in the chemical reaction at the silicon interface also enters as a boundary condition of mixed type. We denote by c_1 the concentration (particles per volume) of oxygen, by c_2 the concentration (particles per volume) of silicon and $\mathbf{v}$ designates the velocity field in the whole layer system. The diffusion-convection equation in the oxide region is given by

$$\frac{\partial}{\partial t} c_1 + \operatorname{div}(c_1 \mathbf{v} + \mathbf{j}) = 0,$$

where the flux is

$$\mathbf{j} = -D\nabla c_1$$

with a constant diffusivity D.

The boundary conditions are

$$\mathbf{j} \cdot \mathbf{n} = h \left(c_1 - c_1^* \right), \quad \text{on the oxide-gas-interface},$$
$$\left(\mathbf{j} + c_1 \left(\mathbf{v} - \mathbf{v}_\Gamma \right) \right) \cdot \mathbf{n} = k c_1 c_2, \quad \text{on the reaction-front}, \tag{2.1}$$
$$\mathbf{j} \cdot \mathbf{n} = 0, \quad \text{on all other boundaries}$$

where c_1^* designates the equilibrium concentration of dissolved oxygen, $\mathbf{v}_\Gamma$ the velocity of the boundary movement, h the transition coefficient of oxygen from gas into oxide, k denotes the reaction coefficient and $\mathbf{n}$ is the outward unit normal to the surface of the oxide region. Alternatively, we could use

$$c_1 = c_1^*$$

as a boundary condition on the oxide-gas-interface.

To describe the mechanical behaviour of the structure, we have to establish mass and momentum balances in the volume together with appropriate interface and boundary conditions. The mass density ϱ of the silicon and the silicon-dioxide layer is the product of the respective particle concentrations c_2 and c_3 with their atomic masses $m_2 = 28.089 m_0$ gr and $m_3 = 60.102 m_0$ gr. The atomic mass unit m_0 is defined as

$$m_0 = \frac{\text{mass of a carbon atom}}{12} = 1.66055 \cdot 10^{-24} \text{ gr}.$$

Thus we have

$$\varrho = m_2 c_2 \quad \text{for silicon}$$

and

$$\varrho = m_3 c_3 \quad \text{for oxide}.$$

In order to distinguish between the different layers we designate these with $+$ and $-$ according to the next figure.

Here $\mathbf{n}^+$ and $\mathbf{n}^-$ denote the outward unit normal vectors to the interface between the different subregions. The vectors differ from each other by the relationship $\mathbf{n}^+ = -\mathbf{n}^-$. Also for the tangent vectors we have $\mathbf{t}^+ = -\mathbf{t}^-$.

The conservation of mass in each of the different subregions is given by

$$\frac{\partial}{\partial t} \varrho + \operatorname{div} \left(\varrho \mathbf{v} \right) = 0.$$

If we assume that there is a complete phase change (i.e. there is no silicon in the oxide-region and no oxide in the silicon-region) the mass densities of the silicon and the oxide depend on the particle concentrations of the layer materials c_2 and c_3.

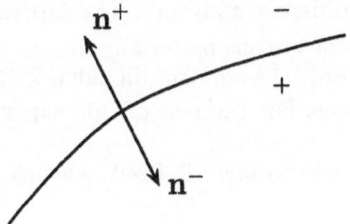

Due to the chemical reaction at the interface between silicon and oxide we have to account the supply and the consumption of layer material to get

$$c_3 \left(\mathbf{v}^+ - \mathbf{v}_\Gamma\right) \cdot \mathbf{n}^+ = -kc_1c_2$$
and   (2.2)
$$c_2 \left(\mathbf{v}^- - \mathbf{v}_\Gamma\right) \cdot \mathbf{n}^- = kc_1c_2 \,.$$

From these we deduce an equation describing the movement of the reaction front and due to the different particle concentrations a jump condition for the normal velocity, namely

$$\mathbf{v}_\Gamma = \left(-kc_1 + \mathbf{v}^- \cdot \mathbf{n}^-\right)\mathbf{n}^-$$ (2.3)

and

$$\mathbf{v}^+ \cdot \mathbf{n}^+ + \mathbf{v}^- \cdot \mathbf{n}^- = -kc_1 \left(\frac{c_2}{c_3} - 1\right).$$ (2.4)

At all the other boundaries there is no supply or consumption of mass. Thus we need no more boundary conditions for the mass balance.

The momentum balance, which holds in each subregion, is described by the Navier-Stokes-equation,

$$\varrho\left(\frac{\partial}{\partial t}\mathbf{v} + (\mathbf{v} \cdot \nabla)\mathbf{v}\right) = \operatorname{div}\sigma \,,$$

where σ_{ij}, $i, j = 1, 2$, is the linearized stress tensor.

For a viscous compressible material we have

$$\sigma_{ij} = -p\delta_{ij} + \lambda(\operatorname{div}\mathbf{v})\delta_{ij} + \mu\left(\frac{\partial v_i}{\partial x_j} + \frac{\partial v_j}{\partial x_i}\right), \quad i, j = 1, 2 \,,$$ (2.5)

p represents the pressure, which has to be related to the mass densities in the different layers by material laws. We assume

$$p = p_0 + p_1 \frac{\varrho - \varrho_1}{\varrho_1}$$ (2.6)

with positive constants p_0 and p_1. The function ϱ_1 represents a known equilibrium density of the different materials. The expression $\frac{1}{2}\left(\frac{\partial v_i}{\partial x_j} + \frac{\partial v_j}{\partial x_i}\right)$ is the deformation velocity of a viscous material.

Next, we have to supply the momentum balance with appropriate boundary- and transition-conditions. For the rest of this paper we use Einstein's summation convention.

The bottom of the silicon layer is fixed, whence the velocity field has to satisfy

$$v_i = 0 \quad i = 1, 2.$$

On the left and right borders we take

$$v_i n_i = 0,$$
$$\sigma_{ij} n_i t_j = 0.$$

The remaining boundaries are free boundaries and have to be determined in the problem. Therefore we have to state kinematic equations for their motion. According to (2.3),

$$\mathbf{v}_\Gamma = \left(-kc_1 + v_i^- n_i^-\right)\mathbf{n}^-$$

describes the movement of the reaction front. The other free boundaries are moving with the phases, why we take

$$\mathbf{v}_\Gamma = v_i n_i \, \mathbf{n}.$$

The tangential velocity at all internal boundaries is assumed to be continuous,

$$v_i^+ t_i^+ + v_i^- t_i^- = 0,$$

and by (2.4), the jump condition of the normal velocity at the reaction front is

$$v_i^+ n_i^+ + v_i^- n_i^- = -kc_1 \left(\frac{c_2}{c_3} - 1\right).$$

On the other internal boundaries without any chemical reaction we assume continuity of the normal velocity,

$$v_i^+ n_i^+ + v_i^- n_i^- = 0.$$

The tangential stress balance at the free surface towards the gas region is taken to be

$$\sigma_{ij} n_i t_j = 0,$$

whereas the normal stresses from the interior are balanced by the external pressure p_A and the surface tension θ, which can be assumed to be proportional to the curvature κ. Indeed, we have

$$\sigma_{ij} n_i n_j = -p_A - \kappa\theta.$$

At all internal boundaries we also have to pose tangential and normal stress balances, namely

$$\sigma_{ij}^+ n_i^+ t_j^+ - \sigma_{ij}^- n_i^- t_j^- = 0,$$
$$\sigma_{ij}^+ n_i^+ n_j^+ - \sigma_{ij}^- n_i^- n_j^- = -\theta\,\kappa^+.$$

From the relative motion between the reaction front and the phases additional inertial forces have to be considered,

$$
\begin{aligned}
&\left(\sigma_{ij}^+ + \varrho^+ v_i^+ \left(v_j^+ - v_{j\Gamma+}\right)\right) n_i^+ n_j^+ - \\
&\left(\sigma_{ij}^- + \varrho^- v_i^- \left(v_j^- v_{j\Gamma-}\right)\right) n_i^- n_j^- = -\kappa^+ \theta.
\end{aligned}
\tag{2.7}
$$

2.2   A Phase Field Model

The numerical treatment of the previous oxidation-model is difficult due to the rapid topological changes of the layer system, since we assume a complete consumption of silicon during the reaction. Techniques, like front adaptation and the generation of new grids in each time step would be necessary. We avoid this difficulty by replacing the reaction front of the given model by an extended reaction zone (see Fig. 2). For this we regard the silicon and the oxide as components of a compressible and viscous mixture in which the oxygen is assumed to be dissolved. The reaction term, which appeared in several boundary conditions, now enters the mass continuity equations as a source or sink term and reads as $\bar{k}c_1 c_2$, $\bar{k}$ a volumetric reaction constant. We consider the partial differential equations in the Si-SiO$_2$-region.

The mass density of the Si-SiO$_2$-mixture is defined as

$$\varrho = m_2 c_2 + m_3 c_3.\tag{2.8}$$

The continuity equations for the different components now read as follows:

$$\frac{\partial}{\partial t}c_1 + \operatorname{div}(c_1 \mathbf{v}) = \operatorname{div}\mathbf{j} - \bar{k}c_1 c_2,\tag{2.9}$$

$$\frac{\partial}{\partial t}c_2 + \operatorname{div}(c_2 \mathbf{v}) = -\bar{k}c_1 c_2,\tag{2.10}$$

$$\frac{\partial}{\partial t}c_3 + \operatorname{div}(c_3 \mathbf{v}) = \bar{k}c_1 c_2,\tag{2.11}$$

where $\mathbf{j} = -D\nabla c_1$.

The boundary condition for equation (2.9) is

$$\mathbf{j}\cdot\mathbf{n} = \begin{cases} h(c_1 - c_1^*), & \text{at the gas boundary } \Gamma_1; \\ 0, & \text{elsewhere.} \end{cases}\tag{2.12}$$

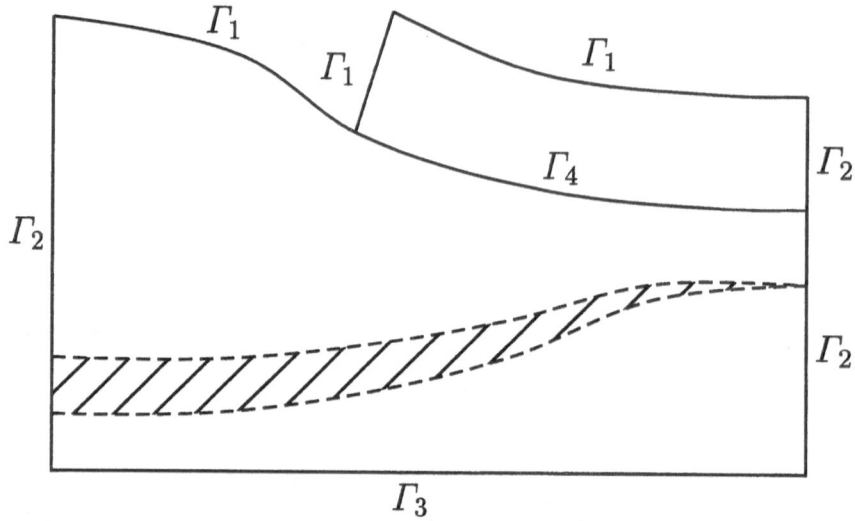

Fig. 2. Extended reaction region

With $m_2 \cdot$ (2.10) $+ m_3 \cdot$ (2.11) we get according to (2.8) the continuity equation of the mixture, i.e.

$$\frac{\partial}{\partial t}\varrho + \mathrm{div}(\varrho \mathbf{v}) = (m_3 - m_2)\,\bar{k}c_1 c_2\,. \tag{2.13}$$

The momentum balance is now related to the Si-SiO$_2$-mixture. If the velocity field of its movement is again denoted by $\mathbf{v} = (v_1, v_2)$, it reads as

$$\varrho \left(\frac{\partial}{\partial t}\mathbf{v} + (\mathbf{v} \cdot \nabla)\,\mathbf{v}\right) = \mathrm{div}\,\sigma + (m_3 - m_2)\,\bar{k}c_1 c_2 \mathbf{v}\,, \tag{2.14}$$

where σ is according to (2.5). The material law for the pressure p as a function of the mass density ϱ was assumed to be

$$p = p_0 + p_1 \frac{\varrho - \varrho_1}{\varrho_1}$$

with positive constants p_0 and p_1. The function ϱ_1 represents a known equilibrium density of the mixture, which is a function of the chemical reaction only. The mixture comes to rest if $\varrho = \varrho_1$. The case of an incompressible material means in our situation $\varrho \equiv \varrho_1$, but not $\varrho \equiv \mathrm{const}$. The viscosities obey the relationship $\lambda, \mu > 0$.

On the boundary sections $\Gamma_1 \cdots \Gamma_4$ we have to establish several conditions, necessary for the momentum equation (2.14). We summarize the previous boundary data:

On Γ_1:

$$\mathbf{v}_\Gamma = v_i n_i \,\mathbf{n} \qquad \text{kinematic equation for the free boundary}$$
$$\sigma_{ij} n_i t_j = 0 \qquad \text{no tangential stress}$$
$$\sigma_{ij} n_i n_j = -p_A - \theta\kappa \quad \text{normal stress balance}$$

On Γ_2:

$$\mathbf{v}_\Gamma = 0 \qquad \text{kinematic equation for the fixed boundary}$$
$$v_i n_i = 0 \qquad \text{no normal flow}$$
$$\sigma_{ij} n_i t_j = 0 \quad \text{no tangential stress}$$

On Γ_3:

$$\mathbf{v}_\Gamma = 0 \quad \text{kinematic equation for the fixed boundary}$$
$$v_i = 0 \quad \text{no slip}$$

On Γ_4:

$$\mathbf{v}_\Gamma = v_i^+ n_i^+ \mathbf{n}^+ \qquad \text{kinematic equation for the free boundary}$$
$$v_i^+ = v_i^- \qquad \text{continuity of the velocity field}$$
$$(\sigma_{ij}^+ n_i^+ + \sigma_{ij}^- n_i^-)\, n_j^+ = -\theta\kappa^+ \quad \text{normal stress balance}$$
$$(\sigma_{ij}^+ n_i^+ + \sigma_{ij}^- n_i^-)\, t_j^+ = 0 \qquad \text{tangential stress balance}$$

The parameters P (e.g. D, h, λ, μ, θ and ϱ_1) significantly depend on the local composition of the mixture. But since only the data of the pure substances are available, we can define some average values for the mixture, e.g.

$$P = \frac{P_2 m_2 c_2 + P_3 m_3 c_3}{\varrho},$$

where P_i, $i = 2, 3$, denote the parameters of the silicon and the oxide, respectively.

3   Mathematical Results

In this section we summarize some results concerning the models stated above. In a bounded and fixed domain we state an existence- and uniqueness result concerning the phase field model. Next, we consider this model in a region which contains the moving transition layer and derive by means of formal asymptotic analysis a sharp reaction front between the phases. An interesting question is how this model agrees with the one in Sect. 2.1.

3.1   Existence and Uniqueness Result

We consider an open and bounded domain $\Omega \subset \mathbb{R}^3$ occupied by a mixture consisting of silicon and silicon-dioxide. Oxygen is supplied from the outside and diffuses inside that mixture. In $Q_T := \Omega \times (0,T)$, $T < +\infty$, we consider the equations (2.9), (2.10), (2.13) and (2.14). For simplicity we assume that the viscosities μ and λ as well as the diffusion coefficient D and the equilibrium density of the mixture ϱ_1 are constant. On the boundary $\Sigma_T = \partial\Omega \times (0,T)$ we assume boundary conditions of Dirichlet type. Thus, we treat the following mathematical model denoted by (P):

$$\varrho\left(\frac{\partial}{\partial t}\mathbf{v} + (\mathbf{v}\cdot\nabla)\mathbf{v}\right) = -\nabla p(\varrho) + \mu\Delta\mathbf{v} + (\mu+\lambda)\nabla\operatorname{div}\mathbf{v} + (m_3 - m_2)\,\bar{k}c_1c_2\mathbf{v}\,,$$

$$\frac{\partial}{\partial t}c_1 + \mathbf{v}\cdot\nabla c_1 + c_1\operatorname{div}\mathbf{v} = -\bar{k}c_1c_2 + D\Delta c_1\,,$$

$$\frac{\partial}{\partial t}c_2 + \mathbf{v}\cdot\nabla c_2 + c_2\operatorname{div}\mathbf{v} = -\bar{k}c_1c_2\,,$$

$$\frac{\partial}{\partial t}\varrho + \mathbf{v}\cdot\nabla\varrho + \varrho\operatorname{div}\mathbf{v} = (m_3 - m_2)\bar{k}c_1c_2\,.$$

The boundary- and initial conditions for the velocity field $\mathbf{v}$ are

$$\mathbf{v} = \bar{\mathbf{v}} \quad \text{on} \quad \Sigma_T\,, \quad \text{with} \quad \bar{\mathbf{v}}\cdot\mathbf{n} = 0\,,$$
$$\mathbf{v}(\cdot,0) = \mathbf{v}_0 \quad \text{in} \quad \Omega\,,$$

where $\mathbf{n}$ denotes the outward unit normal at $\partial\Omega$. For the concentration c_1,

$$c_1 = \bar{c}_1 \quad \text{on} \quad \Sigma_T\,,$$
$$c_1(\cdot,0) = c_1^0 \quad \text{in} \quad \Omega\,,$$

and the initial data for c_2 and c_3 are

$$c_2(\cdot,0) = c_2^0 \quad \text{in} \quad \Omega\,,$$
$$c_3(\cdot,0) = c_3^0 \quad \text{in} \quad \Omega\,.$$

The pressure p appearing in the momentum balance is a linear function of the density ϱ,

$$p(\varrho) = p_1\varrho + p_2\,, \quad p_1 > 0\,, \quad p_2 \geq 0\,,$$

and the viscosities μ and λ satisfy the relationship

$$2\mu + 3\lambda \geq 0\,, \quad \mu > 0\,, \quad \lambda > 0\,.$$

Now we can state the following result:

Theorem 3.1. *Let $\Omega \subset \mathbb{R}^3$ be an open and bounded domain with a sufficiently smooth boundary $\partial\Omega$. For the initial data in Ω we suppose*

$$(\mathbf{v}_0, c_1^0, c_2^0, \varrho_0) \in [H^3(\Omega)]^6$$

as well as $c_1^0, c_2^0 \geq 0$ and $\varrho_0 \geq m_1 > 0$ with a positive constant m_1. The boundary data should satisfy

$$\bar{\mathbf{v}} \in [H^{\frac{7}{2},\frac{7}{4}}(\Sigma_T)]^3 \quad with \quad \bar{\mathbf{v}} \cdot \mathbf{n} = 0$$

and

$$\bar{c}_1 \in H^{\frac{7}{2},\frac{7}{4}}(\Sigma_T).$$

Assume that the necessary compatibility conditions

$$\mathbf{v}_0 = \bar{\mathbf{v}}(0),$$

$$\frac{\partial}{\partial t}\bar{\mathbf{v}}(0) = \frac{1}{\varrho_0}(\mu\Delta\mathbf{v}_0 + (\mu+\lambda)\nabla\,div\,\mathbf{v}_0) - p_1\frac{\nabla\varrho_0}{\varrho_0} - (\mathbf{v}_0 \cdot \nabla)\mathbf{v}_0$$

and

$$c_1^0 = \bar{c}_1(0),$$

$$\frac{\partial}{\partial t}\bar{c}_1(0) = D\Delta c_1^0 - \mathbf{v}_0 \cdot \nabla c_1^0 - c_1^0\,div\,\mathbf{v}_0 - kc_1^0 c_2^0$$

are satisfied on $\partial\Omega$. Then there exists an instant of time $T_f \in (0,T]$, such that problem (P) has a unique solution

$$(\mathbf{v}, c_1, c_2, \varrho) \in [H^{4,2}(Q_{T_f})]^4 \times [L^\infty(0, T_f; H^3(\Omega))]^2$$

and

$$\left(\frac{\partial}{\partial t}c_2, \frac{\partial}{\partial t}\varrho\right) \in [L^\infty(0, T_f; H^2(\Omega))]^2$$

with $c_1, c_2 \geq 0$ and $\varrho \geq \bar{m} > 0$ in $\bar{Q}_T$, where $\bar{m}$ is a positive constant.

In order to prove this result we use the fixed point theorem of Schauder, see for instance [6]. An existence and uniqueness proof of the corresponding free boundary value problem in one spatial dimension can be found in [4].

3.2   Asymptotic Analysis

Starting from the phase field model we now use asymptotic methods to derive a limiting model with a sharp reaction front. In the second section we did it the other way round. We assumed the existence of a mixture consisting of silicon and silicon-dioxide to remove the sharp reaction front and replaced it by a smooth transition layer. Thus we are left with two "sharp front models", and to close the cycle, we finally relate them to each other.

Numerical simulations of the phase field model indicate that the mixture separates into two phases, a silicon phase in which the oxygen concentration is small and an oxide phase in which the silicon concentration is small. The (moving) interfacial layer between these phases becomes sharper as the reaction rate k (we omit the bar) increases, and so this suggests performing an asymptotic analysis with respect to $k \to \infty$.

As an expansion parameter we choose

$$\varepsilon = \frac{1}{\sqrt{k}}$$

which is the order of the thickness of the transition layer between the phases. To motivate this assertion (which is justified by the 'success' of the asymptotic expansion), consider the crosscut through an idealized interface of thickness d given in Fig. 3.

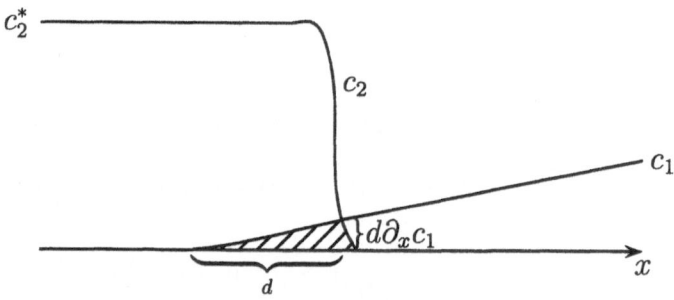

Fig. 3. Crosscut

Suppose that the silicon concentration drops quite suddenly from its 'pure crystal' value c_2^* to zero, that the oxide concentration increases almost linearly with slope $\partial_x c_1$ in most of the interface region, and that the transport velocity $\mathbf{v}$ is small. Then the flow of O_2 into the interface is

$$\mathcal{J} = D \partial_x c_1$$

and the cross interface reaction

$$\mathcal{G} = \int k c_1 c_2 \approx k c_2^* \int_{\text{interface}} c_1 \approx k c_2^* \frac{1}{2} d(d \partial_x c_1) = k d^2 \frac{c_2^*}{2} \partial_x c_1 .$$

As all oxygen is consumed in the reaction, those quantities must be equal, i.e. $\mathcal{J} = \mathcal{G}$. This implies

$$k d^2 \approx 2 \frac{D}{c_2^*} = O(1)$$

and thus the required order relationship

$$d = O(\frac{1}{\sqrt{k}}) = O(\varepsilon),$$

follows.

As the reaction coefficient k tends to infinity and thus ε tends to zero, the domain Ω will decompose into two phases $\Omega^{\pm}$ where either the concentration of oxygen c_1 or the concentration of silicon c_2 vanishes:

$$\Omega^- = \{x \in \Omega : c_1 = 0\} \qquad \text{(silicon-region)}$$

and

$$\Omega^+ = \{x \in \Omega : c_2 = 0\} \qquad \text{(oxide-region)}.$$

The common boundary

$$\Gamma = \overline{\Omega^+} \cap \overline{\Omega^-}$$

is assumed to be smooth. For positive ε, there will be a transition zone of thickness ε around Γ in which c_1, c_2, c_3 and $\mathbf{v}$ change rapidly.

Writing equations of the phase field model in terms of ε we get

$$\varepsilon^2 \left(\frac{\partial}{\partial t} c_1 - D\Delta c_1 + \text{div}(c_1 \mathbf{v}) \right) = -c_1 c_2, \tag{3.1}$$

$$\varepsilon^2 \left(\frac{\partial}{\partial t} c_2 + \text{div}(c_2 \mathbf{v}) \right) = -c_1 c_2, \tag{3.2}$$

$$\varepsilon^2 \left(\frac{\partial}{\partial t} c_3 + \text{div}(c_3 \mathbf{v}) \right) = c_1 c_2, \tag{3.3}$$

$$\varepsilon^2 \left(\varrho \left(\frac{\partial}{\partial t} \mathbf{v} + (\mathbf{v} \cdot \nabla) \mathbf{v} \right) - \text{div}\,\sigma \right) = (m_2 - m_3) c_1 c_2 \mathbf{v}. \tag{3.4}$$

Next we have to formulate the *outer expansion* of equations (3.1)–(3.4) in the regions $\Omega^{\pm}$ and the *inner expansion* in the transition zone around Γ. A detailed description of this formulation can be found in [8]. Anyway, the limiting case results in the following set of equations:

The bulk equations are the zeroth order equations of the outer expansion and the boundary conditions at the reaction front stem from matching the inner to the outer expansion. If we return to our previous notation the resulting two-phase problem reads as follows:
In the Ω^--region

$$c_1 = 0, c_2 = c_2^*, c_3 = 0,$$
$$\text{div}\,\mathbf{v} = 0,$$
$$\varrho \left(\frac{\partial}{\partial t} \mathbf{v} + (\mathbf{v} \cdot \nabla) \mathbf{v} \right) = \text{div}\,\sigma, \tag{3.5}$$

and in the Ω^+-region

$$c_2 = 0,$$

$$\frac{\partial}{\partial t}c_1 - D\Delta c_1 + \text{div}(c_1 \mathbf{v}) = 0,$$

$$\frac{\partial}{\partial t}c_3 + \text{div}(c_3 \mathbf{v}) = 0, \qquad (3.6)$$

$$\varrho\left(\frac{\partial}{\partial t}\mathbf{v} + (\mathbf{v} \cdot \nabla)\mathbf{v}\right) = \text{div}\,\sigma,$$

where $\varrho = m_2 c_2 + m_3 c_3$, $\sigma = \left(-p_3(c_2, c_3) + \lambda\text{div}\mathbf{v}\right)Id + \mu\left(\nabla\mathbf{v} + (\nabla\mathbf{v})^T\right)$ and where $p_3(c_2, c_3)$ is an equivalent representation of (2.6).

We derive from the following set of boundary conditions at the reaction front Γ:

$$c_1^+ = c_1^- = 0,$$

$$D\nabla c_1^+ \cdot \mathbf{n} = -[c_2(\mathbf{v} \cdot \mathbf{n} - v_\Gamma)]_-^+ = [c_3(\mathbf{v} \cdot \mathbf{n} - v_\Gamma)]_-^+,$$

$$[\mathbf{v} \cdot \mathbf{t}]_-^+ = 0, \qquad (3.7)$$

$$[\sigma\mathbf{n}]_-^+ = [\rho(\mathbf{v} \cdot \mathbf{n} - v_\Gamma)\mathbf{v}]_-^+ + (2\lambda + 3\mu)\kappa_\Gamma [\mathbf{v}]_-^+,$$

where v_Γ and κ_Γ are the normal velocity and curvature of Γ, respectively. The first condition says that due to the instantaneous reaction, no oxygen is present at the reaction front. The oxide production and thus the front velocity are governed only by the flow of oxygen into Γ, as can be seen from the second condition. Due to the third condition the tangential component of the velocity field is continuous across the free boundary, while the fourth condition is the normal stress balance, describing the conservation of momentum (see [7]).

It is reasonable to assume that initially the velocity field vanishes in the Ω^- region. This will also hold true during the evolution, and we may thus assume

$$\mathbf{v} = 0 \quad \text{in } \Omega^-.$$

This simplifies the boundary conditions into

$$D\nabla c_1^+ \cdot \mathbf{n} = -c_2^* v_\Gamma = c_3^+(\mathbf{v}^+ \cdot \mathbf{n} - v_\Gamma),$$

$$\sigma^+\mathbf{n} + p^*\mathbf{n} = m_3 D(\nabla c_1^+ \cdot \mathbf{n})\mathbf{v}^+ + (2\lambda + 3\mu)\kappa_\Gamma \mathbf{v}^+, \qquad (3.8)$$

where $p^* = p_3(c_2^*, 0)$ is the equilibrium pressure of the silicon.

Since $c_1 = 0$ on Γ and $c_1 \geq 0$ in Ω^+, we have $\nabla c_1^+ \cdot \mathbf{n} \geq 0$ on Γ and the boundary conditions (3.7) imply

$$v_\Gamma \leq 0 \quad \text{and} \quad \mathbf{v} \cdot \mathbf{n} - v_\Gamma \geq 0,$$

thus the Ω^- region is shrinking. This conclusion reflects the fact that the silicon is consumed in order to produce the oxide.

Now we compare these boundary conditions with the relevant conditions stated in Sect. 2.5. If we combine condition (2.1) with (2.2), use (2.9) and (2.7) we have under the assumption $\mathbf{v}^- = 0$ in the actual terminology

$$D\nabla c_1^+ \cdot \mathbf{n} + c_1^+(\mathbf{v} \cdot \mathbf{n} - v_\Gamma) = c_3^+(\mathbf{v}^+ \cdot \mathbf{n} - v_\Gamma),$$

$$\sigma^+ \mathbf{n} + p^* \mathbf{n} = \kappa_\Gamma \theta, \tag{3.9}$$

$$v_\Gamma = -kc_1^+ \mathbf{n}.$$

Formally taking the limit $k \to \infty$ in (3.9) gives $c_1^+ = 0$, and thus the free boundary models proposed in the present paper emerge.

References

1. Chin, D.: Two Dimensional Oxidation. IEEE Trans. in Electron Devices **30** (1983), 744
2. Collard, D., Baccus, B., Hamonic, B.: A Robust Numerical Procedure for Stress Dependent 2D-Oxidation Simulation. NUPAD-IV, 1992
3. Deal, B. E., Grove, A. S.: General Relationship for the Thermal Oxidation of Silicon. J. Appl. Phys. **36** (1965), 3770–3778
4. Götz, I., Merz, W., Pulverer, K., Zhang, J.: A Fluid Dynamical Free Boundary Problem for the Oxidation Process of Silicon. Nonlin. World **2** (1995), 429–457
5. Merz, W.: Analytische und numerische Behandlung des Oxidationsprozesses von Silizium. Dissertation, Universität Augsburg, 1992
6. Merz, W.: An Existence and Uniqueness Result of a Model Describing the Oxidation of Silicon. Report No. 438, Universität Augsburg, 1993
7. Merz, W., Strecker, N.: The Oxidation Process of Silicon: Modelling and Mathematical Treatment. Math. Meth. in the Appl. Sci. **17** (1994), 1165–1191
8. Merz, W., Pulverer, K., Stoth, B.: Construction of an Asymptotic Model for the Oxidation Process of Silicon. ZAMM **78** (1998), 711–720
9. Poncet, A.: Finite Element Simulation of Local Oxidation of Silicon. IEEE Trans. on CAD **4** (1985), 41
10. Rafferty, C.: Stress Effects in Silicon Oxidation, Simulation and Experiments. PhD Dissertation, Stanford Electronics Laboratories, Stanford University, 1989
11. Rank, E., Weinert, U.: Simulation for Diffusive Oxidation of Silicon: A Two-Dimensional Finite Element Approach. IEEE Transactions on Computer-Aided Design **9** (1990)
12. Tung, T. J., Antoniadis, D. J.: A Boundary Integral Approach to Oxidation Modelling. IEEE Trans. on CAD **4** (1985), 361

The Growth of Vapor Deposited Amorphous $Zr_{65}Al_{7.5}Cu_{27.5}$-Alloy Films: Experiment and Simulation

S. G. Mayr[1], M. Moske[2], and K. Samwer[3]

[1] Institut für Physik, Universität Augsburg, D-86135 Augsburg, Germany
[2] Stiftung caesar, Friedensplatz 16, D-53111 Bonn, Germany
[3] I. Physikalisches Institut, Universität Göttingen, Bunsenstraße 9, D-37073 Göttingen, Germany

Dedicated to Professor Karl-Heinz Hoffmann
on the occasion of his 60th birthday

Abstract. The surface structures of amorphous $Zr_{65}Al_{7.5}Cu_{27.5}$-alloy films prepared by ultra high vacuum physical vapor deposition are investigated with scanning tunneling microscopy to identify the major surface mechanisms for amorphous film growth. A fourier analysis of the STM images shows that curvature induced surface diffusion and a non-smoothing surface mechanism is present. Numerical simulations using a Monte-Carlo and a continuum model which includes adatom concentration triggered surface diffusion as additional coarsing mechanism are in excellent agreement with the experimental results.

1 Introduction

Thin films prepared by various deposition techniques generally do not grow smooth on substrates; instead, they show structures on a mesoscopic and atomic scale [1], which is also true for glassy films prepared by *co*condensation of transition metals [2–5]: Since the experiments of *Buckel* and *Hilsch* [6] it is known, that under appropriate conditions, various material systems can be grown with a disordered atomic structure. This is especially true for the new metallic glass $Zr_{65}Al_{7.5}Cu_{27.5}$ with a high stability concerning crystallisation and an increased temperature interval accessible in the region of the undercooled liquid (the temperature region between the glass temperature and the temperature of crystallisation) [7]. In contrast to crystalline systems, amorphous films can be characterized by a liquid-like structure of statistical character without any lattice constraints, long-range structural order or anisotropy. Concerning surface physics, due to the lack of well-defined steps, the Ehrlich-Schwoebel barrier present in many crystalline systems forming mesa like structures [8] is not known to be present, which offers the possibility to investigate other mechanisms independently, which are normally hidden by step edge barriers. Additionally, the spatial isotropy of the amorphous nature enables a detailed image analysis and thus the possibility to model in a quasi two dimensional way [9]. For a modelling of the extreme non-equilibrium surface dynamics, thermodynamic methods can be applied only with great restrictions, and thus – on an atomic level – kinetic Monte-Carlo methods and – from a mesoscopic point of view – Langevin-type rate equations are usually used, with the help of which (in a deterministic and linearized version) *Herring* [10] already modelled the sintering process: The time evolution (where t corresponds to the overall thickness of the growing film) of the surface $h(\boldsymbol{x}, t)$; $\boldsymbol{x} = (x, y)$ is described by:

$$\frac{\partial h(\boldsymbol{x}, t)}{\partial t} = F\left[h(\boldsymbol{x}, t)\right] + \eta(\boldsymbol{x}, t) + I(\boldsymbol{x}, t) \qquad (1.1)$$

Here, $F\left[h(\boldsymbol{x}, t)\right]$ denotes a functional with various terms to describe the surface processes, $\eta(\boldsymbol{x}, t)$ is a temporal and spatial uncorrelated white noise, i.e.

$$\langle \eta(\boldsymbol{x}, t) \rangle_{\text{ensemble}} = 0 \qquad (1.2)$$
$$\langle \eta(\boldsymbol{x}, t)\, \eta(\boldsymbol{x}', t') \rangle_{\text{ensemble}} = 2\, D\, \delta(\boldsymbol{x} - \boldsymbol{x}')\, \delta(t - t') \qquad (1.3)$$

and the flux term $I(\boldsymbol{x}, t)$ can be removed by an appropriate choice of the coordinate system [1]. The main question therefore is, investigating with scanning tunneling microscopy (STM), how the growth properties and surface morphologies depend on the experimental parameters, such as film thickness and deposition conditions, and how they can be theoretically modelled. In this context, the question of measuring artefacts must also be addressed.

2   Experimental Details

Due to their reactivity, the glassy $Zr_{65}Al_{7.5}Cu_{27.5}$-films are prepared and investigated *in situ* in a three chamber ultra high vacuum (UHV) system: After heat cleaning the thermally oxidized Si wafers (thickness of the SiO_2 layer: 500nm; surface roughness: (0.3 ± 0.1)nm), the films are prepared in the evaporation chamber (base pressure $< 2 \cdot 10^{-10}$mbar) by three independently rate-controlled electron beam evaporators, using a total evaporation rate of 0.79nm/sec, where the geometric arrangement of crucibles and substrate guarantees a particle flux nearly perpendicular to the substrate. During deposition, the substrate rotates (40rpm) to ensure a homogeneous film composition and thickness. After transfer into the analytical chamber, the films are investigated with Auger electron spectroscopy to check for possible surface contamination and to verify the film composition and by STM using electrochemically etched tungsten tips (measurement in constant current mode: typical scanning parameters: $U_T \approx 1.0$ V; $I_T \approx 1.0$ nA). A typical Auger spectrum of a $Zr_{65}Al_{7.5}Cu_{27.5}$ film shortly before STM measurement is depicted in Fig. 1. It verifies the chosen film composition and shows minor contamination of the surface with some remaining oxygen present in the UHV system. *Ex situ* investigations concentrate on the verification of the amorphicity using X-ray diffraction and DSC-measurements. Typical X-ray measurements of amorphous and crystalline films are shown in Fig. 2.

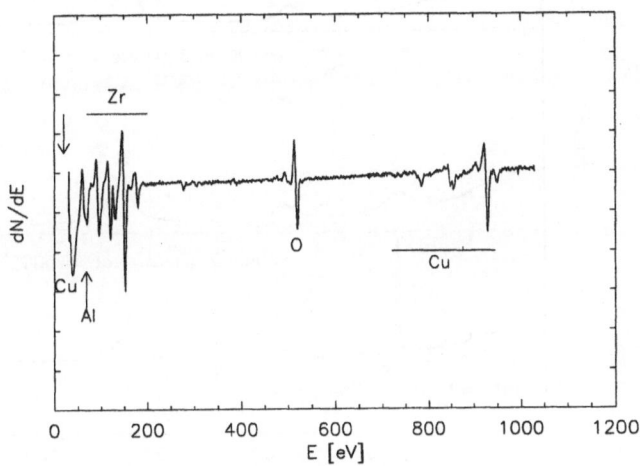

Fig. 1. Typical Auger spectrum for a $Zr_{65}Al_{7.5}Cu_{27.5}$ film on SiO_2 directly after deposition and shortly before a STM measurement

3   Experimental Results

In contrast to their crystalline counterparts, which typically can be distinguished in STM measurements from amorphous films by the formation of atomic steps – Fig. 3, amorphous films measured by STM can be characterized by radially isotropic mesoscopic hill-like structures. The evolution with increasing film thickness is depicted in Fig. 4. Both, the lateral hill size, as well as the hill height, increase with increasing film thickness, where a slowing down is qualitatively obvious from the greyscale images. Quantitatively, a measure of the hill height is the RMS-roughness ($\langle h(x,t)\rangle_x = 0$)

$$\sigma(t) = \sqrt{\langle h(x,t)^2\rangle_x} \tag{3.1}$$

and the height-height-correlation function [11]

$$C(r,t) = \langle h(x+R,t)h(x,t)\rangle_{x,|R|=r} \tag{3.2}$$

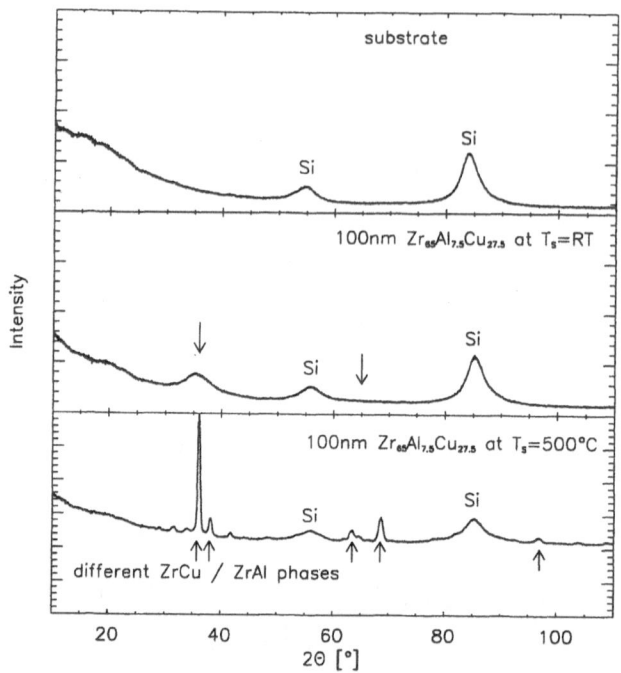

Fig. 2. X-ray diffraction measurements: Typical intensity patterns vs. diffraction angle 2Θ for an untreated Si wafer, a 100nm thick amorphous $Zr_{65}Al_{7.5}Cu_{27.5}$ film and a 100nm thick crystalline $Zr_{65}Al_{7.5}Cu_{27.5}$ film (*co*condensed at a substrate temperature of $500^{\circ}C$)

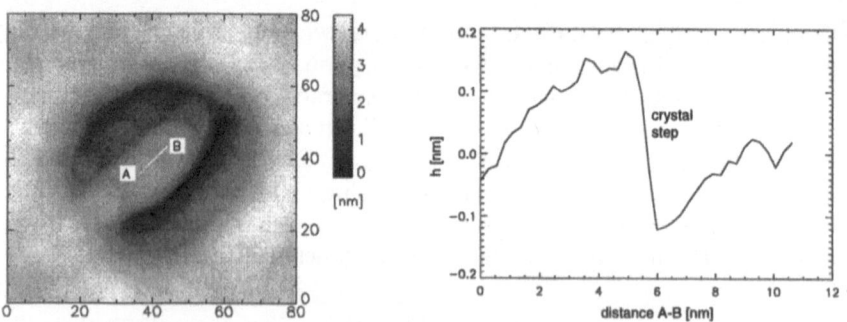

Fig. 3. Crystalline $Zr_{65}Al_{7.5}Cu_{27.5}$ films can be distinguished from amorphous films in STM by the formation of steps. Left: STM topographical image; right: line scan A–B

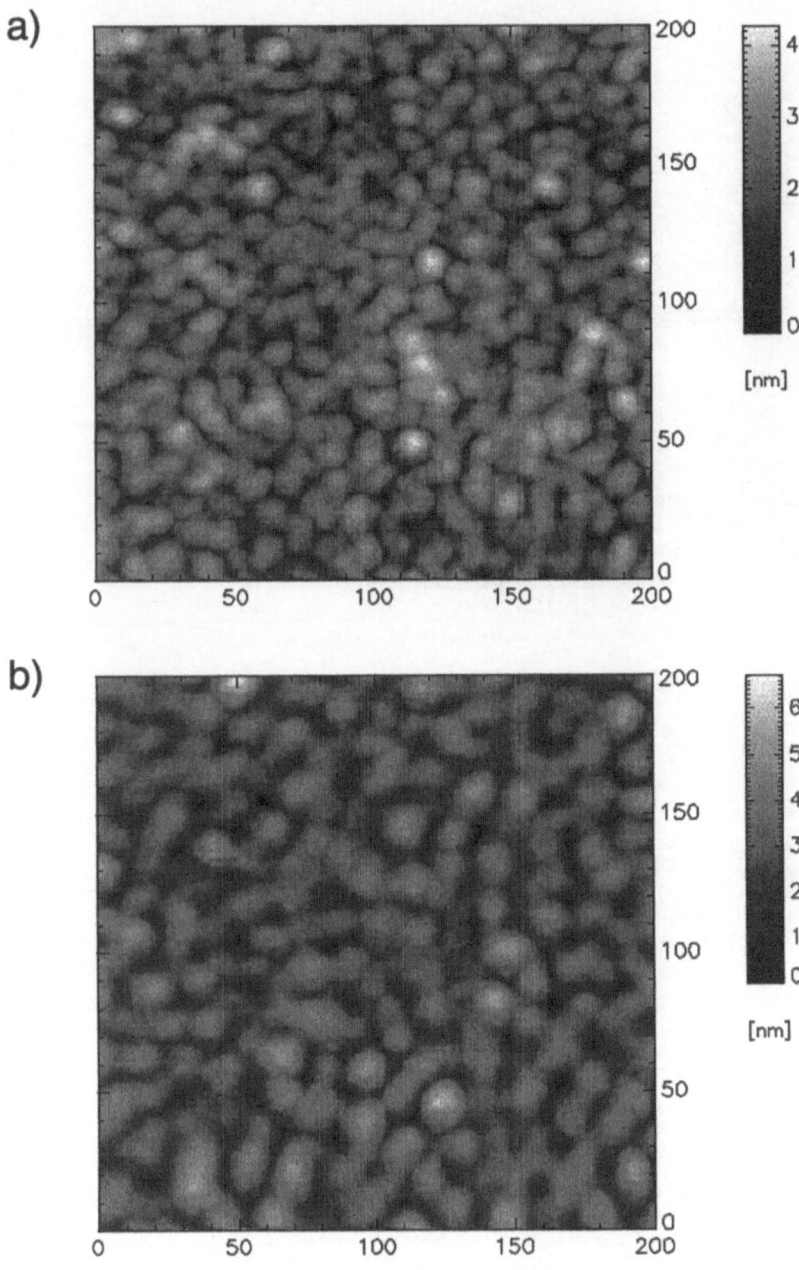

Fig. 4. Film thickness series of amorphous $Zr_{65}Al_{7.5}Cu_{27.5}$: a) 100nm, b) 200nm, c) 360nm and d) 480nm. Scanning parameters: $U_T \approx -1.0$V, $I_T \approx 0.8$nA, 400×400 measurement points; to be continued.

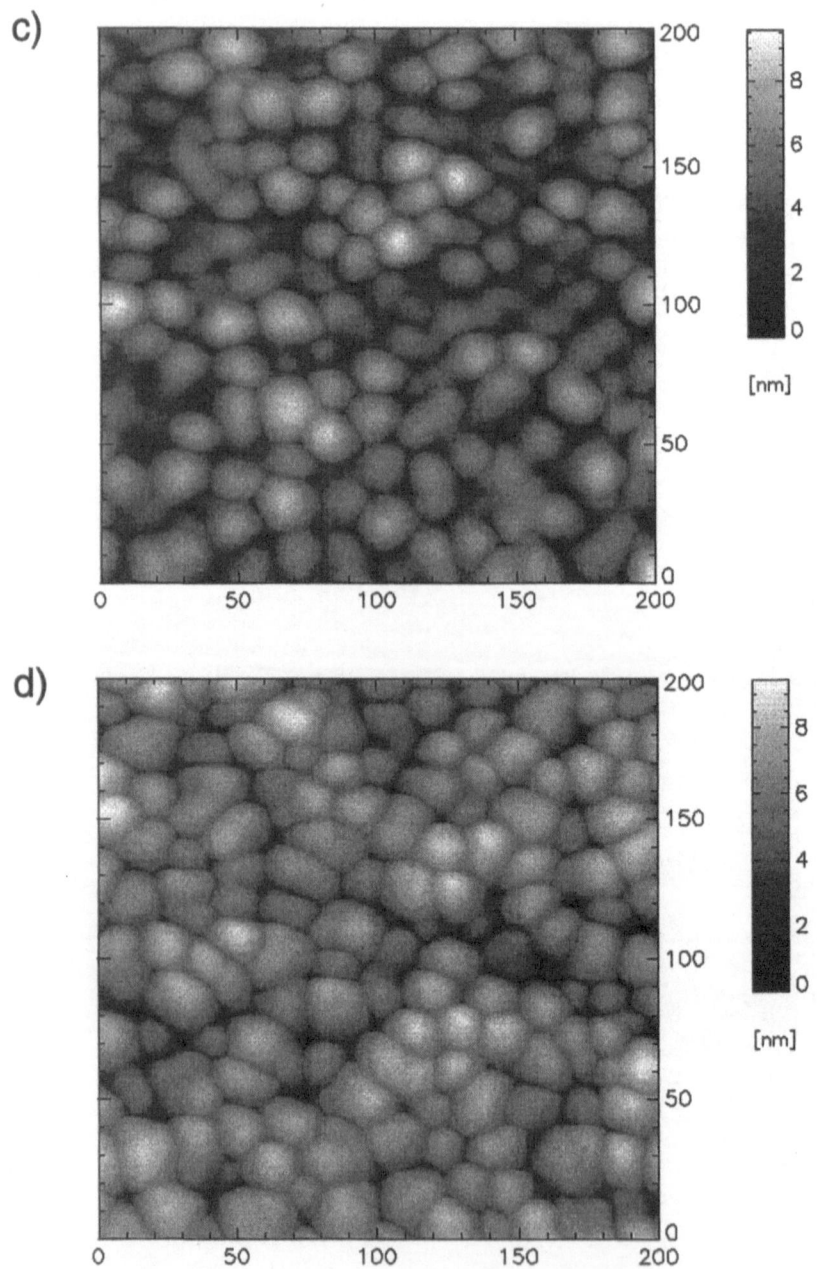

Fig. 4. Film thickness series of amorphous $Zr_{65}Al_{7.5}Cu_{27.5}$: a) 100nm, b) 200nm, c) 360nm and d) 480nm. Scanning parameters: $U_T \approx -1.0V$, $I_T \approx 0.8nA$, 400×400 measurement points

The latter shows – in the case of structure formation – a distinct oscillatory behaviour in dependence of r; the abscissa of the first maximum is a measure for the typical lateral dimension of the mesoscopic hills. $C(r,t)$ is directly linked to the height-difference-correlation function

$$H(r,t) = 2\,\sigma(t)^2 - 2\,C(r,t) \tag{3.3}$$

which typically shows a scaling behaviour following a power law of $r^{2\alpha}$, where α denotes the roughness exponent [1]. The other exponent important in the framework of kinetic roughening is the dynamical exponent β defined by

$$\sigma(t) \propto t^\beta \tag{3.4}$$

The fourier transformation of $C(r)$, $C(q)$, has the exciting property of show-

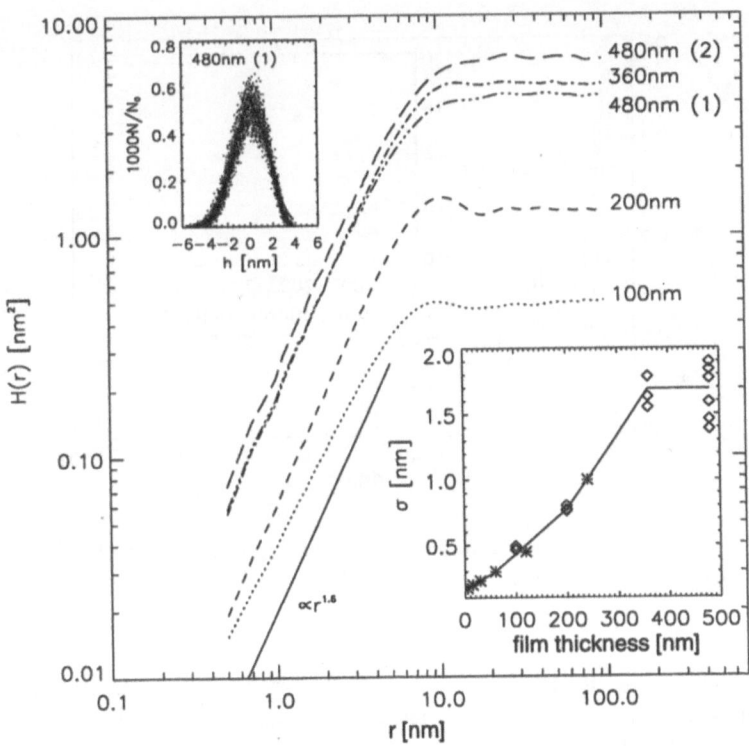

Fig. 5. Height-difference-correlation functions $H(r)$, roughnesses σ and height distribution for various Zr$_{65}$Al$_{7.5}$Cu$_{27.5}$ film thicknesses. The roughness data marked with (*) are taken from [3].

ing a characteristic scaling behaviour in dependence of the major surface relaxation mechanism, as already shown by *Mullins* [12]: In the high frequency range, the damping by surface relaxation mechanisms leads to a decrease following a power law of $q^{-\gamma}$, where γ is found to be 4 in the case of curvature induced surface diffusion in a linear approximation. The roughness σ as well as the height-difference-correlation function $H(r)$ are depicted in Fig. 5. In the early stages of film growth, a dynamical exponent of 0.2 ± 0.05 can be determined [13] and the roughness exponent α is found to be 0.8 ± 0.05 for all measured film thicknesses. The dependence of the roughness on film thickness increases stronger in the medium film thickness range and shows a saturation in the late stages of film growth. The height distribution of the hills for any film thickness is found to be Gaussian distributed. The lateral sizes of the hills determined from the height-height-correlation function are shown in Fig. 6. Similar to the roughness, it saturates in the range of high film thickness.

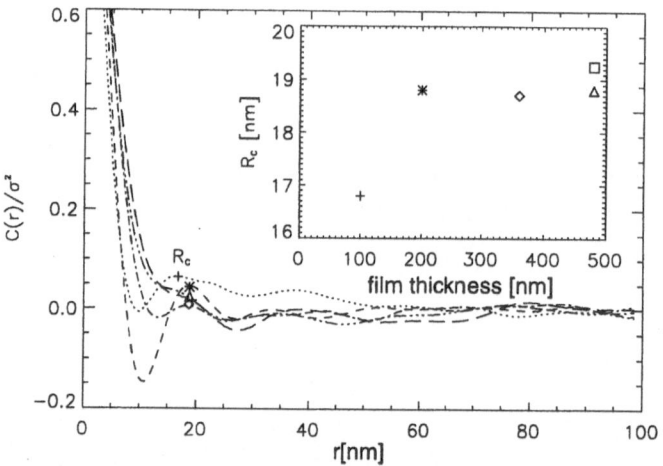

Fig. 6. Height-height-correlation functions $C(r)$ as a function of film thickness with inset of correlation length R_C

The pronounced structure formation for all measured film thickness stages raises the question, which mechanism is responsible for this lateral correlation. Further insight is provided by $C(q)$ in Fig. 7, which follows a power law of q^{-4} in the high frequency range and which is characteristic for curvature induced surface diffusion. The deviation for very high spatial frequencies can be attributed to the discrete nature of the underlying measuring grid, noise effects and limitations of the theoretical assumptions in the range of atomic sizes. A fit also reveals a significant contribution of a q^{-2} term, counteracting the q^{-4}-term (i.e. a non-smoothing term in the medium frequency range). The underlying atomic mechanisms will be discussed in Sect. 6.

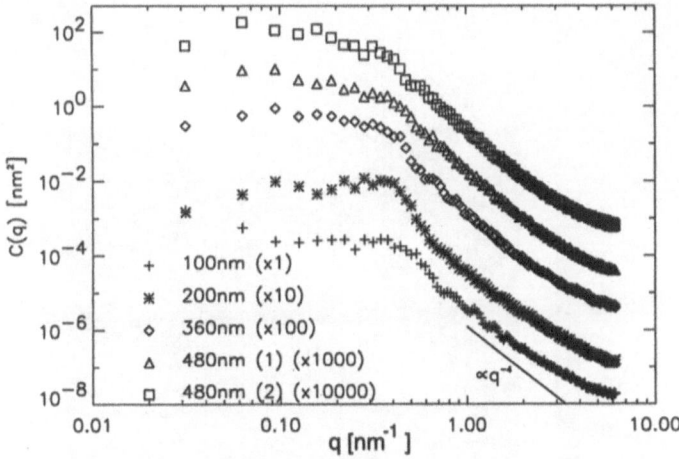

Fig. 7. Radially averaged spectral power densities $C(q)$ for various film thicknesses

4   Structures on the Atomic Scale

In all STM topographs an additional fine structure on a typical distance of
$(0.8 - 2.1)$nm, with a FWHM (full with at half maximum) of $(0.4 - 1.5)$nm
and corrugations of $(0.02 - 0.3)$nm can be resolved, which obviously belong
to the intrinsic atomic structure of amorphous surfaces. Fig. 8 and 9 show a
typical topograph with the corresponding height-height-correlation function.
As an interpretation, the imaging of exposed surface atoms and local clusters
of atoms (in the sense of an atomic short range order) are reasonable [14].

5   Monte-Carlo Model for Amorphous Film Growth

In contrast to continuum models, in which mathematical expressions represent the underlying physical and atomic processes, Monte-Carlo models
investigate film growth processes from an atomistic point of view. In this
sense, they can be understood as a second means to identify mechanisms
in experiments, as a basis for further continuum modelling [15]. As implicitly a crystal lattice underlies the discreteness of the atomic positions, they
fail to account for the correct atomic structure for amorphous films. However,
from a mesoscopic point of view, crystalline and amorphous systems with the
same underlying mechanisms should behave rather similar, if spatial isotropy
is guaranteed. This can be achieved by weighting the diffusional jump with
a diffusion length. Here, particles are deposited ballistically onto the surface,
where they immediately relax isotropically to the energetic most favourable
position in the direct neighbourhood. In addition, surface diffusion acts in

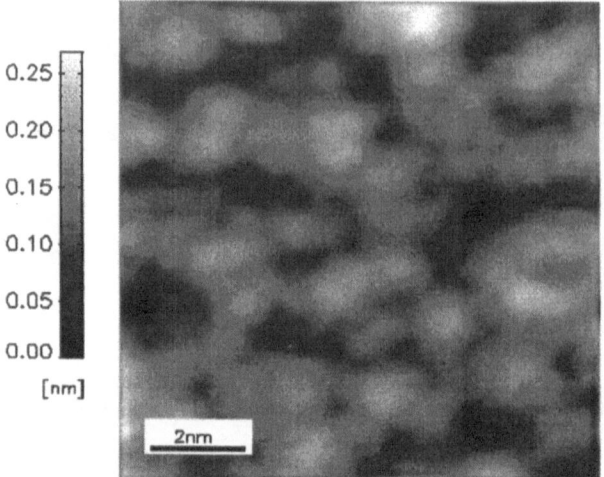

Fig. 8. Structures on the atomic scale: STM-image, 400 × 400 measuring points

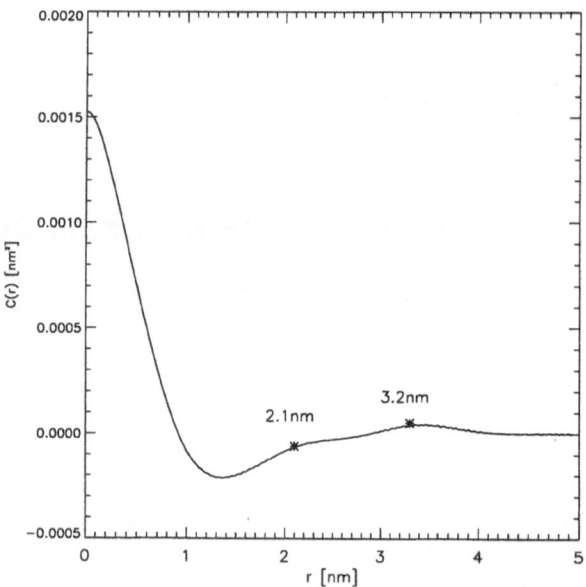

Fig. 9. Structures on the atomic scale: Height-height-correlation function with typical fine structure sizes marked.

smoothing the surface afterwards. Although realistic atomic binding energies and diffusion attempt frequencies are incorporated, the simulations here concentrate more on the behaviour of many particles rather than on simulational details. Fig. 10 shows two example topographs from the early and the late stages of film growth. The corresponding roughnesses and height-difference-correlation functions are depicted in Fig. 11 and 12. From the comparison with the experiments, not only a qualitative agreement (from the greyscale images) can be observed, but also a quantitative agreement concerning the scaling behaviour.

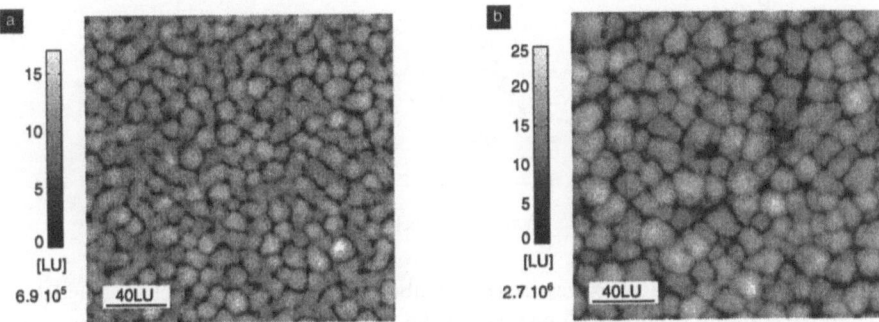

Fig. 10. Two topographs for typical early (a) and late (b) stages, created by the Monte-Carlo model (with tip simulation). The number of deposited particles is given at the margin of the images (LU: length unit).

6   Continuum Equations

The analysis of the experimental results suggests curvature induced surface diffusion as a main surface relaxation process. Already *Mullins* [12] has suggested an expression for modelling in a continuum approximation leading to the fourth order term (where r is the local radius of curvature; D_S =const.):

$$F_D[h] = -D_S \frac{\partial}{\partial x} \left(\left(1 + \left(\frac{\partial h}{\partial x}\right)^2\right)^{-1/2} \frac{\partial}{\partial x}\left(\frac{1}{r}\right) \right) \qquad (6.1)$$

The other mechanism leading to a q^{-2}-term in the Fourier analysis of the experiments (a non-smoothing term) can be attributed to geometrical effects and particle attraction, respectively, as suggested by *Shevchik* [16] and *Mazor et al.* [17] (S =const.):

$$F_S[h] = -S\frac{1}{r} = -S\frac{\frac{\partial^2 h}{\partial x^2}}{\left(1 + \left(\frac{\partial h}{\partial x}\right)^2\right)^{3/2}} \qquad (6.2)$$

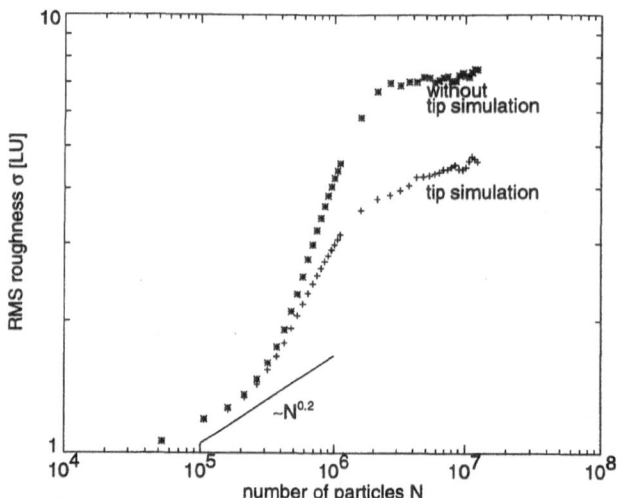

Fig. 11. RMS roughness σ versus the number of deposited particles. The RMS roughness shows a pronounced increase in the medium thickness range, in the early stages it first increases much slower with film thickness, and in the late stages it saturates.

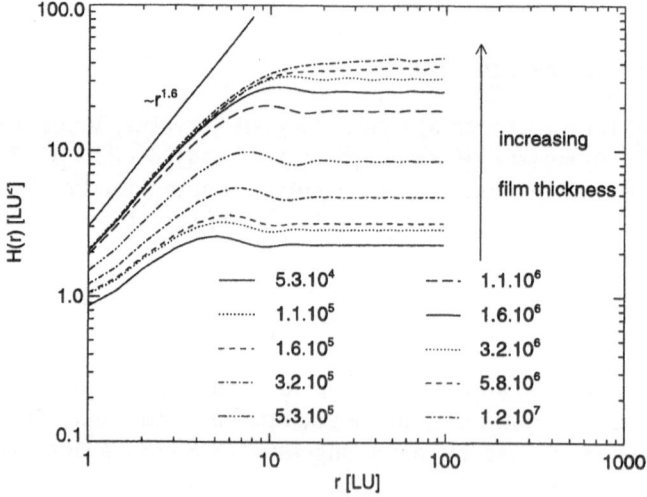

Fig. 12. Height-difference-correlation functions for various film thicknesses. From $H(r)$, $\alpha = 0.8 \pm 0.1$ can be determined.

A continuum equation including both terms from above does not show the experimentally observed structure coarsening, and thus may be attributed to an additional effect as suggested by *Moske* [9]: Due to the local slope, the adatom concentration varies with the surface slope, leading to a total diffusion current from the areas with low slope to those with high slope, $C =$const. (surface mobility assumed to be constant):

$$F_C[h] = C\frac{\partial}{\partial x}\left(\left(1+\left(\frac{\partial h}{\partial x}\right)^2\right)^{-1/2}\frac{\partial}{\partial x}\left(1+\left(\frac{\partial h}{\partial x}\right)^2\right)^{-1/2}\right) \tag{6.3}$$

Thus the total equation considered to compute has the shape:

$$\frac{\partial h}{\partial t} = F_D[h] + F_C[h] + F_S[h] + \eta \tag{6.4}$$

This equation was solved numerically using a simple finite difference scheme for space discretization and the Euler and Heun scheme for time discretization, leading to similar results [18,19,4,20,21]. The parameters were varied for good accordance with the experiments. Greyscale visualizations are depicted in Fig. 13. It can be observed that the height-difference-correlation functions and the roughness as a function of film thickness, shown in Fig. 14 and 15, are in good agreement with the experiments.

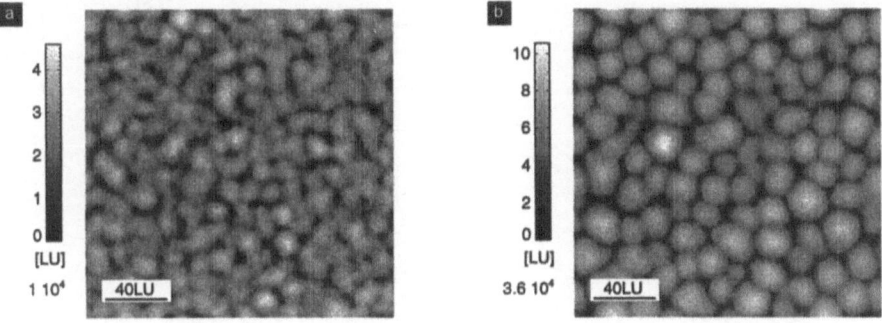

Fig. 13. Topographs generated by numerical integration (Euler method) of the continuum model on a 200×200 grid, $\Delta t = 0.001$, $\Delta x = 0.7$, $D = 0.1$, $D_S = 2$, $C = 5$, $S = 1$ in units of Δx and Δt. The numbers at the margin denote the integration steps.

7   The Problem of Comparing STM Measurements with Simulational Results

For a comparison of STM measurements with surface topographs generated by numerical modelling, and thus for a final evaluation of the quality of the

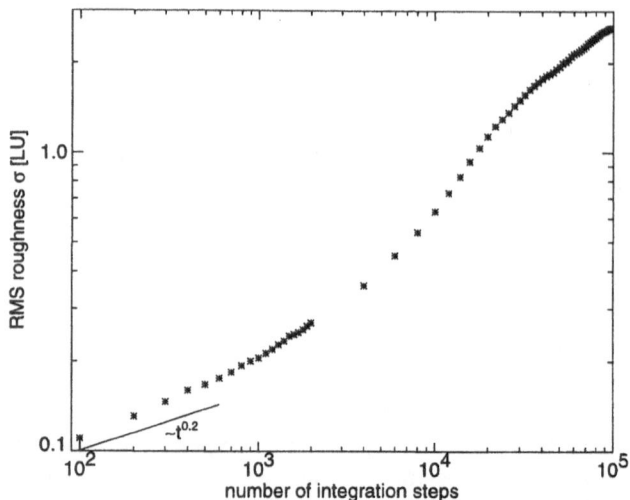

Fig. 14. RMS roughnesses σ using the continuum model (time unit: number of integration steps): A scaling exponent $\beta = 0.2 \pm 0.1$ can be determined.

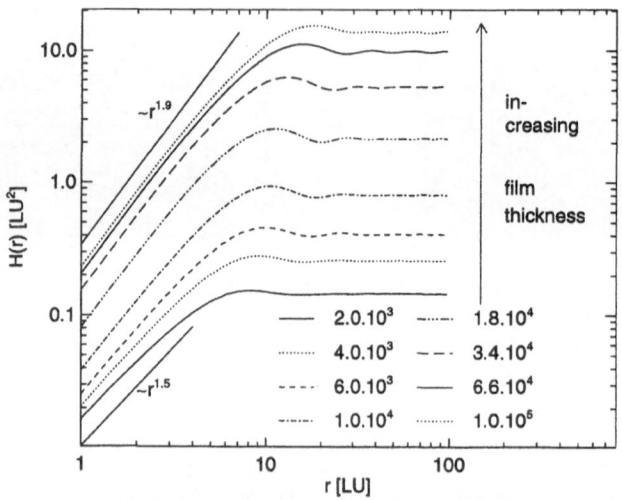

Fig. 15. Height-difference-correlation functions $H(r)$ for various stages of simulated film growth using the continuum model (time unit: number of integration steps): A scaling exponent $\alpha = 0.85 \pm 0.15$ can be determined.

model assumptions, the influence of the measurement method itself, *i.e.* the convolution of the actual surface with the tip geometry has to be considered. This is especially necessary for mesoscopic surfaces, like the hill-like stuctures here. Qualitatively, the overgrowth of the hills in the very late stages observed here experimentally, is an indication for tip imaging. The interpretation of the topographs remains ambiguous in the sense, whether columnar growth is present or not, as the image information lacks for a final decision in that question. A worst-case tip can be estimated by considering the maximum slopes present in the STM topographs, as the maximum accessible slope corresponds to the slope determined by the tip geometry. A systematic method for the estimation of the tip geometry based on this observation, has been suggested in literature [22,23], and was implemented here [4]: Fig. 16 shows the estimated tip geometry determined from the STM topograph in Fig. 4. Using this method, the areas on the surface, which might not be properly

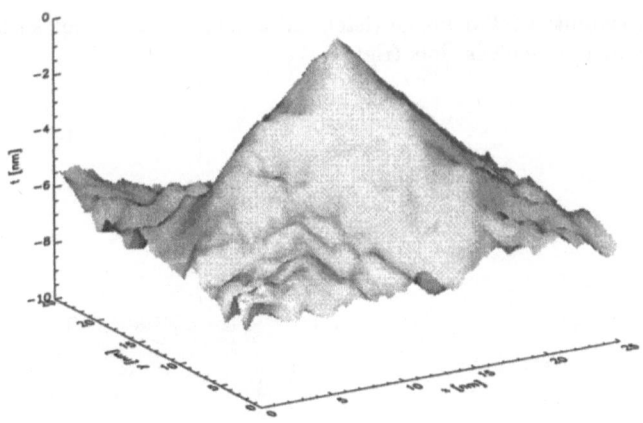

Fig. 16. Estimated worst-case geometry of the STM tip

imaged, have been determined and marked in Fig. 17 as black dots. In Fig. 17 the decovoluted version of the image of Fig. 4d is presented, which shows deeper grooves than the directly measured results. To investigate quantitatively the influence of the tip convolution on the characteristic values, an STM measurement itself is now modelled vice versa on top of the simulated surface growth by the use of a simple cone-shaped tip: Topographs before and after STM tip convolution are presented in Fig. 18. Qualitatively, the roughness is reduced by the effect, that the tip is unable to reach all valleys between the hills, as has been quantitatively evaluated in Fig. 11. The effect of tip convolution on the scaling behaviour of the height-difference-correlation

Fig. 17. Deconvoluted STM image (left) and ambiguous areas for surface reconstruction – depicted as black dots (right)

Fig. 18. Original surface (generated by a MC simulation – left) in comparison with a simulated STM measurement (right) under the assumption of a tip with an opening angle of 60°

function is an increase of the roughness exponent with increasing tip angle, see Fig. 19. However, the main statements, such as the saturation of the lateral hill size growth with increasing film thickness remain undisturbed by the measurement method itself.

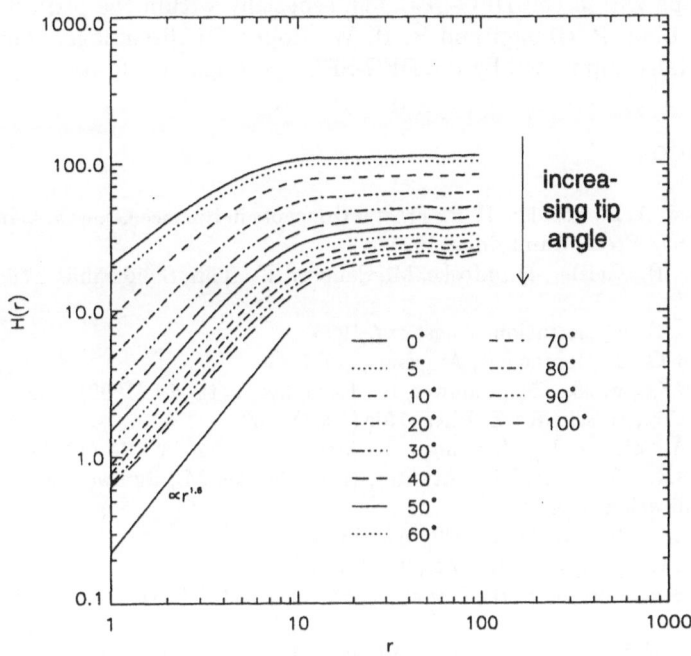

Fig. 19. Dependence of the height-difference-correlation function and its scaling behaviour on the convolution of the original surface with a simulated STM tip of the specified angle. The units of both axes are LU.

8   Conclusion

The results from both numerical treatments – the Monte-Carlo simulation as well as the continuum model – show surface profiles, which are in excellent agreement with the experimental results – not only in a qualitative manner observable from the greyscale images, but also quantitatively. Together with the fact, that the simulated profiles are very parameter sensitive, this is a very strong indication for the fact, that the mechanisms considered in the simulations are also the dominant mechanisms in the experiments.

9 Acknowledgements

The authors acknowledge fruitful discussions with B. Reinker, especially concerning his experience in image processing, STM measurements, and his previous results, on which this work is based. A. Spörhase is acknowledged for technical assistance concerning UHV film preparation. The authors thank the other groups within the DFG-SFB 438, especially within the project A1 M. Raible, S. Linz, P. Hänggi and R. H. W. Hoppe for discussions. This work was financially supported by the DFG-SFB 438 Augsburg-München, TP A1.

References

1. Barabasi, A. L., Stanley, H. E.: Fractal Concepts in Surface Growth. Cambridge University Press, Cambridge, 1995
2. Reinker, B., Geisler, H., Moske, M., Samwer, K.: Thin Solid Films **275** (1996), 240
3. Reinker, B.: Dissertation, Augsburg, 1996
4. Mayr, S. G.: Diplomarbeit, Augsburg, 1997
5. Mayr, S. G., Moske, M., Samwer, K.: Europhys. Lett. **44** (1998), 465
6. Buckel, W., Hilsch, R.: Z. Phys. **138** (1954), 109
7. Inoue, A., Zhang, T., Masumoto, T.: Mater. Sci. Eng. A **134** (1991), 1125
8. Mayr, S. G., Taylor, M. E., Atwater, H. A., Moske, M., Samwer, K.: accepted for publication in APL
9. Moske, M.: Habilitationsschrift, Augsburg, 1997
10. Herring, C.: J. Appl. Phys. **21** (1950), 301
11. Sinha, S. K., Sirota, E. B., Garoff, S., Stanley, H. B.: Phys. Rev. B **38** (1988), 2297
12. Mullins, W. W.: J. Appl. Phys. **30** (1959), 77
13. Reinker, B., Moske, M., Samwer, K.: Phys. Rev. B **56** (1997), 9887
14. Schaub, T. M., Bürgler, D. E., Schmidt, C. M., Güntherodt, H.-J.: J. Non-Cryst. Solids **205–207** (1996), 748
15. Maksym, P. A.: Semicond. Sci. Technol. **3** (1988), 594
16. Shevchik, J.: J. Non-Cryst. Solids, **12** (1973), 141
17. Mazor, A., Srolovitz, D. J., Hagan, P. S., Bukiet, B. G.: Phys. Rev. Lett. **60** (1988), 424
18. Amar, J. G., Family, F.: Phys. Rev. A **41** (1990), 3399
19. Kloeden, P. E., Platen, E., Schurz, H.: Numerical Solution of SDE through Computer Experiments. Springer Verlag, Berlin, Heidelberg, New York, 1991
20. Mayr, S. G., Moske, M., Samwer K.: accepted for publication in Phys. Rev. B **60** (1999)
21. Raible, M., Mayr, S. G., Linz, S., Moske, M., Samwer, K., Hänggi, P.: submitted to EPL
22. Villarrubia, J. S.: Surf. Sci. **321** (1994), 287
23. Williams, P. M., Shakesheff, K. M., Davies, M. C., Jackson, D. E., Roberts, C. J., Tendler, S. J. B.: J. Vac. Sci. Technol. B **14** (1996), 1557

Modelling, Simulation, and Control of Electrorheological Fluid Devices

R. H. W. Hoppe[1], G. Mazurkevitch[1], U. Rettig[2], and O. von Stryk[2]

[1] Universität Augsburg, Institut für Mathematik, D-86159 Augsburg, Germany
[2] Technische Universität München, Zentrum Mathematik, D-80290 München, Germany

Dedicated to Professor Karl-Heinz Hoffmann
on the occasion of his 60th birthday

Abstract. The new generation of electrorheological fluids (ERFs) offers a wide range of applicability in fluid mechatronics with automotive ERF devices such as ERF shock absorbers mentioned at first place. The optimal design of such tools requires the proper modelling and simulation both of the operational behaviour of the device itself as well as its impact on the dynamics of the complete vehicle. This paper addresses these issues featuring an extended Bingham fluid model and its numerical solution as well as substitutive models of viscoelastic-plastic system behaviour. Also control issues for optimal semi-active suspension of vehicles with controllable ERF shock absorbers are discussed.

1   Introduction

Electrorheological fluids (ERFs) are microstructured fluids consisting of a dispersion of electrically polarizable particles in a nonconducting liquid. The characteristic feature is that, under the influence of an outer electric field, the initially unordered particles get oriented and stick together to form particle chains in the fluid. On the macroscopic scale this process results in a significant increase of the dynamic viscosity of the ERF yielding a considerable increase of the shear stress (cf., e.g., [9,10,32,38]). Therefore, ERFs are highly amenable to an efficient control of the transmission of forces. They are thus potentially useful for applications in hydraulic systems and automotive devices such as clutches, engine bearings, and shock absorbers (cf., e.g., [15–17,20,21,39]). The new generation of ERFs which has been developed in recent years features high ER effects by direct activation through electrical signals with an adaptation of the forces in the range of milliseconds that goes along with low abrasive wear, good redispersibility, and high shear stress and sedimentation stability (see [5,3,37]). Despite this progress, ERFs are not yet used in mass production due to still existing problems, as for instance a stable operational behaviour over a wide range of temperatures or their interaction with other components of the devices such as the power supply. Nevertheless, it can be foreseen that the rapid development of ongoing research will put them into the marketplace within only a few years. The optimal design and layout of individual ERF devices is one important aspect. Another equally important one is to study their behaviour, in particular their controllability, as an integral part of a complete system.

It is the purpose of this paper to provide mathematical tools for modelling and simulation of both the fluid flow in specific ERF devices within a continuum mechanical framework and the dynamical behaviour of such devices as integral parts in mechatronical vehicle systems based on the description of the vehicle as a multibody system. The issue of an efficient control of ERF dampers for active suspension of vehicles is also addressed.

The paper is organized as follows: After these introductory remarks, in Sect. 2.1, we will begin with a brief overview on the basic modes of fluid flows in ERF devices. As far as the modelling aspect is concerned, in Sect. 2.2, we present an extension of the classical Bingham fluid model that goes beyond pure shear or flow modes and is thus able to describe the flow pattern in complete devices. In Sect. 2.3, we elaborate in some detail on the method of augmented Lagrangians combined with operator splitting techniques that is based on a primal-dual formulation of the problem and that can efficiently handle the inherent nonlinearities. This is followed by a documentation of numerical simulation results for shear and flow modes in ERF devices given in Sect. 2.4. Geometric design aspects of ERF shock absorbers are mentioned in Sect. 3.1 and followed in Sect. 3.2 by a discussion of substitutive models for the viscoelastic-plastic system behaviour. Optimal and robust optimal feedback controllers for optimal active suspension of a quarter car model based on LQR and H^∞ techniques are investigated in Sect. 3.3. Numerical results are presented in Sect. 3.4.

2   Modelling and Simulation of ERF flows

2.1   Modes of Fluid Flows in ERF devices

In applications of ERFs in automotive devices there are basically three different types of fluid modes: the shear mode, the flow mode, and the squeeze mode.

The shear mode occurs when the electrodes are sheared against each other, as for instance in clutches. A simple model of an ER clutch is shown in Fig. 1. The ERF is located in a housing consisting of two coaxial cylinders with the inner one being driven at some constant angular speed. The inner cylinder hosts a high voltage lead supplying the lateral surface serving as the electrode whereas the lateral surface of the outer cylinder acts as the output. If voltage is applied to the electrodes, an electric field builds up which in the gap between the cylinders is perpendicular to the Couette type flow. The viscosity of the ERF increases with increasing electric field strength as does the torque exerted on the outer cylinder.

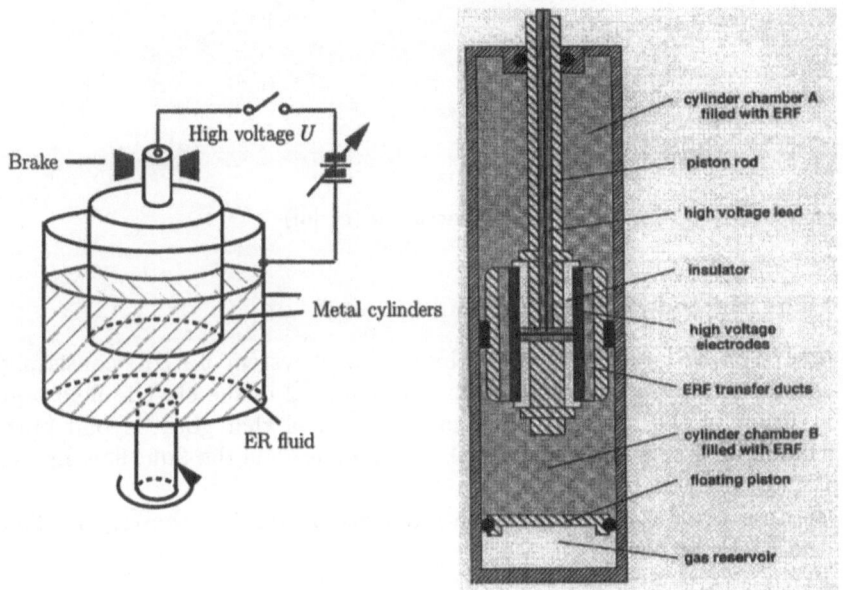

Fig. 1. Schematical representation of an ERF clutch (left) and ERF shock absorber (right) (cf. [7])

The flow mode is predominant in devices where the ERF passes through ducts with fixed electrodes such as shock absorbers or engine mounts. Fig. 1 displays the longitudinal section of a cylindrical shock absorber featuring two ERF-filled chambers with a piston in between that contains two tranfer ducts. There is a third gas-filled chamber separated from the others by a floating piston. The inner walls of the ducts serve as electrodes that are supplied by a voltage lead within the piston rod. In the ducts the generated electric field is perpendicular to the Poiseuille type flow of the ERF.

A third type of mode, the so-called squeeze mode, prevails in ERF actuators used, for instance, as vibration dampers. In this case the electrodes are subjected to vibrations so that the ERF is either pressed out of the gap between the electrodes or sucked into it (cf. Fig. 3 for a schematic representation of an industrially produced ERF actuator).

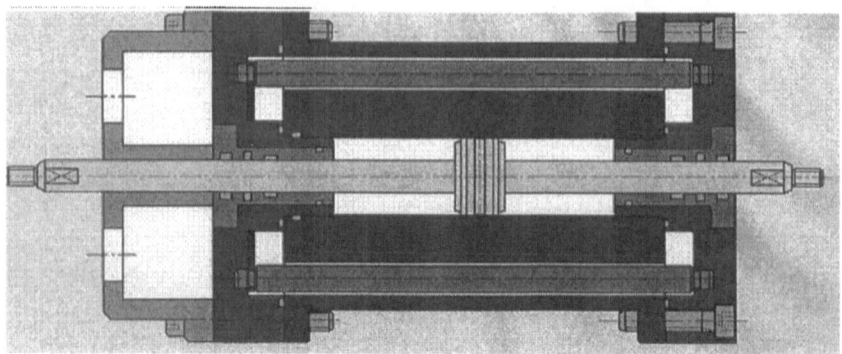

Fig. 2. ERF actuator (cf. [6])

2.2   The Extended Bingham Fluid Model

The flow of an ERF can be modelled by a coupled system of PDEs consisting of a second order elliptic equation for the potential of the electric field and an extension of the classical Bingham fluid model that goes beyond pure shear mode or flow mode and thus allows to determine the fluid flow in the complete device (cf., e.g., [14]).

Denoting by $\Omega \subset \mathbf{R}^d$, $d \in \mathbf{N}$ the physical domain, i.e., the region filled with the ERF, we have

$$\rho(\frac{\partial \mathbf{u}}{\partial t} + (\mathbf{u} \cdot \nabla)\mathbf{u}) = \nabla \cdot \sigma(\mathbf{u}, \mathbf{E}) \quad \text{in } \Omega, \qquad (2.1)$$

$$\nabla \cdot \mathbf{u} = 0 \quad \text{in } \Omega, \qquad (2.2)$$

where the stress tensor $\sigma = \partial \mathcal{D}$ is the subgradient of the (local) energy functional

$$\mathcal{D}(\mathbf{u}, \mathbf{E}) := -\nabla p + \gamma \, | \, \mathbf{E} \, || \, \mathbf{D}(\mathbf{u})\mathbf{E} \, | + \frac{\eta}{2} \|\mathbf{D}(\mathbf{u})\|_F^2 \ . \qquad (2.3)$$

Here, $\mathbf{u} := (u_1, ..., u_d)^T$ stands for the velocity field, $\mathbf{D}(\mathbf{u}) := \frac{1}{2}(\nabla \mathbf{u} + (\nabla \mathbf{u})^T)$ is the rate of deformation tensor, $\mathbf{E} := (E_1, E_2, E_3)^T$ is the electric field, $-p\mathbf{I}$ denotes the spherical part of the stress tensor so that p can be interpreted as the pressure and $\| \cdot \|_F$ refers to the Frobenius norm of a matrix. Moreover, ρ, η and γ are material parameters with ρ and η being the density and zero field viscosity of the ERF, respectively, whereas γ is the proportionality factor in the quadratic dependence ($\sigma_E := \gamma \, | \, \mathbf{E} \, |^2$) of the yield limit on the electric field.

The stress tensor is not well defined if $| \, \mathbf{D}(\mathbf{u})\mathbf{E} \, | = 0$ which indicates the rigid zone. For an increasing electric field $\mathbf{E}$, the rigid zones grow and can completely block the flow for $| \, \mathbf{E} \, |$ sufficiently large.

Note that (2.1), (2.2) have to be interpreted in a weak sense. For instance, if we prescribe the tangential component $\mathbf{t} \cdot \mathbf{u}$ of the velocity on some part

$\Gamma_t \subset \Gamma := \partial\Omega$ and impose no-slip boundary conditions $\mathbf{n} \cdot \mathbf{u} = 0$ on all of Γ, an appropriate variational setting is provided by the function spaces

$$\mathbf{V} := \{\mathbf{v} \in H^1(\Omega)^d \mid \mathbf{t} \cdot \mathbf{v} \mid_{\Gamma_t} = u_t , \; \mathbf{n} \cdot \mathbf{v} \mid_{\Gamma} = 0\} ,$$

$$\mathbf{V}^0 := \{\mathbf{v} \in \mathbf{V} \mid \nabla \cdot \mathbf{v} = 0\}, \quad \mathbf{V}_0^0 := \{\mathbf{v} \in H_0^1(\Omega) \mid \nabla \cdot \mathbf{v} = 0\} .$$

In particular, the variational formulation of the flow problem amounts to the solution of the parabolic variational inequality:

Find $\mathbf{u} \in L^2(0, T; \mathbf{V}_0^0) \cap L^\infty(0, T; \mathbf{V}_0)$, $\partial\mathbf{u}/\partial t \in L^2(0, T; (\mathbf{V}_0^0)^*)$ such that for all $\mathbf{v} \in \mathbf{V}_0^0$ and almost all $t \in (0, T]$

$$\int_\Omega \rho(\frac{\partial\mathbf{u}}{\partial t} + (\mathbf{u} \cdot \nabla)\mathbf{u}) \cdot (\mathbf{v} - \mathbf{u}) dx$$

$$+ \int_\Omega \mathcal{D}(\mathbf{v}, \mathbf{E}) dx - \int_\Omega \mathcal{D}(\mathbf{u}, \mathbf{E}) dx \geq 0 . \tag{2.4}$$

We remark that (2.4) admits a unique solution if $d = 2$ (cf., e.g., [12,19]).

The electric field $\mathbf{E}$ is generated by some voltage U applied to an electrode Γ_e that is part of the boundary whereas the opposite counterelectrode Γ_c is voltage free. Observing $\mathbf{E} = -\varepsilon\nabla\varphi$ where φ is the electric potential and ε refers to the permeability of the ERF, the potential can be obtained as the solution of the elliptic boundary value problem

$$-\nabla \cdot \varepsilon\nabla\varphi = 0 \quad \text{in } \Omega , \tag{2.5}$$

$$\varphi = U \text{ on } \Gamma_e , \quad \varphi = 0 \text{ on } \Gamma_c ,$$
$$\mathbf{n} \cdot \varepsilon\nabla\varphi = 0 \text{ on } \Gamma \setminus (\Gamma_e \cup \Gamma_c) . \tag{2.6}$$

Setting $V := \{\psi \in H^1(\Omega) \mid \psi \mid_{\Gamma_e \cup \Gamma_c} = 0\}$, the variational formulation of (2.4,2.5) is to find $\varphi \in H^1(\Omega)$, $\varphi \mid_{\Gamma_e} = U$, $\varphi \mid_{\Gamma_c} = 0$ such that

$$\int_\Omega \sigma\nabla\varphi \cdot \nabla\psi dx = 0 , \quad \psi \in V . \tag{2.7}$$

2.3   The Method of Augmented Lagrangians

The difficulty associated with the inherent nonlinearity in (2.4) can be circumvented by applying the method of augmented Lagrangians. Combined with an appropriate operator splitting technique this confines the nonlinearity to local, low-dimensional problems (cf., e.g., [18,19]).

We discretize (2.4) implicitly in time by the backward Euler scheme with respect to a partition $0 =: t_0 < t_1 < ... < t_M := T$ of the time interval

$[0, T]$ resulting in elliptic type variational inequalities that have to be solved at each time level $t_m := t_{m-1} + (\Delta t)_m$, $1 \le m \le M$:

Find $\mathbf{u}^m \in \mathbf{V}^0$ such that for all $\mathbf{v} \in \mathbf{V}_0^0$

$$
\rho \int_\Omega \mathbf{u}^m \cdot (\mathbf{v} - \mathbf{u}^m) \, dx + \rho(\Delta t)_m \int_\Omega (\mathbf{u}^m \cdot \nabla)\mathbf{u}^m \cdot (\mathbf{v} - \mathbf{u}^m) \, dx
$$

$$
+ (\Delta t)_m \cdot \left(\int_\Omega \mathcal{D}(\mathbf{v}, \mathbf{E}) \, dx - \int_\Omega \mathcal{D}(\mathbf{u}^m, \mathbf{E}) \, dx \right) \tag{2.8}
$$

$$
\ge \rho \int_\Omega \mathbf{u}^{m-1} \cdot (\mathbf{v} - \mathbf{u}^m) \, dx .
$$

The method of augmented Lagrangians is applied to (2.8) based on a primal-dual formulation by introducing $\mathbf{p}_i := \nabla u_i^m$, $1 \le i \le d$, as additional variables and by coupling the constraints $\mathbf{p}_i - \nabla u_i^m = 0$, $1 \le i \le d$, as well as the incompressibility constraint $\nabla \cdot \mathbf{u}^m = 0$ both by Lagrangian multipliers and penalty terms.

Setting $\mathbf{f} := \rho \mathbf{u}^{m-1} - (\Delta t)_m \nabla p$, introducing the matrix-valued functions $\mathbf{P} := (\mathbf{p}_1 \mid ... \mid \mathbf{p}_d)$, $\boldsymbol{\lambda} := (\lambda_1 \mid ... \mid \lambda_d)$, where $\mathbf{p}_i, \lambda_i \in L^2(\Omega)^d$, $1 \le i \le d$, and given penalization parameters $\boldsymbol{\kappa} := (\kappa_1, \kappa_2)^T$ with $\kappa_\nu > 0, 1 \le \nu \le 2$, we are thus led to the saddle point problem

Find $(\mathbf{u}^m, \mathbf{P}, \boldsymbol{\lambda}, \theta) \in \mathbf{V} \times L^2(\Omega)^{d \times d} \times L^2(\Omega)^{d \times d} \times L^2(\Omega)$ such that

$$
L_\kappa(\mathbf{u}^m, \mathbf{P}, \boldsymbol{\lambda}, \theta) = \inf_{\mathbf{v}, \mathbf{Q}} \sup_{\mu, \tau} L_\kappa(\mathbf{v}, \mathbf{Q}, \boldsymbol{\mu}, \tau) . \tag{2.9}
$$

The Lagrangian L_κ is given by

$$
L_\kappa(\mathbf{v}, \mathbf{Q}, \boldsymbol{\mu}, \tau) :=
$$

$$
\rho(\Delta t)_m \int_\Omega (\mathbf{Q} - \frac{1}{2} \mathrm{diag}\, \mathbf{Q}) \mathbf{v} \cdot \mathbf{v} \, dx + \gamma(\Delta t)_m \int_\Omega |\mathbf{E}| |\mathbf{Q}_s \mathbf{E}| \, dx
$$

$$
+ \frac{\eta}{2}(\Delta t)_m \int_\Omega \|\mathbf{Q}_s\|_F^2 \, dx + \frac{\rho}{2} \int_\Omega |\mathbf{v}|^2 \, dx - \int_\Omega \mathbf{f} \cdot \mathbf{v} \, dx + \sum_{i=1}^d \int_\Omega \mu_i \cdot (\mathbf{q}_i - \nabla v_i) \, dx
$$

$$
+ \int_\Omega \tau \nabla \cdot \mathbf{v} \, dx + \frac{\kappa_1}{2} \sum_{i=1}^d \int_\Omega |\mathbf{q}_i - \nabla v_i|^2 \, dx + \frac{\kappa_2}{2} \int_\Omega |\nabla \cdot \mathbf{v}|^2 \, dx ,
$$

where $\mathbf{Q}_s := \frac{1}{2}(\mathbf{Q} + \mathbf{Q}^T)$.

As far as the discretization in space is concerned, we assume a simplicial triangulation $\mathcal{T}_h$ of Ω and approximate the velocities $u_i, 1 \le i \le d$, by continuous, piecewise linear finite elements augmented by cubic bubble functions associated with each element of $\mathcal{T}_h$. We denote by $S_1(\Omega; \mathcal{T}_h)$ the conforming

P1 finite element space and by $B_3(\Omega; \mathcal{T}_h) := \mathrm{span}\{\prod_{i=1}^{d+1} \lambda_i^T \mid T \in \mathcal{T}_h\}$ the linear space of cubic bubbles where $\lambda_i^T, 1 \le i \le d+1$, are the barycentric coordinates associated with $T \in \mathcal{T}_h$. We set

$$\mathbf{V}_h := V_h^d \quad , \quad V_h := S_1(\Omega; \mathcal{T}_h) \oplus B_3(\Omega; \mathcal{T}_h) \ . \tag{2.10}$$

The components $p_{ij}, \lambda_{ij}, 1 \le i, j \le d$, of the variables $\mathbf{p}_i, 1 \le i \le d$, and the multipliers $\lambda_i, 1 \le i \le d$, as well as the multiplier θ will be approximated by elementwise constants, i.e., we define

$$\mathbf{W}_h := W_h^d \ , \quad W_h := \{w_h \in L^2(\Omega) \mid w_h \mid_T \in P_0(T), T \in \mathcal{T}_h\} \ . \tag{2.11}$$

Then, the discretized saddle point problem reads as follows:

Find $(\mathbf{u}_h, \mathbf{P}_h, \boldsymbol{\lambda}_h, \theta_h) \in \mathbf{V}_h \times \mathbf{W}_h^d \times \mathbf{W}_h^d \times W_h$ such that

$$L_\kappa(\mathbf{u}_h, \mathbf{P}_h, \boldsymbol{\lambda}_h, \theta_h) = \inf_{\mathbf{v}_h, \mathbf{Q}_h} \sup_{\boldsymbol{\mu}_h, \tau_h} L_\kappa(\mathbf{v}_h, \mathbf{Q}_h, \boldsymbol{\mu}_h, \tau_h) \ . \tag{2.12}$$

Note that (2.12) satisfies the Ladyzhenskaja-Babuska-Brezzi (LBB-) condition and is thus well-defined.

We will solve (2.12) by an iterative procedure that is based on operator splitting techniques. Each iteration consists of two steps. The first one requires the subsequent solution of global linear problems whereas the second one involves the solution of local, low-dimensional nonlinear problems.

We assume that we are given initial values $u_i^0 \in V_h, 2 \le i \le d$, $\mathbf{p}_i^0 \in \mathbf{W}_h$, $\lambda_i^1 \in \mathbf{W}_h, 1 \le i \le d, \theta^1 \in W_h$ and sequences $(\kappa_\nu^n)_{\mathbf{N}}, (\rho_\nu^n)_{\mathbf{N}}, 1 \le \nu \le 2$, of penalization parameters $\kappa_\nu^n > 0$ and update parameters $\rho_\nu^n > 0, 1 \le \nu \le 2, n \in \mathbf{N}$. For notational convenience we have dropped the super- resp. subscripts h and m for the independent variables. Then, for $n \in \mathbf{N}$ we perform the following iteration steps:

Step 1: For $i = 1, .., d$ compute $u_i^n \in V_h$ as the solution of the minimization problem

$$
\begin{aligned}
L_\kappa(u_1^n, ..., u_{i-1}^n, u_i^n, u_{i+1}^{n-1}, ..., u_d^{n-1}, \mathbf{P}^{n-1}, \boldsymbol{\lambda}^n, \theta^n) &= \\
= \inf_{v_i \in V_h} L_\kappa(u_1^n, ..., u_{i-1}^n, v_i, u_{i+1}^{n-1}, ..., u_d^{n-1}, \mathbf{P}^{n-1}, \boldsymbol{\lambda}^n, \theta^n)
\end{aligned}
\tag{2.13}
$$

and update the multiplier λ_i according to

$$\lambda_i^{n+\frac{1}{2}} = \lambda_i^n + \rho_1^n(\nabla u_i^n - \mathbf{p}_i^{n-1}) \ . \tag{2.14}$$

Finally, update the multiplier θ

$$\theta^{n+1} = \theta^n + \rho_2^n \nabla \cdot \mathbf{u}^n \ . \tag{2.15}$$

Step 2: For $i = 1, .., d$ compute $\mathbf{p}_i^n \in \mathbf{W}_h$ by solving the elementwise non-linear minimization problems

$$
\begin{aligned}
L_\kappa^T(\mathbf{u}^n, \mathbf{p}_1^n, ..., \mathbf{p}_{i-1}^n, \mathbf{p}_i^n, \mathbf{p}_{i+1}^{n-1}, ..., \mathbf{p}_d^{n-1}, \lambda^{n+\frac{1}{2}}, \theta^{n+1}) \; &= \\
\inf_{\mathbf{q}_i \in \mathbf{W}_h} L_\kappa^T(\mathbf{u}^n, \mathbf{p}_1^n, ..., \mathbf{p}_{i-1}^n, \mathbf{q}_i, \mathbf{p}_{i+1}^{n-1}, ..., \mathbf{p}_d^{n-1}, \lambda^{n+\frac{1}{2}}, \theta^{n+1}) \; &,
\end{aligned} \tag{2.16}
$$

$T \in \mathcal{T}_h$, where L_κ^T denotes the Lagrangian L_κ restricted to $T \in \mathcal{T}_h$.

Update the multiplier λ_i according to

$$
\lambda_i^{n+1} \; = \; \lambda_i^{n+\frac{1}{2}} + \rho_1^n (\nabla u_i^n - \mathbf{p}_i^n) \; . \tag{2.17}
$$

Remarks: The computation of $u_i^n \in V_h, 1 \le i \le d$, in the first step requires the solution of the following variational equations

$$
\sum_{j=1}^d \int_\Omega (\kappa_1^n \frac{\partial u_i^n}{\partial x_j} + \kappa_2^n \frac{\partial u_i^n}{\partial x_j} \delta_{ij}) \frac{\partial v}{\partial x_j} dx + \rho \int_\Omega u_i^n v dx + \Delta t \rho \int_\Omega p_{ii}^{n-1} u_i^n v dx =
$$

$$
= \int_\Omega f_i v dx - \Delta t \rho \int_\Omega \left(\sum_{j=1}^{i-1} (p_{ij}^{n-1} + p_{ji}^{n-1}) u_j^n + \sum_{j=i+1}^d (p_{ij}^{n-1} + p_{ji}^{n-1}) u_j^{n-1} \right) dx
$$

$$
+ \int_\Omega \lambda_i^n \cdot \nabla v dx + \kappa_1^n \int_\Omega \mathbf{p}_i^{n-1} \cdot \nabla v dx - \int_\Omega \theta^n \frac{\partial v}{\partial x_i} dx -
$$

$$
- \kappa_2^n \int_\Omega \left(\sum_{j=1}^{i-1} \frac{\partial u_j^n}{\partial x_j} \frac{\partial v}{\partial x_i} + \sum_{j=i+1}^d \frac{\partial u_j^{n-1}}{\partial x_j} \frac{\partial v}{\partial x_i} \right) dx \; , \quad v \in V_h \; .
$$

For the computation of $\mathbf{p}_i^n \mid_T \in P_0(T)^d, 1 \le i \le d, T \in \mathcal{T}_h$, in the second step we have to solve d-dimensional minimization problems in the unknowns $p_{ij}^n \mid_T \in P_0(T), 1 \le j \le d$, which can be done by Newton's method applied to the first order optimality conditions.

2.4   Simulation Results

Based on the extended Bingham fluid model and its numerical solution by the method of augmented Lagrangians as described in the previous sections we have performed simulations of an ERF clutch and an ERF shock absorber (cf. Fig. 1).

In particular, for a clutch with the radii $r_i = 3.5\,cm$, $r_e = 7.0\,cm$ and lengths $\ell_i = 25.0\,cm$, $\ell_e = 30.0\,cm$ of the inner resp. outer cylinder, an angular velocity of $125\,rads^{-1}$ of the inner cylinder, and an ERF with $\eta = 0.9 \cdot 10^{-1}\,kgm^{-1}s^{-1}$, $\gamma = 1.0 \cdot 10^{-9}\,PaV^{-2}$ the Figs. 3 and 4 display the adaptively generated computational grids and velocity distributions (see Fig. 4) and the angular velocity profiles between the inner and outer cylinder (see Fig. 3) for different applied voltages. For a detailed discussion and more results as e.g. the torque on the outer cylinder as a function of the applied voltage we refer to [14].

On the other hand, for an ERF shock absorber with a geometry of the fluid chamber as shown in Fig. 1 and an ERF with the same characteristics as for the ERF clutch, Fig. 5 (left) shows the equipotential lines of the electrostatic potential in case of an applied voltage of $U = 1000\ V$, whereas Fig. 5 (right) represents the velocity distribution in the chamber for a velocity $u_t = 0.001\ ms^{-1}$ of the piston and a pressure difference of $\Delta p = 0.0\ Nm^{-2}$ between the inflow and outflow boundaries.

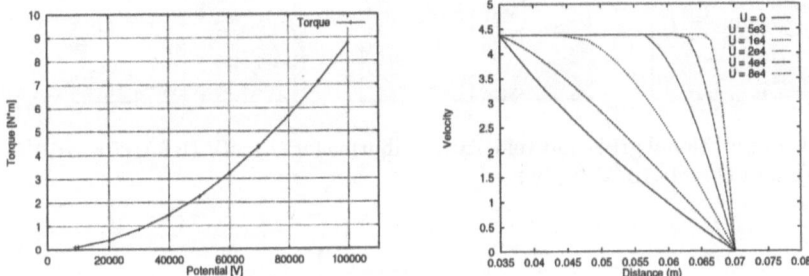

Fig. 3. Torque as a function of applied voltage (left), angular velocity profiles for different values of potential (right)

3   Modelling, Simulation, and Control of ERF Dampers for Semi-Active Vehicle Suspension

Models of ERF shock absorbers suitable for control purposes are investigated in this section. However, due to their qualitatively similar behaviour phenomenological models of electro- and magnetorheological (MR) fluid devices can mostly be applied to either material [23].

3.1   Design Aspects of Electrorheological Fluid Dampers

Typical geometric designs of dampers with controllable electrorheological or magnetorheological fluids for use as automotive shock absorbers are displayed in Fig. 6. The piston rod which moves up and down in a chamber filled with the ERF may be one- or double-sided (Figs. 6(b), 6(a)). The cylinder may be equipped with a bypass to enable fluid flow between the separated upper and the lower parts of the chamber (Fig. 6(c)). However, ERF dampers with double sided piston rod are longer and require more space than dampers equipped with one sided piston rods and equivalent performance. Thus, they

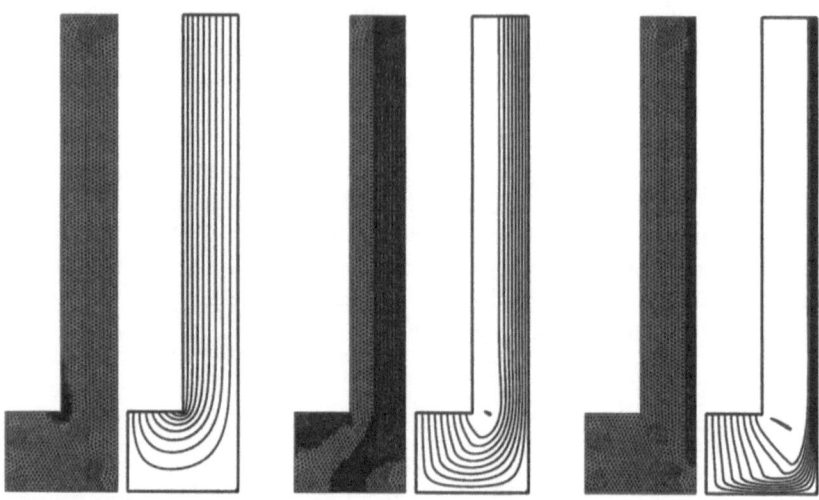

Fig. 4. Computational grids and velocity distribution for $U = 0\text{V}$ (left), $U = 1\cdot10^4\text{V}$ (middle), and $U = 4\cdot10^4\text{V}$ (right)

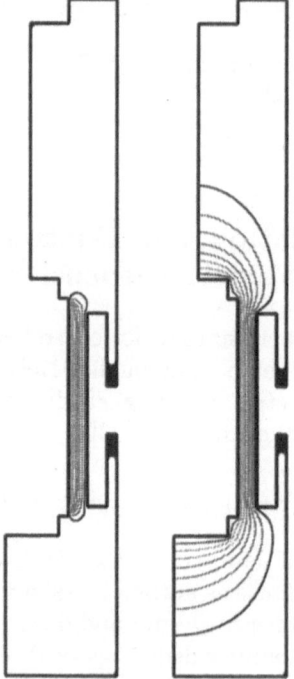

Fig. 5. Potential and z-component of velocity

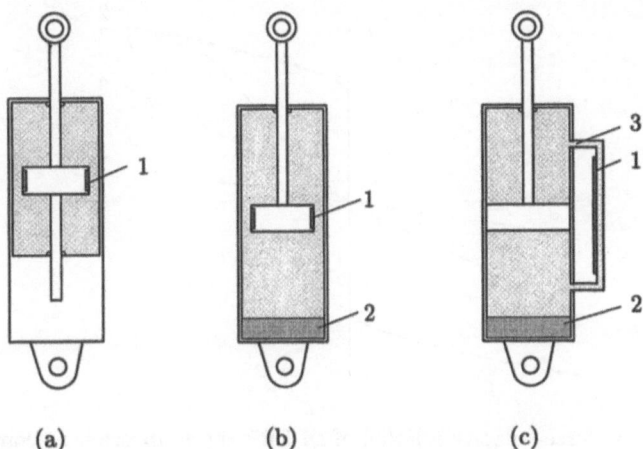

(a)   (b)   (c)

Fig. 6. ERF damper geometries (1: field generator, 2: accumulator, 3: bypass)

are less suited for automotive applications. Dampers designed with a bypass also require more space than without one.

The accumulator (Figs. 6(b), 6(c)) compensates the change in volume of the fluid caused by movements of the piston rod and by thermal expansion. It prevents cavitation while the pressure amounts about 20 bar. Hence, the accumulator affects the dynamical behaviour of the ERF damper – from a phenomenological point of view – like a spring.

For the optimal design and application of ERF dampers for active suspension of vehicles mathematical models are needed taking into account the nonlinear dynamic behaviour of the damper and the vehicle. The purpose of numerical simulation is to reproduce and to predict the behaviour of the damper to investigate the performance of various damper designs and to design the best possible control for a specific application. The characteristic damper properties include the *force-velocity* (Fig. 7) as well as the *force-displacement* relations.

Numerical fluid simulation in complex three-dimensional geometries is computationally too expensive for control purposes where hundreds or thousands of situations must be investigated quickly during the simulated or experimentally conducted ride of a vehicle. However, the accurate simulation of the ERF flow inside the chambers is needed for investigating the performance of various ERF damper designs and for providing reference data to calibrate the parameters of a phenomenological model of the ERF damper dynamics.

The performance of an ERF damper depends on the geometric design as well as on the properties of the used ERF. The spectrum of the yield point of the fluid depends on the electric field strength and is of significant importance for shock absorbing and control properties. Figure 7 shows typical force-velocity relations for a conventional and an ERF shock absorber. The force-

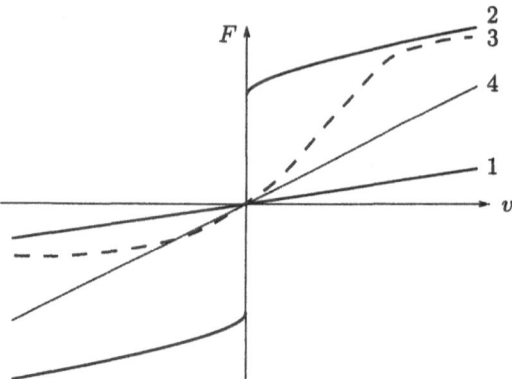

Fig. 7. Typical force-velocity relation of an ERF shock absorber at constant minimum (1) and constant maximum (2) field strengths and for varying field strengths (3) compared to a conventional, passive shock absorber (4)

velocity relation of the ERF damper depends on the shear stress versus shear strain rate relation of the ERF. The design of an adaptively controllable ERF damper must take into account that the curves for constant minimum and maximum field strengths must span the region of all desired damping rates (Fig. 7). The larger the enclosed region of possible damping curves the higher the ERF requirements. While the ERF properties are mainly responsible for the width of the area of possible force-velocity curves and for the slopes of the envelopes, the geometric design determines the relation between bandwidth and slope.

3.2   Phenomenological Models of ERF Dampers

The geometry to be considered for the computation of the fluid flow within an ERF shock absorber without bypass and with a one-sided piston rod (Fig. 6(b)) is less complicated than in a damper with bypass (cf. Sect. 2.1). The fluid flow in the region between the inner piston head and the main cylinder can be computed almost analytically if a Poiseuille type flow excited by a pressure difference between the lower and upper chambers is assumed. For a radial symmetric design, a constant electric field strength, perpendicular directions of the both fields, a constant flow velocity u within the gap and using radial symmetric coordinates, the PDEs investigated in Sect. 2.2 reduce to a steady state, one dimensional ordinary differential equation (ODE)

$$\frac{d\tau}{dr} + \frac{\tau}{r} = \frac{\Delta P}{L} \qquad (3.1)$$

with a simple Bingham model

$$\tau = \tau_0 \operatorname{sgn}\left(\frac{du}{dr}\right) + \eta \frac{du}{dr} \qquad \text{if} \quad \tau > \tau_0,$$

$$\frac{du}{dr} = 0 \qquad \text{if} \quad \tau < \tau_0.$$

Here τ denotes the shear stress, τ_0 the yield point corresponding to the applied electric field strength, η the plastic viscosity as the slope of shear stress versus shear strain rate, r the radius with origin at the center of the cylinder. A linear pressure gradient along the piston head is denoted by $\Delta P/L$.

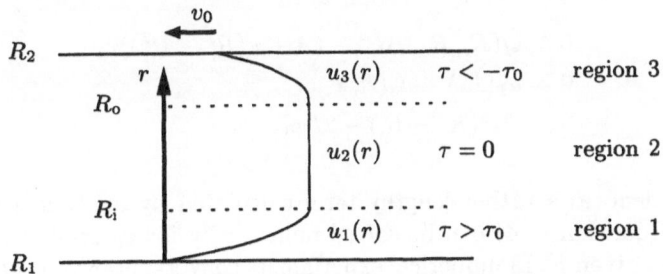

Fig. 8. Velocity field within the gap, $u(R_1) = -v_0$, $u(R_2) = 0$

The analytic solution of the ODE requires a separate treatment of the three regions of evolution of the shear stress (Fig. 8). Inside the inner region 2, i. e., $R_\mathrm{i} \leq r \leq R_\mathrm{o}$, the velocity gradient vanishes and we obtain a plug flow region. If R_i and R_o would reach the boundary of the interval $[R_1, R_2]$ the gap would operate as a locked valve. If we consider the boundary conditions at $r = R_1, R_2$ and the transition conditions at $r = R_\mathrm{i}, R_\mathrm{o}$ we obtain (cf. [2,22])

$$u_1(r) = \frac{\Delta P}{4\eta L}\left[r^2 - R_1^2 - 2R_\mathrm{i}^2 \ln\left(\frac{r}{R_1}\right)\right]$$
$$- \frac{\tau_0}{\eta}\left[r - R_1 - R_\mathrm{i} \ln\left(\frac{r}{R_1}\right)\right] - v_0 \, ,$$

$$u_3(r) = \frac{\Delta P}{4\eta L}\left[r^2 - R_2^2 - 2R_\mathrm{o}^2 \ln\left(\frac{r}{R_2}\right)\right] \qquad (3.2)$$
$$- \frac{\tau_0}{\eta}\left[r - R_2 - R_\mathrm{o} \ln\left(\frac{r}{R_2}\right)\right] \, ,$$

$$u_2(r) = u_1(R_\mathrm{i}) \quad (= u_3(R_\mathrm{o})) \, .$$

The variables R_i, R_o and the piston head velocity v_0 define the velocity field $u(r)$. Thus, the volume flow through the electrode gap is

$$Q(R_i, R_o, v_0) = \int_{R_1}^{R_2} 2\pi r u(r) \, \mathrm{d}r \ . \tag{3.3}$$

On the other hand, the volume replacement of the piston head determines the volume flow according to $Q = Av_0$, where A denotes the area of the piston head. This relation and the two transient conditions at $r = R_i, R_o$ lead to a system of three nonlinear equalities that define F, R_i, and R_o for a given velocity v_0. Considering the relation $F = -\Delta P/L$ we obtain

$$
\begin{aligned}
0 &= \bar{Q}(R_i, R_o, \Delta P) - (A + \pi(R_i^2 - R_1^2))v_0 \,, \\
0 &= u_1(R_i) - u_3(R_o) \,, \\
0 &= \Delta P(R_o - R_i) - 2L\tau_0 \,,
\end{aligned}
\tag{3.4}
$$

where $\bar{Q}$ denotes a rather lengthy term calculated by analytical evaluation of (3.3). Equation (3.4) can be solved numerically for v_0 applying Newton's method for given F. In numerical experiments convergence was obtained from feasible initial values within a few iterations. However, the iterates must be prevented from leaving the valid reagion, as a singularity occurs at $v_0 = 0$ corresponding to the "locking" of the valve.

The approach described so far relies on several strong assumptions. Inertia terms have been neglected, i. e., an oscillating excitation leads to a transient PDE. Furthermore, viscoelasticity of the ERF occurs at small shear strain rates, resp. small velocities of the piston rod. Model equations considering this kind of fluid flow are currently not available and difficult to derive. Viscoelasticity causes hysteresis not only in the force-displacement relation, but also in the force-velocity relation. This effect is related to elastic properties at small shear strain rates, which are kept in "memory" during the elastic phase. Complex considerations assume a "fading memory" effect [8]. In addition, the presence of the accumulator and other mechanical details of the shock absorber lead to further mutation of the input-output behaviour.

A substitutive model of viscoelastic fluids which may be combined with the Bingham model is investigated next. Maxwell recognized that the equation

$$\tau + \frac{\eta}{G} \frac{\partial \tau}{\partial t} = \mu \dot{\gamma} \tag{3.5}$$

where $\dot{\gamma}$ specifies the shear strain rate, η and G denote the viscosity resp. the modulus of elasticity, contains both the ideas of Newtonian liquids and

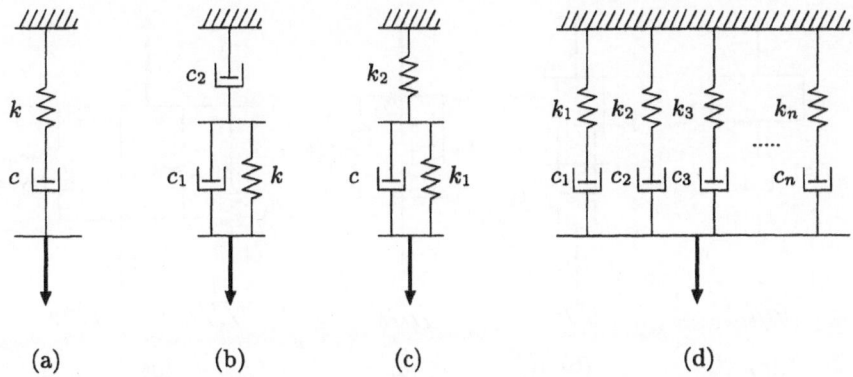

Fig. 9. Substitutive models for viscoelastic behaviour: (a) conventional Maxwell model, (b) 3-parametric fluid model, (c) 3-parametric solid model, (d) general Maxwell model

Hookian solids [8]. More general types of the Maxwell model are

$$\left(1 + \sum_{n=1}^{\infty} a_n \frac{\partial^n}{\partial t^n}\right) \tau_{ij} = \eta_0 \left(1 + \sum_{n=1}^{\infty} b_n \frac{\partial^n}{\partial t^n}\right) \gamma_{ij} \qquad (3.6)$$

$$\tau = \sum_k \tau_k, \qquad \tau_k + \lambda_k \frac{\partial}{\partial t} \tau_k = \eta_k \dot{\gamma}_k \qquad (3.7)$$

with appropriate constants. Equation (3.7) represents a linear combination of several Maxwell models. A special case of (3.6) follows with

$$\tau + p_1 \dot{\tau} = q_1 \dot{\gamma} + q_2 \ddot{\gamma} . \qquad (3.8)$$

If we compare with the substitutive models depicted in Figs. 9(b) and 9(c) we obtain for

$$\text{case (b)} \qquad p_1 = \frac{c_1 + c_2}{k}, \qquad q_1 = c_2, \qquad q_2 = \frac{c_1 c_2}{k},$$

$$\text{resp. case (c)} \qquad p_1 = \frac{c}{k_1}, \qquad q_1 = \frac{k_1 + k_2}{k_1 k_2}, \qquad q_2 = \frac{c}{k_1 k_2}.$$

The constants c_i, k_i or p_i cannot directly be derived from material constants. They must be fitted to reference data, e. g., from experiments or ERF flow simulations. Both the 3-parametric fluid and the 3-parametric solid models are popular substitutive models for viscoelastic materials such as polymers.

These models may represent the behaviour of the ERF at small shear strain rates, i. e., for shear stresses less than the yield point. When combined with suitable models representing the phase of large shear stresses, e. g., the Bingham model, we expect a substitutive model of the damping behaviour of

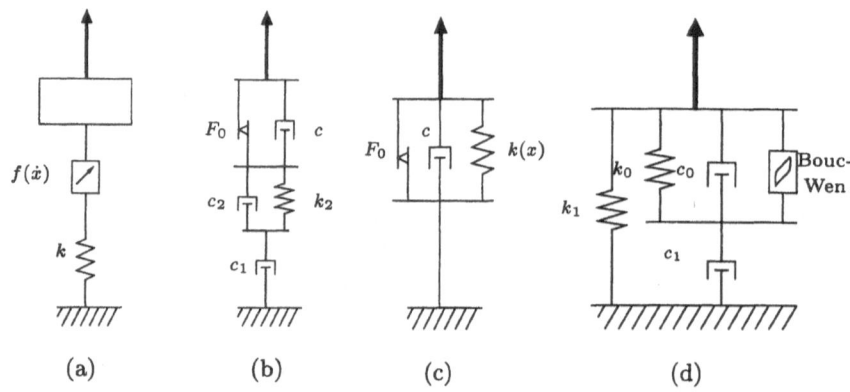

Fig. 10. Substitutive models of viscoelastic-plastic system behaviour: (a) Peel, Stanway, Bullough; (b) Powell; (c) Ehrgott, Masri resp. Kamath, Wereley; (d) Spencer et al.

an ERF shock absorber over the whole range of operation. Stanway et al. [35] (Fig. 10(a)) use a simple spring according to elastic properties of the fluid and the accumulator. The damping characteristics are calculated by the simplified flow model mentioned earlier in this section. However, this approach obviously neglects the so-called viscous damping at the viscoelastic phase.

As depicted in Fig. 10(b) this problem is tackled straightforward with the model of Ehrgott and Masri [13,24]. Both the Bingham model and the 3-parametric fluid model have been arranged serially with satisfying results comparing simulation and measurement data. Another phenomenological model has been described by Powell [31] with a model containing a nonlinear spring and with parameters fitted to measurement data.

The approach of Spencer et al. [33] is a completely phenomenological model. It takes into account the hysteresis within the force-velocity relation and is called the *augmented Bouc-Wen model* (Fig. 10(d)). It includes the largest number of parameters. The augmented Bouc-Wen model and the model of Ehrgott and Masri have been most well accepted. Transcribing the mechanical system into mathematical terms we obtain

$$(c_0 + c_1)\dot{s} = c_1\dot{x} - \alpha z - k_0 s,$$
$$\dot{z} = (A - \beta(1 + \mathrm{sgn}(\dot{s}z))z^2)\dot{s}, \tag{3.9}$$

which is a hysteresis operator with the inner variables s, z and constants c_0, c_1, k_0, α, β, and A. The acting force is

$$F = c_1(\dot{x} - \dot{s}) + k_1(x - x_0) \tag{3.10}$$

with further constants k_1 and x_0.

The augmented Bouc-Wen Model has been tested in the context of field dependence [34]. The characteristic parameters c_0, c_1 and α are modelled as linearly depending on the field strength including a time delay, i. e.,

$$p = p_a + v\,p_b$$
$$\dot{v} = \eta(u - v)$$

(3.11)

where p is either one of c_0, c_1 or α. Furthermore, u denotes the actual field strength, while v stands for a virtual time delayed field strength depending on a constant η according to the known time delay of acting ERF devices.

In summary, all the described models include a priori unknown parameters which account for the uncertainty of unknown behaviour of ERFs. Consequently some authors applied general functional approximation techniques, e. g., Chebyshev polynomials [16,27] or neural nets [29], to fit measurement data. However, it seems to be advantageous to use a priori knowledge about the system behaviour for deriving model equations whose solutions behave similar to the fluid properties, i. e., taking the main properties of the physical system into account within the describing equations. One property is the stiff behaviour of the dynamic equations [36] caused by the phase changes between viscoelastic and plastic mode. These phase changes occur very often at typical frequencies of operating automotive shock absorbers (about 50 Hz [25]).

However, to some extent ERF devices exhibit even more properties which are not included in the models described so far as a temperature dependence of the fluid's yield point and a dropping of the yield point with an increase in the frequency acting on the ERF device. Furthermore, some authors [29] believe that elastic properties of the ER fluids do not significantly affect the system behaviour of an ERF shock absorber contrary to the results of other authors for magnetorheological fluids [33].

3.3 Control of ERF Dampers for Active Vehicle Suspension

An active suspension can be defined as a suspension layout which controls the forces acting in the shock absorber by control of energy dissipation, or, if required by the control law, by generating additional force [25]. A suspension based on control of the previously described ERF shock absorbers cannot input energy into the system and is thus called "semi-active".

The regulator problem for an ERF damper as part of a semi-active vehicle suspension is to control the electric field in such a way that comfort and safety are *maximized* during the vehicle ride. For this purpose, *optimal* control methods must be investigated that take into account the nonlinear dynamical behaviour of the ERF damper as well as of the vehicle itself.

Conventional regulator techniques for active suspension of vehicles are based on controlling the pressure of gas inside a cylinder or the flow of a viscous fluid by a regulator valve. Commonly used control algorithms are only semi-active: The driver or a control unit selects the actual damper characteristic from several force-velocity relations of fixed damper settings corresponding to various ride levels of comfort or safety.

The technology of ER (or MR) fluids offers new control strategies. As already mentioned the electric field regulates the viscosity of the fluid within milliseconds. Hence, the flow through a gap acts as the flow through a continuously adjustable valve which can be opened or closed very fast and with almost no wear. It is possible in principle to adapt the damping rate within one cycle of a damper movement thus enabling fully active vehicle suspensions. In the sequel, an *optimal* damping force is investigated for active suspension of a quarter vehicle. The resulting ideal damping rate serves as a setpoint trajectory for the ERF damper.

The essential gains in shock absorber regulation are comfort and safety of the vehicle ride which are, to some extent, antagonistic gains. If, as usual, the objective is modeled as a weighted sum of both objectives then the proper selection of the weights is an ambiguous problem that must be addressed.

Technically speaking the regulation is concerned with eliminating or decreasing the effect of street disturbances. Especially the excitations close to eigenfrequencies of the vehicle body have to be avoided as well as frequencies which lead to unhealthy vibrations of the vehicle passengers. These effects are observed by the history of the sprung mass acceleration $\ddot{x}_a$ if a quarter car model is considered for the vehicle dynamics (Fig. 11). Furthermore the priority and, thus, the weights of comfort and safety depend on the current driving situation. E. g., driving safety has absolute priority in dangerous situations. This comes along with ensuring large values of the vertical wheel load forces which is necessary for a good maneuverability of a vehicle [1].

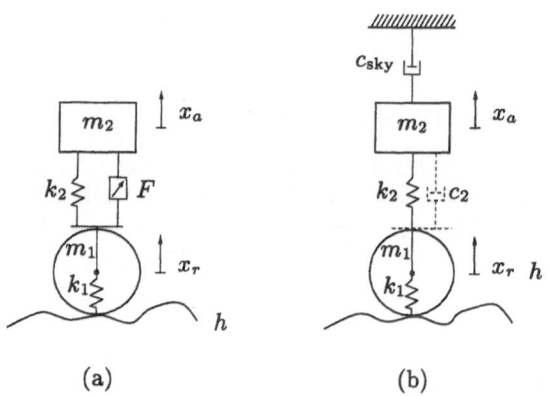

(a)   (b)

Fig. 11. Quarter car model with (a) a controllable damping force F and (b) the sky-hook assumption

Fig. 11(b) depicts a possibility to avoid sprung mass oscillations with eigenfrequencies of a quarter car model (cf. [30]). The damping rate c_2 is adjusted as if a virtual damper c_{sky} would act with a constant proper damping rate. The resulting damping law is technically not realizable, but approximately included in common damping strategies. The outcome of this approach are different compression and decompression damping rates.

Since the *sky-hook* regulator is based on a heuristic approach, a better performance can be expected if *optimal* control techniques are applied based on a general problem formulation: A linear system behaviour of the differential equations for the state variables $\mathbf{x}$ of the vehicle is considered:

$$\dot{\mathbf{x}} = \mathbf{A}\mathbf{x} + \mathbf{B}\mathbf{u}, \quad \mathbf{x}(t_0) = \mathbf{x}_0 \tag{3.12}$$

with $\mathbf{x} \in \mathbb{R}^{n_x}$, $\mathbf{u} \in \mathbb{R}^{n_u}$, $\mathbf{A} \in \mathbb{R}^{n_x \times n_x}$, $\mathbf{B} \in \mathbb{R}^{n_x \times n_u}$. For the linear quadratic Bolza type objective

$$J[\mathbf{u}] = \mathbf{x}_f{}^{\mathrm{T}}\mathbf{Q}_f\mathbf{x}_f + \int_{t_0}^{t_f} \mathbf{x}^{\mathrm{T}}\mathbf{Q}\mathbf{x} + 2\mathbf{u}^{\mathrm{T}}\mathbf{S}\mathbf{x} + \mathbf{u}^{\mathrm{T}}\mathbf{R}\mathbf{u}\,\mathrm{d}t \tag{3.13}$$

with $\mathbf{x}_f = \mathbf{x}(t_f)$, a solution of the problem can be calculated straightforward solving the corresponding Riccati differential equation for $\mathbf{P} \in \mathbb{R}^{n_x \times n_x}$

$$\dot{\mathbf{P}} + \mathbf{A}^{\mathrm{T}}\mathbf{P} + \mathbf{P}\mathbf{A} + \mathbf{Q} - (\mathbf{P}\mathbf{B} + \mathbf{S})\mathbf{R}^{-1}(\mathbf{B}^{\mathrm{T}}\mathbf{P} + \mathbf{S}^{\mathrm{T}}) = \mathbf{0}\,,$$
$$\mathbf{P}(t_f) = \mathbf{Q}_f, \tag{3.14}$$

if no inequality constrains are active. A steady state problem, i. e., $\dot{\mathbf{P}} = \mathbf{0}$, $\mathbf{Q}_f = \mathbf{0}$ and $t_f \to \infty$ in (3.14), leads to an algebraic Riccati equation. Thus, the *optimal* feedback control $\mathbf{u}^*(\mathbf{x})$ is determined by

$$\mathbf{u}^* = -\mathbf{R}^{-1}(\mathbf{B}^{\mathrm{T}}\mathbf{P} + \mathbf{S}^{\mathrm{T}})\mathbf{x}^* \tag{3.15}$$

where $\mathbf{x}^*$ denotes the solution of

$$\frac{\mathrm{d}}{\mathrm{d}t}\mathbf{x}^* = (\mathbf{A} - \mathbf{B}\mathbf{R}^{-1}(\mathbf{B}^{\mathrm{T}}\mathbf{P} + \mathbf{S}^{\mathrm{T}}))\mathbf{x}^*\,,$$
$$\mathbf{x}^*(t_0) = \mathbf{x}_0. \tag{3.16}$$

However, it must be assumed that the system is stabilizable, $\mathbf{R}$ is positive definite, $\mathbf{Q} - \mathbf{S}\mathbf{R}^{-1}\mathbf{S}^{\mathrm{T}}$ is positive semidefinite, and there are no limitations on the observability of $\mathbf{x}$. This method is well known as the *linear quadratic regulator* (LQR) and is popular in many fields of applications (cf. [26,28]).

A drawback of the previous approach is that the solution is only optimal with regard to a step disturbance at initial time. Although the LQR can be extended to a linear, time varying road disturbance $\mathbf{w}$ acting in the following way with $\mathbf{D} \in \mathbb{R}^{n_x \times n_w}$

$$\dot{\mathbf{x}} = \mathbf{A}\mathbf{x} + \mathbf{B}\mathbf{u} + \mathbf{D}\mathbf{w} \tag{3.17}$$

a desirable property of the regulator is to compensate *all* possible disturbances $\mathbf{w}$, particularly those which lead to unstable total systems.

With the state, control and disturbance variables $\mathbf{x}, \mathbf{u}$ and $\mathbf{w}$ as elements of suitable Hilbert spaces H_x, H_u, resp. H_w we now consider a given feedback control $\mathbf{u} = \mu(\mathbf{x}, t)$. The control law μ determines an operator

$$T_\mu : H_w \to H_x \tag{3.18}$$

which maps disturbances $\mathbf{w}$ onto solutions $\mathbf{x}$ of the state equations. With appropriate norms of the according Hilbert spaces a maximum norm of the operator T_μ is defined [4]

$$\|T_\mu\|_\infty = \sup_{w \in H_w} \frac{\|T_\mu\|_x}{\|w\|_w} \tag{3.19}$$

which is related to the values of disturbances ($\|.\|_x$ and $\|.\|_w$ are defined according to (3.22) and (3.23)). Decreasing the maximum value means to attenuate disturbance influence on the total system, so we have to find the control law μ^*, which yields an infimum for the worst case disturbance

$$\inf_\mu \|T_\mu\|_\infty = \phi^* , \tag{3.20}$$

i. e., the worst case is bounded simultaneously. Here the feasible controls μ are restricted implicitly, as a stable system is required. If there exists a solution μ^* of (3.20) we obtain the inequality

$$\|T_{\mu^*}(\mathbf{w})\|_x{}^2 \le \phi^{*2}\|\mathbf{w}\|_w{}^2 \quad \text{for all} \quad \mathbf{w} \in H_w \tag{3.21}$$

as well as the uniqueness of μ^* (cf. [4]). Using the definition of the objective

$$J_\phi[\mu, \mathbf{w}] = \|T_\mu(\mathbf{w})\|_x{}^2 - \phi^2\|\mathbf{w}\|_w{}^2 = J[\mu, \mathbf{w}] - \phi^2\|\mathbf{w}\|_w{}^2 \tag{3.22}$$

and replacing $J[\mu, \mathbf{w}]$ by the objective from (3.13) leads to a zero-sum differential game with the expanded linear quadratic objective

$$J_\phi[\mathbf{u}, \mathbf{w}] = \mathbf{x}_f{}^T\mathbf{Q}_f\mathbf{x}_f + \\ + \int_0^{t_f} \mathbf{x}^T\mathbf{Q}\mathbf{x} + 2\mathbf{u}^T\mathbf{S}\mathbf{x} + \mathbf{u}^T\mathbf{R}\mathbf{u} - \phi^2\mathbf{w}^T\mathbf{w} \, dt . \tag{3.23}$$

The first player chooses the control $\mathbf{u}$ to minimize the objective while the second player chooses the road disturbances $\mathbf{w}$ to maximize it. The solution defines the *best worst case*, i. e., what can be obtained at least with optimal control under the worst possible circumstances.

The solution of the H^∞ problem again reduces to the solution of the corresponding Riccati equation for $\mathbf{P}_\phi \in \mathbb{R}^{n_x \times n_x}$

$$\dot{\mathbf{P}}_\phi + \mathbf{A}^T\mathbf{P}_\phi + \mathbf{P}_\phi\mathbf{A} + \mathbf{Q} - \\ - [\mathbf{P}_\phi\,[\mathbf{B}\,\mathbf{D}] + [\mathbf{S}\,\mathbf{0}]] \begin{bmatrix} \mathbf{R}^{-1} & \mathbf{0} \\ \mathbf{0} & -\frac{1}{\phi^2} \end{bmatrix} \left[\begin{bmatrix} \mathbf{B}^T \\ \mathbf{D}^T \end{bmatrix} \mathbf{P}_\phi + \begin{bmatrix} \mathbf{S}^T \\ \mathbf{0}^T \end{bmatrix} \right] = \mathbf{0} \tag{3.24}$$

$$\mathbf{P}_\phi(t_f) = \mathbf{Q}_f$$

with a chosen $\phi > \phi^*$, and a zero matrix $\mathbf{0}$ of proper dimensions. The resulting control laws are

$$\mathbf{u}^* = \boldsymbol{\mu}^*(t, \mathbf{x}_0) = -\mathbf{R}^{-1}(\mathbf{B}^T\mathbf{P}_\phi + \mathbf{S}^T)\mathbf{x}^* , \tag{3.25}$$

$$\mathbf{w}^* = \boldsymbol{\nu}^*(t, \mathbf{x}_0) = \frac{1}{\phi^2}\mathbf{D}^T\mathbf{P}_\phi\mathbf{x}^* \tag{3.26}$$

with the corresponding optimal state trajectory

$$\dot{\mathbf{x}}^* = (\mathbf{A} - \mathbf{B}\mathbf{R}^{-1}(\mathbf{B}^T\mathbf{P}_\phi + \mathbf{S}^T) + \frac{1}{\phi^2}\mathbf{D}\mathbf{D}^T\mathbf{P}_\phi)\mathbf{x}^* ,$$

$$\mathbf{x}^*(t_0) = \mathbf{x}_0 . \tag{3.27}$$

However, the existence of a solution to the Riccati equation is not guaranteed for the infimum ϕ^* if $\dot{\mathbf{P}} = \mathbf{0}$ [4]. Then only a suboptimal solution can be obtained for a $\phi > \phi^*$. Suboptimal solutions exist for sufficiently large ϕ because the solution of the LQR problem is obtained as $\phi \longrightarrow \infty$.

The numerical solution is obtained iteratively by solving a sequence of problems with decreased values of ϕ while all iterates must satisfy that the algebraic Riccati equation (3.24) is solvable and the controlled and disturbed systems are stable.

The H^∞ approach ensures *robust optimality* and is suitable for linear dynamical vehicle models such as quarter car models. However, several limitations exist: The calculation of the optimal damping force is unconstrained, although it is limited by the shock absorber properties in practice. For problems with a *nonlinear* objective or *nonlinear* dynamics or active state and control *constraints*, the numerical computation of the optimal feedback controller $\mathbf{u}^*(\mathbf{x})$ is only possible in very special cases. However, a nonlinear optimization technique may be applied to compute an approximation of the optimal open loop control $\mathbf{u}^*(t)$ of a nonlinear dynamical system subject to constraints in case of a given road disturbance [26].

3.4   Numerical Results

Optimal active suspension of the quarter car model depicted in Fig. 11(a) with $\mathbf{x} = (x_r, x_a, \dot{x}_r, \dot{x}_a)^T$, $n_x = 4$, and $\mathbf{u} = F$, $n_u = 1$, is investigated using a controllable rheological fluid damper and comparing LQR- and H^∞-regulators.

The vehicle data consists of the spring constants $k_1 = 190\,\text{kN/m}$, $k_2 = 16.812\,\text{kN/m}$, the masses $m_1 = 59\,\text{kg}$, $m_2 = 290\,\text{kg}$. An uncontrolled conventional damper with the constant damping rate $c_2 = 3.0\,\text{kNs/m}$ is used for comparison (cf. [1]). The controllable rheological fluid damper is described by the Bouc-Wen model of Fig. 10(d) and (3.9,3.11) with constants $c_{1a} = 1.0\,\text{kNs/m}$, $c_{1b} = 20.0\,\text{kNs/m}$, $c_{2a} = 1.2\,\text{kNs/m}$, $c_{2b} = 21.5\,\text{kNs/m}$, $\alpha_a = 1.0\,\text{kN/m}$, $\alpha_b = 200\,\text{kN/m}$, $k_0 = 50\,\text{kN/m}$, $k_1 = 100\,\text{kN/m}$, $A = 47.2$,

$\beta = 3.93 \cdot 10^6 \, \text{cm}^{-2}$ and $\eta = 100 \, \text{s}^{-1}$. The constants have been chosen according to the results of Spencer et al. [34] and adapted to damping rates of common automotive shock absorbers.

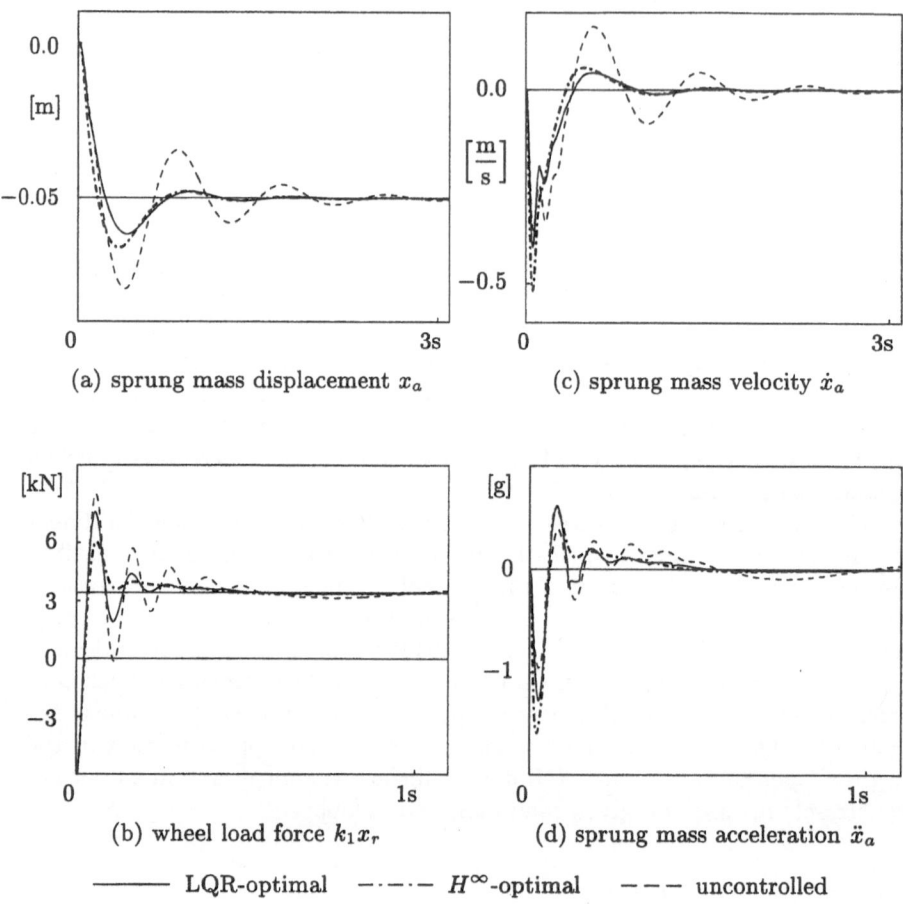

(a) sprung mass displacement x_a

(c) sprung mass velocity $\dot{x}_a$

(b) wheel load force $k_1 x_r$

(d) sprung mass acceleration $\ddot{x}_a$

———— LQR-optimal   —·—·— H^∞-optimal   — — — uncontrolled

Fig. 12. State histories for the actively, LQR- and H^∞-controlled systems and for the passive suspension during the simulated ride with an initial fall off a step of 5 cm

The objective to be minimized by the optimal damping force $F^* = \mathbf{u}^*$ is a weighted quadratic sum of the scaled sprung mass acceleration $\ddot{x}_a/g$ (cf. Fig. 11(a)), the scaled wheel load force $k_1 x_r/g(m_1+m_2)$, the force F/gm_2 and the state variables $x_i/x_{i,\text{max}}$, $\dot{x}_i/\dot{x}_{i,\text{max}}$, $i = a, r$ (cf. [26]). Uniform weights ($\mu = \mu_a = 1$) are chosen except for the states of the wheel, which are weighted by $\mu_r = 1/10$ of the sprung mass states resulting in

$$J[F] = \int_0^\infty \mu \left(\frac{\ddot{x}_a}{g}\right)^2 + \mu \left(\frac{k_1 x_r}{g(m_1 + m_2)}\right)^2 + \mu \left(\frac{F}{m_2 g}\right)^2 +$$

$$+ \sum_{i=a,r} \mu_i \left(\frac{x_i}{x_{i,\max}}\right)^2 + \mu_i \left(\frac{\dot{x}_i}{\dot{x}_{i,\max}}\right)^2 dt. \tag{3.28}$$

The histories of sprung mass displacement x_a and velocity $\dot{x}_a$ (Fig. 11(a)) for the uncontrolled system with a conventional passive damper and the actively, LQR- and H^∞-controlled systems for an initial step disturbance of 5 cm are depicted in Fig. 12. The numerical solutions have been obtained using MATLAB. A comparison of the wheel load forces is also given which are indicators for driving safety. If the wheel load becomes less or equal to zero then the vehicle is not maneuverable. On the other hand, the sprung mass acceleration $\ddot{x}_a$ is a measure of driving comfort. Obviously, an increase in safety, i.e., of the wheel load forces, comes for the price of an increased sprung mass acceleration, i.e., a decrease in comfort.

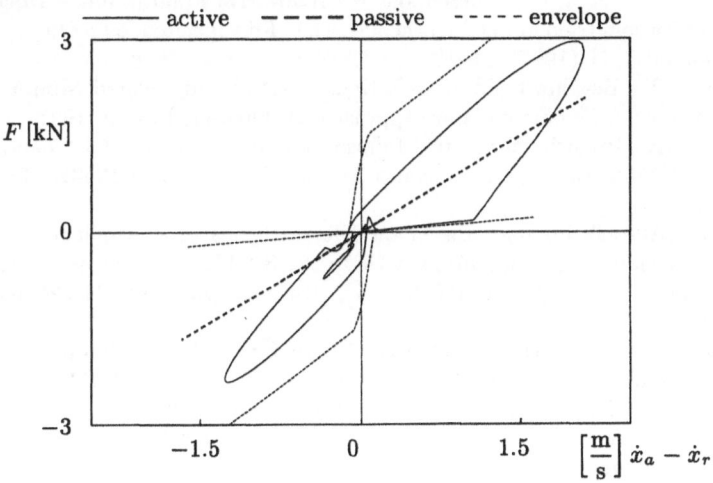

Fig. 13. Force-velocity graph of the actively LQR-controlled rheological fluid damper and the conventional passive damper during the simulated ride

Actually, we are interested in the strength of the electric field u of $(3.11)^1$, which affects the viscosity of the damper. For this purpose, an extension of the approach of [33] to adjust the current damping force F to the computed

---

[1] Please note the difference between u and $\mathbf{u}$.

optimal damping force F^* is applied

$$u = \begin{cases} \min \left(u_{\max}, \left| \frac{F^*-F}{F_{\text{scal}}} \right| \left| \frac{F}{F_{\text{scal}}} \right| \right), & \text{if } |F| < |F^*|, FF^* > 0, \\ u_{\min}, & \text{otherwise.} \end{cases} \qquad (3.29)$$

with a proper scaling force (here $F_{\text{scal}} = 10^2 \text{N}$) and the scaled values $u_{\max} = 1$ and $u_{\min} = 0$.

The force-velocity relation of the LQR-controlled rheological fluid damper model during the simulated ride is depicted in Fig. 13 in comparison with the passive damper demonstrating the innovative potential of optimally controlled electrorheological fluid dampers for semi-active suspension of vehicles.

References

1. Alleyne, A., Hedrick, J. K.: Nonlinear Adaptive Control of Active Suspensions. IEEE Transactions on Control Systems Technology **3** (1995), 94–101
2. Atkin, R. J., Shi, X., Bullough, W. A.: Solutions of the Constitutive Equations for the Flow of an Electrorheological Fluid in Radial Configuations. J. Rheology **35** (1991), 1441–1461
3. Backé, W., Fees, G., Murrenhoff, H.: Innovative Fluidtechnik - Hochdynamischer Servoantrieb mit elektrorheologischen Flüssigkeiten. o+p Ölhydraulik und Pneumatik **41** (1997), 11–12
4. Başar, T., Bernhard, P.: H^∞-Optimal Control and Related Minimax Design Problems: A Dynamic Game Approach. Birkhäuser, Boston, 1991
5. Bayer AG: Provisional Product Information. Rheobay TP AI 3565 and Rheobay TP AI 3566. Bayer AG, Silicones Business Unit, No. AI 12601e, Leverkusen, 1997
6. Bayer AG: Technology Based on ERF. Rheobay for Applications in Fluid Mechatronics (in cooperation with IFAS, RWTH Aachen, and Carl Schenck AG, Darmstadt). Bayer AG, Silicones Business Unit, No. AI 12666d+e, Leverkusen, 1997
7. Bayer AG and Carl Schenck AG: Active ERF-Vibrationdamper (a joint development by Carl Schenck AG and Bayer AG). Bayer AG, Silicones Business Unit, No. AI 12668d+e, Leverkusen/Darmstadt, 1998
8. Bird, B., Armstrong, R., Hassager, O.: Dynamics of Polymeric Liquids. J. Wiley and Sons, New York, 1987
9. Bonnecaze, R., Brady, J.: Dynamic Simulation of an Electrorheological Fluid. J. Chem. Phys. **96** (1992), 2183–2202
10. Bonnecaze, R., Brady, J.: Yield Stresses in Electrorheological Fluids. J. Rheol. **38** (1992), 73–115
11. Butz, T., von Stryk, O.: Modelling and Simulation of Rheological Fluid Devices. Preprint SFB-438-9911, Sonderforschungsbereich 438, Technische Universität München – Universität Augsburg, 1999
12. Duvaut, G., Lions, J.: Inequalities in Mechanics and Physics. Springer, Berlin-Heidelberg-New York, 1976
13. Ehrgott, R. C., Masri, S. F.: Modelling the Oscillatory Dynamic Behaviour of Electrorheological Materials in Shear. Smart Mat. Struct. **1** (1992), 275–285

14. Engelmann, B., Hiptmair, R., Hoppe, R.H.W., Mazurkevitch, G.: Numerical Simulation of Electrorheological Fluids Based on an Extended Bingham Model. Computing and Visualization in Science, submitted 1999
15. Filisko, F.: Overview of ER Technology. In: Progress in ER Technology, Havelka, K., ed., Plenum Press, New York, 1995
16. Gavin, H. P., Hanson, R. D., Filisko, F. E.: Electrorheological Dampers, Part I: Analysis and Design. J. Appl. Mech. **63** (1996), 669–675
17. Gavin, H. P., Hanson, R. D., Filisko, F. E.: Electrorheological Dampers, Part II: Testing and Modeling. J. Appl. Mech. **63** (1996), 676–682
18. Glowinski, R., Le Tallec, P.: Augmented Lagrangian and Operator-Splitting Methods in Nonlinear Mechanics. SIAM Studies in Applied Mathematics **9** (1989)
19. Glowinski, R., Lions, J.L., Trémolières, R.: Numerical Analysis of Variational Inequalities. North-Holland, Amsterdam, 1981
20. Hartsock, D., Nowak, R., Chaundy, G.: ER Fluid Requirements for Automotive Devices. J. Rheol. **35** (1991), 1305–1326
21. Janocha, H., Rech, B., Bölter, R.: Practice-Relevant Aspects of Constructing ER Fluid Actuators. Int. J. Modern Physics B **10** (1996), 3243–3255
22. Kamath, G. M., Hurt, M. K., Wereley, N. M.: Analysis and Testing of Bingham Plastic Behaviour in Semi-Active Electrorheological Fluid Dampers. J. Appl. Mech. **63**c (1996), 676–682
23. Kamath, G. M., Wereley, N. M.: System Identification of ER Fluid Dampers Using a Nonlinear Mechanisms-Based Model. Paper No. SPIE-2717-46, 1996 SPIE Conference on Smart Materials and Structures, 25-29 February, San Diego, CA, 1996
24. Kamath, G. M., Wereley, N. M.: A Nonlinear Viscoelastic-Plastic Model for Electrorheological Fluids. Smart Mat. Struct. **6** (1997), 351–359
25. Kortüm, W., Lugner, P.: Systemdynamik und Regelung von Fahrzeugen. Springer-Verlag, Berlin, 1994
26. Koslik, B., Rill, G., von Stryk, O., Zampieri, D.: Active Suspension Design for a Tractor by Optimal Control Methods. Preprint SFB-438-9801, Sonderforschungsbereich 438, Technische Universität München – Universität Augsburg, 1998
27. McGlamroch, N. H., Gavin, P. G.: Closed Loop Structural Control Using Electrorheological Dampers. Proc. American Control Conference (1995), 4173–4177
28. Levine, W. S. (ed.): The Control Handbook. CRC Press, Boca Raton (1996)
29. Makris, N., Burton, S. A., Taylor, D. P.: Modelling the Response of ER Damper: Phenomenology and Emulation. J. Eng. Mech. (1996), 897–906
30. Mitschke, M.: Dynamik der Kraftfahrzeuge, Band B: Schwingungen. 3rd ed., Springer-Verlag, 1993
31. Powell, J. A.: Modelling the Oscillatory Response of an Electrorheological Fluid. Smart Mat. Struct. **3** (1994), 416–438
32. Rajagopal, K., Wineman, A.: Flow of Electrorheological Materials. Acta Mechanica **91** (1992), 57–75
33. Spencer Jr., B. F., Dyke, S. J., Sain, M. K., Carlson, J. D.: Modeling and Control of Magnetorheological Dampers for Seismic Response Reduction. Smart Mat. Struct. **5** (1996), 565–575
34. Spencer Jr., B. F., Dyke, S. J., Sain, M. K., Carlson, J. D.: Phenomenological Model of a Magnetorheological Damper. J. Engrg. Mech., ASCE. **123** (1997), 230–238

35. Stanway, R., Peel, D. J., Bullough, W. A.: Applications of Electro-Rheological Fluids in Vibration Control: A Survey. Smart Mat. Struct. **5** (1995), 464–482
36. Stoer, J., Bulirsch, R.: Introduction to Numerical Analysis. 2nd ed., Springer-Verlag, Berlin, 1993
37. Wendt, E., Büsing, K.W.: Properties of a New Generation of Non-Abrasive and Water-Free Electrorheological Fluids. Preprint. Bayer AG, Silicones Business Unit, Leverkusen, 1997
38. Whittle, M.: Computer Simulation of an Electrorheological Fluid. J. Non-Newton. Fluid Mech. **37** (1990), 233–263
39. Whittle, M., Atkin, R.J., Bullough, W.: Fluid Dynamic Limitations on the Performance of an Electrorheological Clutch. J. Non-Newton. Fluid Mech. **57** (1995), 61–81

Direct Numerical Simulation of Turbulent Channel Flow of a Viscous Anisotropic Fluid

M. Manhart and R. Friedrich

Lehrstuhl für Fluidmechanik, Technische Universität München, D-85747
Garching, Germany

Dedicated to Professor Karl-Heinz Hoffmann
on the occasion of his 60th birthday

Abstract. Direct Numerical Simulations (DNS) of turbulent channel flow are performed with a viscous anisotropic model to account for the effect of dilute polymer solutions in water. The model is based on the assumption that the polymer molecules behave like rigid elongated rods. In addition it is assumed that the average orientation angle of the molecules is parallel to the instantaneous flow vector. The results of the DNS with two different model parameters show a drag reduction compared to the Newtonian case. The evaluation of the averaged flow variables shows a change in turbulence structure that is in line with published data of turbulent channel or pipe flows of dilute polymer solutions.

1 Introduction

The turbulent drag reduction potential of dilute polymer solutions in water is known for several decades now. The technological importance of drag reduction has drawn much attention on this effect. The way, how the polymer molecules interact with the turbulence structure is subject of intense fundamental research. A number of experimental investigations have shown that the turbulence is not just suppressed, but altered in a specific way. For an overview on this subject of drag reduction by additives, the reader is referred to e.g. Tiedermann [26] or Gyr and Bewersdorff [9]. Recent experimental work has been published e.g. by Pinho and Whitelaw [19], Harder and Tiedermann [10], Wei and Willmarth [27] and Toonder [3] and Toonder et al. [4].

The basic mechanism of drag reduction by polymer additives still remains unclear because of two main reasons. First, measurements can only provide results on a macroscopic and statistical level. Second, there is still no uniquely accepted mathematical model available describing the dynamics of embedded polymer molecules and their interaction with the surrounding fluid. De Gennes [8] and Joseph [12] e.g. explain the drag reduction by elastic effects introduced by the polymers. On the other hand, Landahl [14] concluded in

his theoretical study that the anisotropic stress caused by extended polymeric molecules seems to play an essential role for polymeric drag reduction.
This conclusion is strongly supported by the combined numerical and experimental study of Toonder et al. [4] who used a constitutive model based on
Batchelor's theory of suspensions of elongated particles in a direct numerical
simulation (DNS) of turbulent pipe flow. He showed that his rigid-rod type
model is capable of reproducing most of the effects that occur in a turbulent
pipe flow under drag reducing conditions.

Toonder was one of the first to study drag reduction of turbulent pipe flow
with DNS. Another DNS study with a non-Newtonian constitutive equation
has been performed in a so-called 'minimum channel' by Orlandi [18]. The
problem of DNS is that all relevant turbulent length and time scales have
to be resolved by the computational mesh and the time integration. This
enforces the use of low Reynolds numbers and relatively simple constitutive
models.

The purpose of the present paper is the DNS of turbulent channel flow
with the viscous anisotropic constitutive model used by Toonder. The results of the DNS are used to validate the findings of Toonder and to gain
further insight into the structural changes of turbulence under drag reducing
conditions.

2   Mathematical Model

2.1   Basic Equations

The dynamics of an incompressible fluid consisting of a Newtonian solvent
with a minute amount of added polymers can be decribed by the conservation
of mass and momentum:

$$\nabla \cdot \mathbf{u} = 0, \tag{2.1}$$

$$\rho \frac{D\mathbf{u}}{Dt} = -\nabla p + \nabla \cdot (\tau^N + \tau^{NN}). \tag{2.2}$$

Here, $\mathbf{u}$ is the velocity vector, ρ is the density and p is the pressure. The part
of the stress tensor attributed to the Newtonian solvent is τ^N and τ^{NN} is
the non-Newtonian part of the stress tensor due to the polymeric molecules.
For the Newtonian part of the stress tensor τ^N the following constitutive
equation is generally accepted:

$$\tau^N = 2\mu\mathbf{D}, \tag{2.3}$$

where μ is the dynamic viscosity and $\mathbf{D}$ is the rate-of-strain tensor

$$\mathbf{D} = (\nabla \mathbf{u} + \nabla \mathbf{u}^T)/2. \tag{2.4}$$

For the contribution of the polymeric molecules to the stress tensor τ^{NN}, a
non-Newtonian constitutive relation has to be supplied.

2.2   The Viscous Anisotropic Model

The approach used here starts from the assumption that the effect of polymer molecules on the flow can be modelled by considering macromolecules. Neglecting the effect of elasticity and assuming rigid-rod type molecules, we arrive at a class of constitutive equations, which are believed to capture the essence of the anisotropic stresses that may be caused by stretched polymer molecules. Following the theory of transversely isotropic fluids (TIF, Ericksen [6,7]) the additional stress caused by one single macromolecule according to Fig. 1 can be expressed by the following equation:

$$\tau_{TIF} = 2\mu_0 \mathbf{D}$$
$$+ [\mu_1 + \mu_2(\mathbf{n} \cdot \mathbf{D} \cdot \mathbf{n})]\,\mathbf{nn} \tag{2.5}$$
$$+ 2\mu_3 [\mathbf{n}(\mathbf{D} \cdot \mathbf{n}) + (\mathbf{D} \cdot \mathbf{n})\cdot\mathbf{n}]\,,$$

where $\mathbf{n}$ is the orientation vector of the molecule and $\mu_0 \ldots \mu_3$ are material constants. The term involving μ_1 leads to a stress persisting indefinitely in the absence of a flow and is dropped therefore.

In a suspension of particles, each fluid element contains a large number of particles with varying orientations. The macroscopic stress due to this number of particles is obtained by averaging (2.5):

$$\tau^{NN} = N_P < \tau_{TIF} > \tag{2.6}$$
$$= 2\mu_0 \mathbf{D}$$
$$+\mu_2 \mathbf{D} :< \mathbf{nnnn} >$$
$$+2\mu_3 (< \mathbf{nn} > \cdot\mathbf{D} + \mathbf{D}\cdot < \mathbf{nn} >)\,, \tag{2.7}$$

involving the second moment $< \mathbf{nn} >$ and the fourth moment $< \mathbf{nnnn} >$ of the distribution function of the orientation vector. An equation for the evolution of $\mathbf{n}$ can be found for an isolated particle by:

$$\frac{D\mathbf{n}}{Dt} = \mathbf{\Omega} \cdot \mathbf{n} + \kappa \left[\mathbf{D} \cdot \mathbf{n} - (\mathbf{n} \cdot \mathbf{D} \cdot \mathbf{n})\mathbf{n}\right], \tag{2.8}$$

where Ω is the vorticity tensor. The material constant κ indicates the shape of the macromolecule. If $\kappa = 0$, the particles have a spherical shape and rotate with the average rotation rate of the fluid as indicated by the term $\mathbf{\Omega} \cdot \mathbf{n}$. If $\kappa \to 1$, the particles are slender rods, which will be considered in the present paper.

By pre- and post-multiplying (2.8) by $\mathbf{n}$ and averaging, we arrive at a transport equation for the second moment $< \mathbf{nn} >$ with $(\kappa = 1)$

$$\frac{D< \mathbf{nn} >}{Dt} = \mathbf{L}\cdot < \mathbf{nn} > + < \mathbf{nn} > \cdot\mathbf{L^T} - 2\mathbf{D} :< \mathbf{nnnn} >, \tag{2.9}$$

where $\mathbf{L} = (\nabla \mathbf{u})^T$ is the velocity gradient tensor. As in (2.6) here again, the fourth moment $< \mathbf{nnnn} >$ appears, which leads to a closure problem. One way to solve it, is to consider the orientation distribution function $\Psi(\mathbf{n}, t)$ which describes the fraction of particles in a fluid element which have orientation vectors that reside around $\mathbf{n}$ at time t. Possibilities to solve for the orientation distribution function can be found in Bird et al. [2] or Szeri and Leal [23], [24]. Another possibility to solve the closure problem for the fourth moment is to apply a suitable closure approximation, i.e. to express it in terms of the second moment. In the present paper we are using

$$< \mathbf{nnnn} > = < \mathbf{nn} > < \mathbf{nn} >, \qquad (2.10)$$

which preserves the symmetry and the trace of the fourth moment. In the case of a turbulent channel flow, the solution of (2.1), (2.2) and (2.9) on a sufficiently fine grid for a large number of time steps would require an enormous amount of computational resources. Therefore, we follow in this paper the approach of Toonder [3] and Toonder et al. [4] and further simplify the expression for $< \mathbf{nn} >$. This simplification is based on the observation that large-aspect-ratio fibres ($\kappa \to 1$) often align quickly with the flow direction:

$$< \mathbf{nn} > = \frac{\mathbf{uu}}{\mathbf{u} \cdot \mathbf{u}}. \qquad (2.11)$$

In fact, it can be shown that alignment with the flow direction is a steady solution of (2.8), if $\kappa = 1$. Equation (2.7) can then be solved using (2.11) and (2.10):

$$\begin{aligned}
\tau^{NN} = {} & 2\mu_0 \mathbf{D} + \mu_2 \mathbf{D} : \frac{< \mathbf{uuuu} >}{(\mathbf{u} \cdot \mathbf{u})^2} \\
& + 2\mu_3 \left(\frac{< \mathbf{uu} >}{\mathbf{u} \cdot \mathbf{u}} \cdot \mathbf{D} + \mathbf{D} \cdot \frac{< \mathbf{uu} >}{\mathbf{u} \cdot \mathbf{u}} \right).
\end{aligned} \qquad (2.12)$$

The remaining task is the determination of the material constants μ_i in (2.7). This is done by considering the behaviour of the stresses (2.7) in special flow situations like pure shear flow and uniaxial as well as biaxial extensional flow. These stresses can then be compared with measured non-Newtonian values for shear- and extensional viscosities for the considered fluid. In dilute solutions of polymer molecules in water, the shear viscosity η remains nearly at the Newtonian value. Equation (2.12) gives for pure shear flow a shear viscosity of

$$\eta = \mu + \mu_0 + \mu_3. \qquad (2.13)$$

Therefore, we choose $\mu_0 = 0$ and $\mu_3 = 0$ in order to keep the shear viscosity at the Newtonian value $\eta = \mu$. The final form of our model then is:

$$\tau^{NN} = \mu_2 \mathbf{D} : \frac{< \mathbf{uuuu} >}{(\mathbf{u} \cdot \mathbf{u})^2}. \qquad (2.14)$$

This model is called a viscous anisotropic model by Toonder [3] and Toonder et al. [4], because it diplays purely viscous and neglects elastic effects. The only material constant that remains to be determined in this model is μ_2. Together with the Newtonian dynamic viscosity μ it gives a uniaxial extensional viscosity of

$$\eta_E = 3\mu + \mu_2. \qquad (2.15)$$

From experiments we know, that in dilute polymer solutions this extensional viscosity can be increased by several orders of magnitude compared to the Newtonian value (see e.g. Gyr and Bewersdorff [9]). In the simulations presented here, we choose two values, i.e. $\mu_2 = 18\mu$ and $\mu_2 = 36\mu$, respectively. According to (2.15) the uniaxial extensional viscosity equals $\eta_E = 21\mu$ and $\eta_E = 39\mu$. This is smaller than the values reported by e.g. Metzner [16], but stability requirements of the numerical scheme prohibit the use of much larger values.

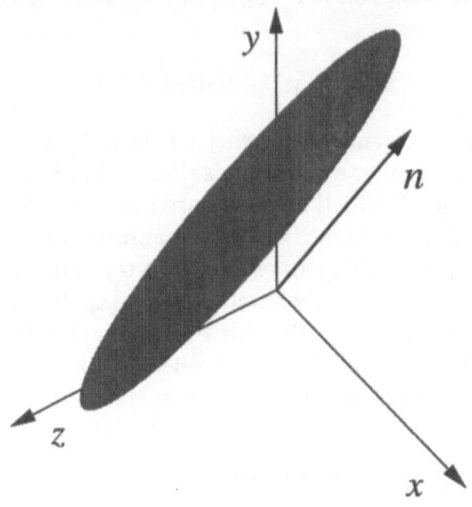

Fig. 1. Configuration of a rigid-rod-type macromolecule

3   Numerical Method

3.1   The Code MGLET

MGLET, is based on a finite volume formulation of the Navier-Stokes equations on a staggered Cartesian non-equidistant grid. The spatial discretization is of second order (central) for the convective and diffusive terms. For the time advancement of the momentum equations an explicit second-order time step (leapfrog with time-lagged diffusion term) is used, i.e.:

$$\mathbf{u}^{n+1} = \mathbf{u}^{n-1} + 2\Delta t \left[C\left(\mathbf{u}^n\right) + D\left(\mathbf{u}^{n-1}\right) - G\left(p^{n+1}\right) \right] \qquad (3.1)$$

where C, D and G represent the discrete convection, diffusion and gradient operators, respectively.

The pressure at the new time level p^{n+1} is evaluated by solving the Poisson equation

$$\nabla \cdot G\left(p^{n+1}\right) = \frac{1}{2\Delta t} \nabla \cdot \mathbf{u}^* \qquad (3.2)$$

where $\mathbf{u}^*$ is an intermediate velocity field, calculated by omitting the pressure term in equation (3.1). By applying the velocity correction

$$\mathbf{u}^{n+1} = \mathbf{u}^* - 2\Delta t G\left(p^{n+1}\right) \qquad (3.3)$$

we arrive at the divergence-free velocity field u^{n+1} at the new time level.

The solution of the Poisson equation is done either in a direct way or iteratively. The direct solver, which has been used for the calculations presented in this paper, uses FFTs in homogeneous streamwise and spanwise directions resulting in independent tridiagonal systems in the inhomogeneous wall normal direction for each wavenumber combination (k_x, k_y), (see Schmitt [22]). The code has been tested together with two different other 2nd order finite volume codes (Manhart et al. [15]) and selected, because it was the most efficient for the DNS of turbulent channel flow.

3.2   Parameters of the Simulations

The flow considered here is a fully turbulent flow between two parallel plates that extend to infinity in the streamwise x- and spanwise y-directions which therefore can be considered as homogeneous directions. The Reynolds number based on the bulk velocity $u_b = \int_{2h} <u>(z)dz$ ($<.>$ denotes spatial and temporal averaging) and the channel half width is $Re_h = 2817$, based on the friction velocity, it is $Re_\tau = 180$. It is therefore comparable to the DNS of Kim et al. [13] which has been performed by a spectral scheme and is generally accepted as a reference case in the literature. As in the DNS of Kim et al., we used $N_x, N_y, N_z = 192, 160, 128$ grid points in streamwise, spanwise and wall-normal direction, respectively. The grid is non-equidistant in wall-normal direction in order to achieve a finer resolution near the wall. This leads to a resolution in wall units of $\Delta x^+, \Delta y^+, \Delta z_{min}^+ = 9.0, 6.75, 1.35$. The time step was $\Delta t \cdot u_b/h = 0.005$, which corresponds to $\Delta t^+ = 3.2 \cdot 10^{-4}$ in wall units. The simulations have been performed on a Fujitsu VPP/700 using one processor. The CPU-time spent during one time step was about 7.5 seconds.

At the wall, no-slip boundary conditions are used and in the homogeneous directions periodic boundary conditions are applied. The flow is driven by a constant pressure gradient in streamwise direction which balances the flow resistance due to the wall shear stress. It is adjusted in such a way that the flow rate through the channel is $u_b = 1.0$ in the Newtonian case. We have run three different simulations: one Newtonian case ($\mu_2 = 0$), one with $\mu_2 = 18\mu$ and one with $\mu_2 = 36\mu$, respectively. We started our simulations from a coarse grid preliminary solution (Newtonian) and waited about $t \cdot u_b/h = 250$ time units ($N_t = 50000$ time steps) until a statistically stationary state has been developed. This has been detected by looking at the bulk velocity which reached its equilibrium value after this time.

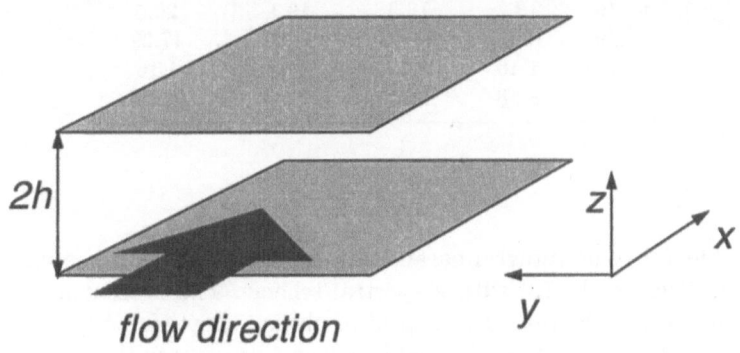

Fig. 2. Geometry of the channel flow simulation

4   Results

4.1   Integral Parameters

In the initial coarse grid simulation, the flow rate (bulk velocity) was about 4% too low. For all three parameters μ_2, it takes about $200 - 300t \cdot u_b/h$ until the final flow rate is approximately reached (see Fig. 3). After that transient period only small oscillations around the long term average occur, that are due to turbulent fluctuations. Note, that the pressure gradient has been adjusted using an empirical formula in order to get a flow rate of unity in the Newtonian case. Starting averaging at $t \cdot u_b/h = 250$, the long term averaged values of u_b are listed in table 4.1. The viscous anisotropic model leads to a drag reduction $DR = (1 - u_{bN}/u_{bP}) \cdot 100\%$ of 6.2% and 9.6% if $\mu_2 = 18\mu$ and $\mu_2 = 36\mu$, respectively. This result is consistent with the findings reported by Toonder et al. [4] where the viscous anisotropic model was applied in a turbulent pipe flow.

Table 4.1. Long term averaged bulk velocity u_b

	$\mu_2 = 0$	$\mu_2 = 18\mu$	$\mu_2 = 36\mu$
u_b	1.0008	1.066	1.106
DR	0.0%	6.2%	9.6%

Table 4.2. Integral parameters compared with the results of Kim et al. [13]

	Kim et al.	$\mu_2 = 0$	$\mu_2 = 18\mu$	$\mu_2 = 36\mu$
u_{cl}/u_τ	18.2	18.3	19.1	20.0
u_b/u_τ	15.63	15.63	16.66	17.29
u_{cl}/u_b	1.16	1.17	1.14	1.16
$c_f \cdot 10^3$	8.18	8.17	7.20	6.69

In table 4.2 some integral parameters are compared with the values obtained by Kim et al. [13] with a spectral scheme. The agreement with our Newtonian case indicates that the grid resolution is satisfactory in our case. One can see that with increasing model parameter μ_2 both the centerline velocity u_{cl} as well as the bulk velocity u_b are increased at nearly the same rate. This implies that the main characteristics of the averaged velocity profile remain untouched during drag reduction. The increase of u_b/u_τ is equivalent to the decrease of the skin friction coefficient $c_f = \tau_w/\frac{1}{2}\rho u_b^2$, i.e. the drag reduction.

4.2   Statistical Evaluation

Averaged velocity profiles. The averaged velocity profiles $< u >$ are obtained by averaging in the homogeneous directions and in time, starting at $t \cdot u_b/h = 250$. In Fig. 4, $< u >$ is plotted for both channel half widths. Although there are slight asymmetries, the main trend can be clearly seen. An increase in μ_2 leads to an overall increase of $< u >$ except in the near wall layer. This effect can be better illustrated in a semi-logarithmic plot of the velocity profile in inner coordinates, i.e. normalized by the wall shear stress. The inner coordinates are defined by

$$u^+ = u/u_\tau \quad \text{and} \quad z^+ = z \cdot u_\tau/\nu, \tag{4.1}$$

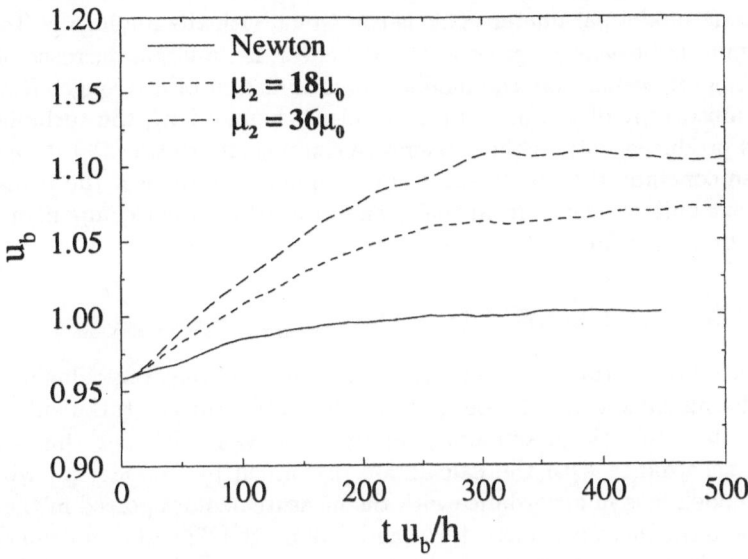

Fig. 3. Temporal development of u_b

where $u_\tau = \sqrt{(\tau_w/\rho)}$. For Newtonian turbulent wall bounded flows, the near wall behaviour is universal and can be divided into a linear layer and a logarithmic layer (e.g. Tennekes and Lumley [25]):

$$u^+ = z^+ \quad \text{for} \quad z^+ < 5, \tag{4.2}$$

$$u^+ = A \cdot ln(z^+) + B \quad \text{for} \quad z^+ > 30. \tag{4.3}$$

Kim et al. [13] estimated for the case of a low Reynolds number turbulent channel flow (our flow case) the constants with $A = 2.5$ and $B = 5.5$. The effect of the viscous anisotropic model starts at $z^+ = 6$ (Fig. 5), i.e. beyond the viscous sublayer. The profiles are shifted to higher values of B in the log layer, whereas the slope remains nearly unchanged. This is consistent with many observations made in the literature for dilute polymer solutions (e.g. Harder and Tiedermann [10]; Wei and Willmarth [27]). Moreover, the results are fully consistent with those of Toonder [3].

Turbulence intensities. The root-mean-square (rms) profiles of the three velocity components can be used as a first indicator of how the turbulence structure is changed. The streamwise fluctuations u_{rms} are increased by the model throughout the whole channel (Fig. 6) and the peak is shifted away from the wall. With increasing model parameter μ_2 this effect is more pronounced. The spanwise v_{rms} and the vertical w_{rms} are both decreased with increasing model parameter. This is consistent with the findings of Toonder [3]. Again, increasing μ_2 gives a stronger effect, i.e. a larger decrease of v_{rms} and w_{rms}. It seems that the model acts in direction of a stronger Reynolds stress anisotropy of the flow. In a turbulent channel flow, the turbulent energy is produced only in the streamwise component (Rotta [21]). From this, one can conclude that the viscous anisotropic model reduces the transfer of turbulent kinetic energy from the streamwise to the other components. The production term for $< u'^2 >$ is defined as

$$P = -2 < u'w' > \frac{\partial < u >}{\partial z}. \tag{4.4}$$

We plotted this term in Fig. 9 for the Newtonian and the drag reducing cases. The viscous anisotropic model reduces the production of turbulent kinetic energy and shifts the maximum away from the wall. Although this result is not in accordance with the calculation performed by Toonder [3] with the same model, it is in accordance with the measurements reported in the same paper and the measurements of Wei and Willmarth [27] and complements the physical picture. It can be concluded that, when the production of streamwise fluctuation is reduced, while the level of these fluctuations is increased, the transfer of energy to the other velocity components must be reduced.

Stresses. The basic mechanism of the model is the introduction of additional anisotropic stresses into the flow. In a fully developed turbulent channel flow, the pressure gradient $\partial < p > /\partial x$ balances the gradient of the total shear stress:

$$\frac{1}{\rho} \frac{\partial < p >}{\partial x} = \frac{\partial}{\partial z} \tau_{13}. \tag{4.5}$$

The total shear stress τ_{13} is the sum of the turbulent part ($\tau_{13}^T = -\rho < u'w' >$), the Newtonian viscous part ($\tau_{13}^N = \mu \partial < u > /\partial z$) and the non-Newtonian viscous part (τ_{13}^{NN}). We evaluated the three stress contributions and plotted them in Fig. 10. Due to the presence of the model, the magnitude of the turbulent shear stress $- < u'w' >$ is reduced and the maximum is shifted away from the wall. The Newtonian part of the viscous shear stress is increased due to the larger gradient of the average velocity profile in the buffer domain which is connected with the larger center line velocity (compare Fig. 4). The overall contribution of the polymeric part of the shear stress is fairly small. In the literature this part is called "Reynolds stress deficit" denoting the deficit in the stress balance (4.5) if one considers only the turbulent and the Newtonian stress. A significant Reynolds stress deficit has been found e.g. in the experiments of Wei and Willmarth [27] where the drag reduction was between 30% and 40%. Toonder et al. [4] also found a stress deficit in their experiments, while in the simulations the observed stress deficit was very small. In Fig. 11 the non-Newtonian part of the stress tensor is plotted in inner coordinates. It can be seen that with increasing model parameter, the maximum of the polymeric stress is increasing. We could speculate that higher model parameters would further increase τ_{13}^{NN} until values comparable to the experiment could be reached. It may be also of interest to note that the effect of the non-Newtonian stress is highest where most of the turbulent kinetic energy is produced, namely in the buffer layer. This also happens to be the zone where coherent (streaky) structures are observed.

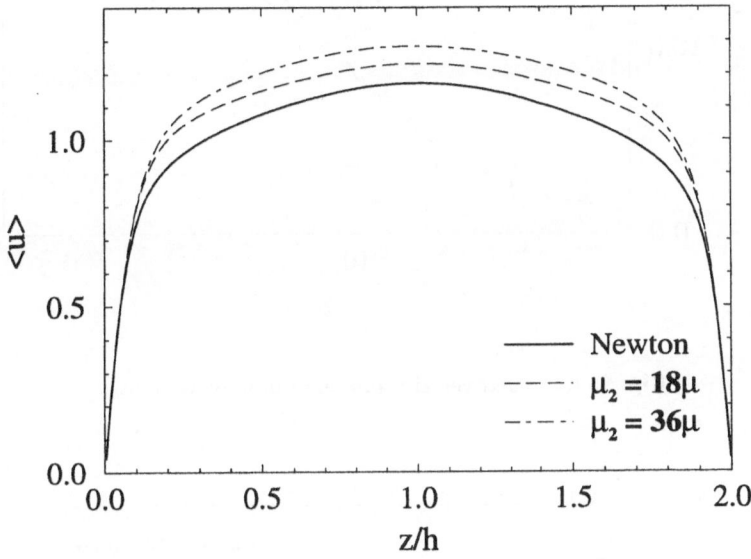

Fig. 4. Averaged velocity profile across the channel

4.3   Instantaneous Velocity Fields

In order to estimate the influence of the viscous anisotropic model on the turbulent large scale structures, we are looking at instantaneous flow fields. In Figs. 12-14 we have plotted contour lines of instantaneous spanwise velocity components for all three runs. In the Newtonian case, we can see large scale structures emanating from the wall and reaching the middle of the channel at an average angle of 45° to the x-axis. These structures are typical for wall-bounded flows and reflect the existence of coherent quasi-vortices in streamwise direction (see e.g. Moin and Moser [17]; Robinson [20]). If the viscous anisotropic model is switched on (Fig. 13), the structures become larger and move away from the wall. This explains why the maxima of the turbulent intensities (rms-values) are moving away from the wall. In addition

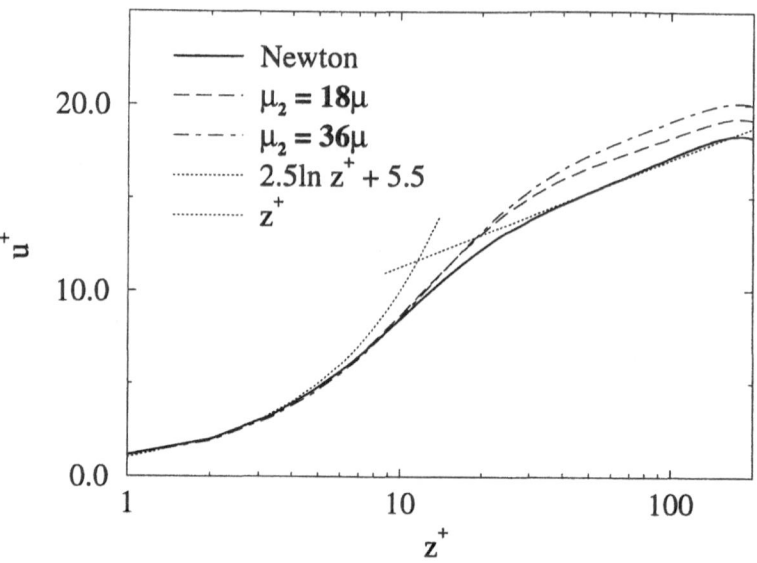

Fig. 5. Averaged velocity profile in inner coordinates

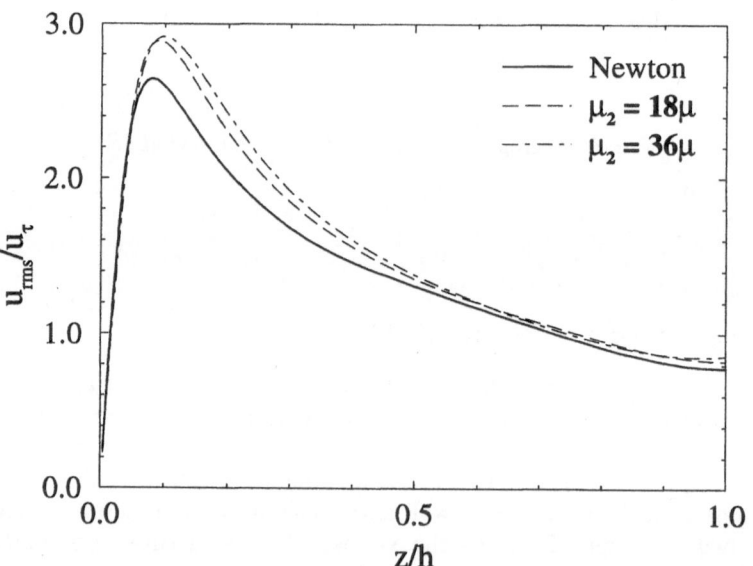

Fig. 6. RMS of streamwise velocity fluctuations

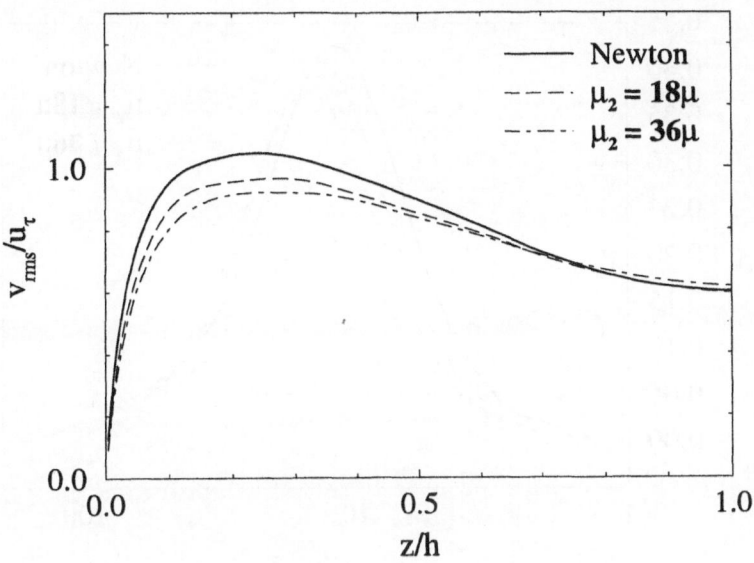

Fig. 7. RMS of spanwise velocity fluctuations

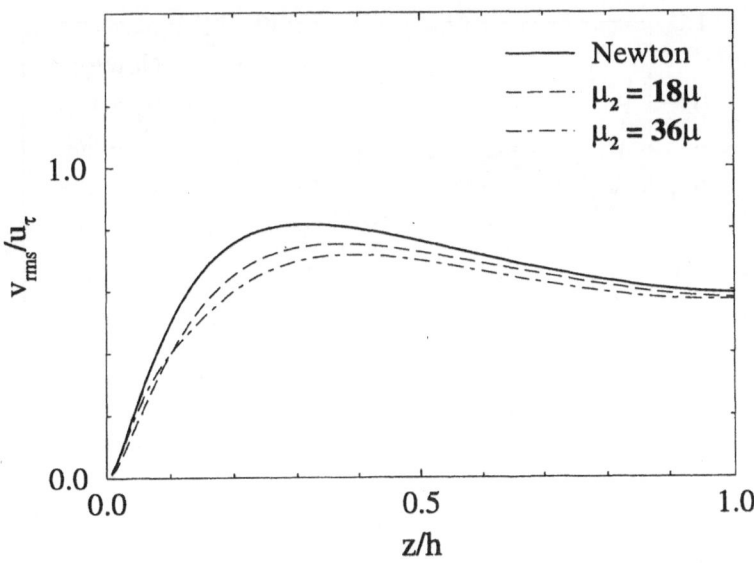

Fig. 8. RMS of wall-normal velocity fluctuations

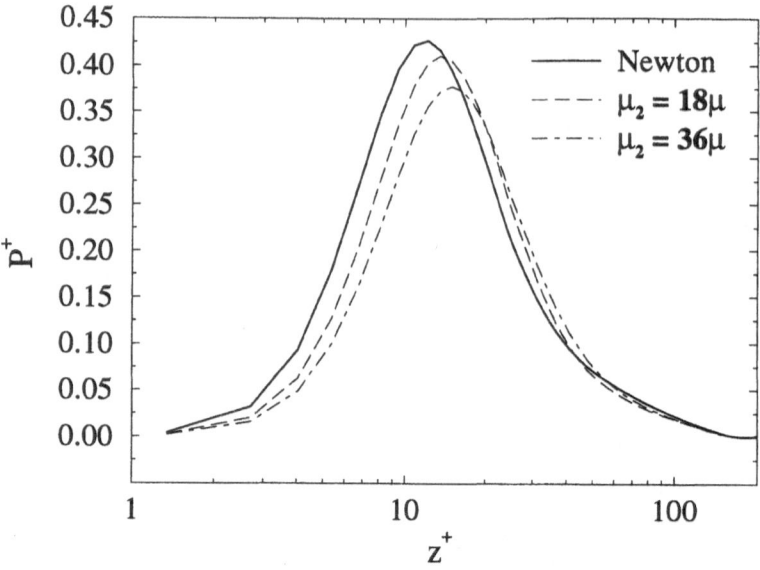

Fig. 9. Production of turbulent kinetic energy

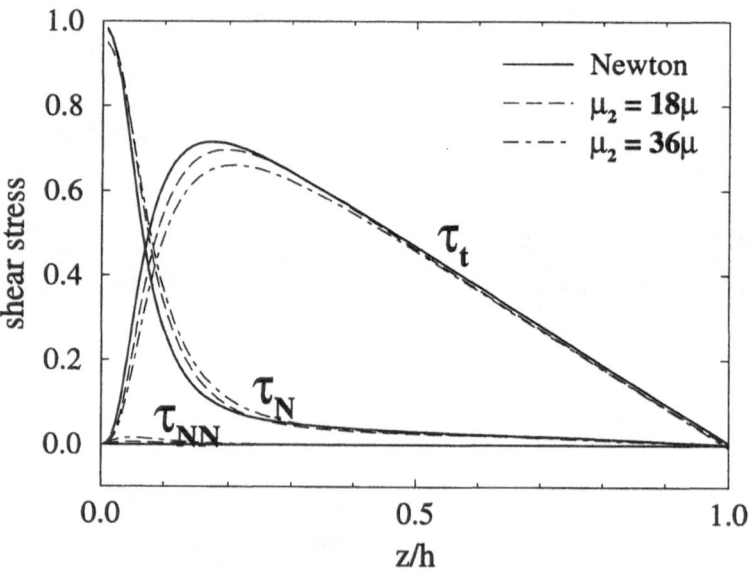

Fig. 10. Components of shear stress τ_{13}

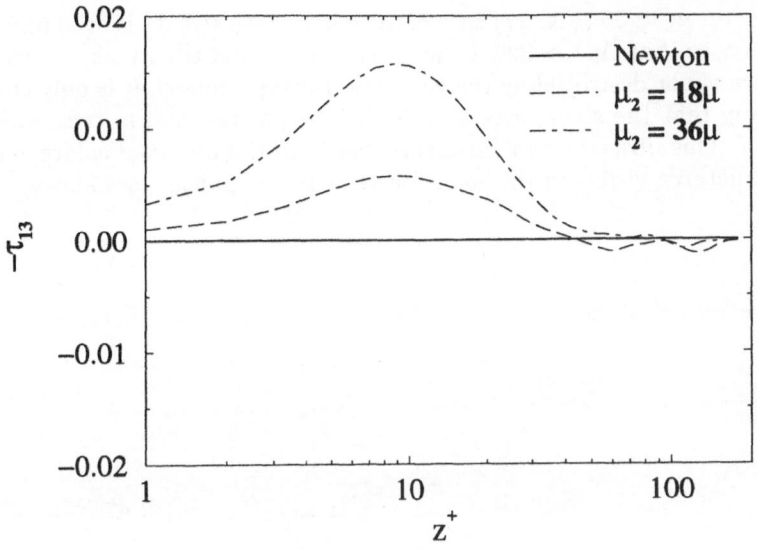

Fig. 11. Averaged polymeric shear stress τ_{13}^{NN}

it seems that the orientation angle of the structures is reduced. Increasing the model parameter to $\mu_2 = 36$ (Fig. 14) results in a further modification of the structures. First, the occurence of structures is reduced and second, the orientation angle is further changed. The angle of 45° can hardly be observed any more.

The dominant feature of wall-bounded shear flows is the occurrence of streaky structures in the streamwise velocity fluctuations at a certain distance from the wall. It is generally believed that these structures play a crucial role in the recreation cycle of wall turbulence (Jiménez and Pinelli [11]). Achia and Thompson [1] found in an experimental study that the spanwise spacing of the streaks is increased during drag-reducing conditions. We made the streaks visible by plotting isolines of the u-fluctuations in a wall-parallel plane at $z^+ = 13$ (see Figs. 15-17). In the Newtonian case (Fig. 15), elongated structures with periodically changing sign in spanwise (y-) direction can clearly be identified as streaks. The average number of streaks (maxima or minima) in spanwise direction is 10, corresponding to a streak spacing of $\Delta y^+ \approx 100$ wall units, which is in line with observations made in the literature (e.g. Eggels et al. [5]). If the model is switched on to $\mu_2 = 18$ (Fig. 16), the average streak spacing is increased to about $\Delta y^+ \approx 180$ (6 structures). In addition, the streamwise coherence of these structures is increased to more than the length of the computational box ($\Delta x^+ = 1728$). In the higher model parameter $\mu_2 = 36$ (Fig. 17) the enlargement of the streaks is even more pronounced, so that $\Delta y^+ \approx 220$. It can be concluded that the streaky structure of the flow is not destroyed by the viscous anisotropic model. It is only changed in a way that the structures are enlarged in spanwise and in streamwise direction. This observation is in accordance with the observed enlargement of the structures visible in the spanwise velocity component (see above).

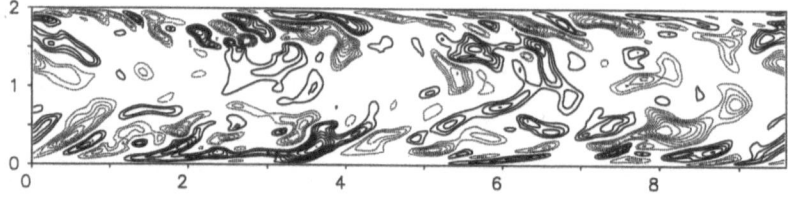

Fig. 12. Instantaneous spanwise velocity component for the Newtonian case. ——— : positive; - - - : negative

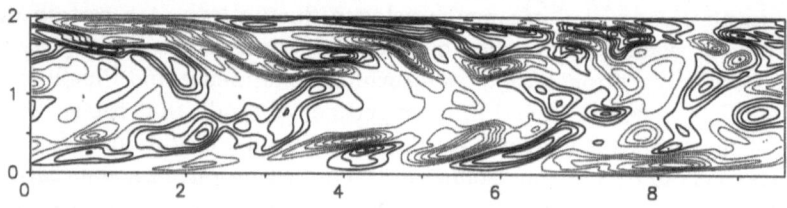

Fig. 13. Instantaneous spanwise velocity component for $\mu_2 = 18\mu$. ——— : positive; - - - : negative

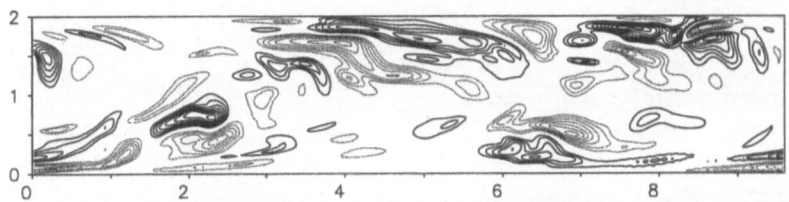

Fig. 14. Instantaneous spanwise velocity component for $\mu_2 = 36\mu$. ——— : positive;
- - - : negative

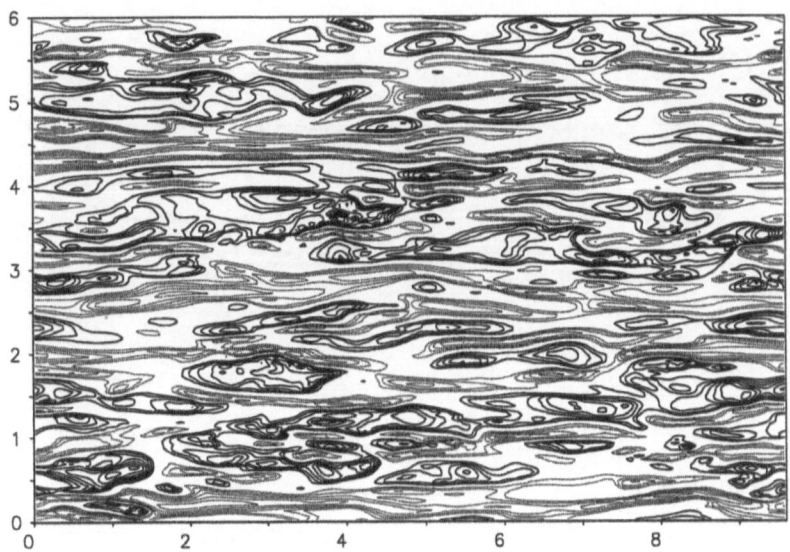

Fig. 15. Instantaneous streamwise velocity fluctuations for the Newtonian case.
——— : positive; - - - : negative

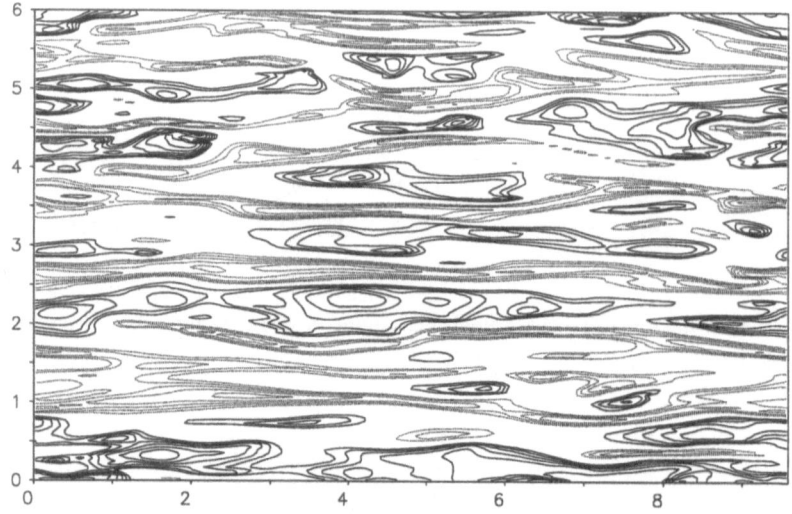

Fig. 16. Instantaneous streamwise velocity fluctuations for $\mu_2 = 18\mu$. ———— : positive; - - - : negative

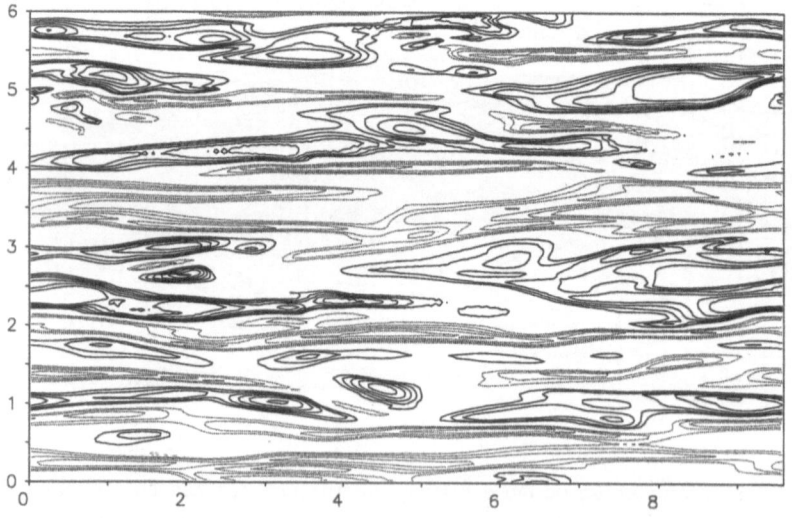

Fig. 17. Instantaneous streamwise velocity fluctuations for $\mu_2 = 36\mu$. ———— : positive; - - - : negative

5 Conclusions

We have presented direct numerical simulations of turbulent channel flow with a viscous anisotropic model for the effect of polymer molecules on the stress tensor in turbulent wall-bounded flow. The model assumes the polymer chains as rigid rods with an elongated shape. The average orientation angle of these rods is assumed to be in line with the instantaneous flow vector. Two different model parameters have been evaluated and compared with Newtonian turbulent channel flow. The model leads to an increase of the flow rate and therefore to a drag reduction compared to the Newtonian case of about 10%. Although this reduction is not as high as it has been observed in experiments, it is associated with a change of the turbulence structure that is in line with measurements in turbulent channel and pipe flows with dilute solutions of polymer in water. The streamwise fluctuations are enhanced at the same time as the spanwise and wall-normal fluctuations are attenuated. The production of turbulent kinetic energy and the Reynolds shear stress are decreased. The observed stress deficit, i.e. the non-Newtonian shear stress, is weak but it is increasing with increasing model parameter. The observable instantaneous turbulent structures are drastically enlarged by the action of the model. The streak spacing is increased from $\Delta y^+ \approx 100$ in the Newtonian case to $\Delta y^+ \approx 220$ in the case with the larger model parameter.

Notes and Comments. We gratefully acknowledge the support of the HLRS in Stuttgart and the LRZ in Munich. The work has been supported by the Deutsche Forschungsgemeinschaft in the framework of the Munich Collaborative Research Center, SFB 438.

References

1. Achia, B. U, Thompson, D. W.: Structure of the Turbulent Boundary Layer in Drag Reducing Pipe Flow. J. Fluid Mech. **81** (1977), 439–464
2. Bird, R. B., Armstrong, R. C., Hassager, O.: Dynamics of Polymeric Liquids, Vol. 1, Fluid Mechanics. John Wiley and Sons, 2nd edition , 1987
3. den Toonder, J. M. J.: Drag Reduction by Polymer Additives in a Turbulent Pipe Flow: Laboratory and Numerical Experiments. PhD thesis, Technische Universiteit Delft, Netherlands, 1995
4. den Toonder, J. M. J., Hulsen, M. A., Kuiken, G. D. C., Nieuwstadt, F. T. M.: Drag Reduction by Polymer Additives in a Turbulent Pipe Flow: Numeric al and Laboratory Experiments. J. Fluid Mech. **337** (1997), 193–231
5. Eggels, J. G. M., Unger, F., Weiss, M. H., Westerweel, J., Adrian, R. J., R. Friedrich, R., Nieuwstadt, F. T. M.: Fully Developed Turbulent Pipe Flow: A Comparison between Direct Numerical Simulation and Experiment. J. Fluid Mech. **268** (1994), 175–209
6. Ericksen, J. L.: Anisotropic fluids. Arch. Ration. Mech. Anal. **4** (1960), 231–237
7. Ericksen, J. L.: Transversely Isotropic Fluids. Kolloid-Z. **173** (1960), 117–122

8. de Gennes, P. G.: Introduction to Polymer Dynamics. Cambridge University Press, 1990

9. Gyr, A., Bewersdorff, H.-W.: Drag Reduction of turbulent flows by additives, volume 32 of Fluid Mechanics and its Applications. Kluwer Academic Publishers, Dordrecht, 1995

10. Harder, K. J., Tiedermann, W. G.: Drag Reduction and Turbulent Structure in Two-Dimensional Channel Flows. Phil. Trans. R. Soc. Lond. A **336** (1991), 19–34

11. Jiménez, J., Pinelli, A.: The Autonomous Cycle of Near-Wall Turbulence. J. Fluid Mech. **389** (1999), 335–359

12. Joseph, D. D.: Fluid Dynamics of Viscoelastic Liquids. Springer, 1990

13. Kim, J., Moin, P., Moser, R.: Turbulence Statistics in Fully Developed Channel Flow at Low Reynolds Number. J. Fluid Mech. **17** (1987), 133–166

14. Landahl, M. T.: Drag Reduction by Polymer Addition. In E. Becker and G.K. Mikhailov, editors, Theoretical and Applied Mechanics, Proc. 13th Intl. Congr. Theor. and Appl. Mech., Springer, 1973, 177–199

15. Manhart, M., Deng, G. B., Hüttl, T. J., Tremblay, F., Segal, A., Friedrich, R., Piquet, J., and Wesseling, P.: The Minimal Turbulent Flow Unit as a Test Case for Three Different Computer Codes. In E.H. Hirschel, editor, Vol. 66, Notes on numerical fluid mechanics. Vieweg-Verlag, Braunschweig, 1998

16. Metzner, A. B., and Metzner, A.P.: Stress Levels in Rapid Extensional Flows of Polymeric Fluids. Rheol. Acta, **9** (1970), 174–181

17. Moin, P., Moser, R.: Characteristic-Eddy Decomposition of Turbulence in a Channel. J. Fluid Mech. **200** (1989), 471–509

18. Orlandi, P.: A Tentative Approach to the Direct Simulation of Drag Reduction by Polymers. J. Non-Newtonian Fluid Mech., **60** (1995), 277–301

19. Pinho, F. T., Whitelaw, J. H.: Flow of Non-Newtonian Fluids in a Pipe. J. Non-Newtonian Fluid Mech., **34** (1990), 129–144

20. Robinson, S. K.: Coherent Motions in the Turbulent Boundary Layer. Annu. Rev. Fluid Mech., **23** (1991), 601–639

21. Rotta, J.: Turbulente Strömungen. Teubner, Stuttgart, 1972

22. Schmitt, L., Richter, K., and Friedrich, R.: Large-Eddy Simulation of Turbulent Boundary Layer and Channel Flow at High Reynolds Number. In U. Schumann and R. Friedrich, editors, Direct and Large Eddy Simulation of Turbulence, Vieweg Braunschweig (1986), 161–176

23. Szeri, A. J., Leal, L. G.: A New Computational Method for the Solution of Flow Problems of Microstructured Fluids. Part 1. Theory. J. Fluid Mech., **242** (1992), 549–576

24. A.J. Szeri, A. J.,Leal, L. G.: A New Computational Method for the Solution of Flow Problems of Microstructured Fluids. Part 2. Inhomogeneous Shear Flow of a Suspension. J. Fluid Mech., **262** (1994), 171–204

25. Tennekes, H., Lumley, J. L.: A First Course in Turbulence. MIT Press, Cambridge, Massachusetts, 1972

26. Tiedermann, W. G.: The Effect of Dilute Polymer Solutions on Viscous Drag and Turbulence Structure. In A. Gyr, editor, Structure of Turbulence and Drag Reduction, IUTAM Symp., Springer (1990), 187–200

27. Wei, T., and Willmarth, W. W.: Modifying Turbulent Structure with Drag-Reducing Polymer Additives in Turbulent Channel Flows. J. Fluid Mech. **245** (1992), 619–641

Numerical Simulation and Experimental Studies of the Fluid-Dynamic Behaviour of Rising Bubbles in Stagnant and Flowing Liquids

A. Lucic[1], F. Meier[2], H.-J. Bungartz[2], F. Mayinger[1], and C. Zenger[2]

[1] Lehrstuhl A für Thermodynamik, Technische Universität München, D-85747 Garching, Germany
[2] Lehrstuhl V für Informatik, Technische Universität München, D-80290 München, Germany

Dedicated to Professor Karl-Heinz Hoffmann
on the occasion of his 60th birthday

Abstract. This paper presents numerical simulations and experimental studies on the fluid-dynamic behaviour of rising bubbles in stagnant and flowing liquids in a vertical straight duct of rectangular cross-section. The bubbles were generated by injecting air through an orifice of the side-wall. Experimental data were obtained for various liquid flow velocities in order to assess the impact of the stream conditions on the bubble characteristics. The bubble size, their rising behaviour as well as the bubble velocity were investigated systematically using high-speed cinematogaphy. The velocity field in the vicinity and the wake of the rising bubbles was measured by means of the Laser Doppler velocimetry to quantify the local structure of the liquid flow in the presence of rising bubbles. According to the experimental configuration and the obtained data numerical calculations have been performed. For these numerical computations, a model for the bubbles was added to a solver for the incompressible Navier-Stokes equations. This model captures the motion of the bubbles, yet omits details of their shape. The results of the experimental study show the complexity of the bubble dynamics in superimposed liquid flow. Experimental data obtained for stagnant liquid compare well with the numerical predictions.

1 Introduction

The generation and movement of bubbles as well as the interaction and transport processes between the gas and liquid phases in forced convection nucleate boiling occur in a great multitude of industrial applications. They can be found in thermal power systems, power and chemical processing plants, as well as in cooling and heating systems. For an efficient and safe operation of these devices, the knowledge of the fluid- and thermo-dynamic behaviour of the bubbles in boiling liquids is of major importance. Compared to single-phase convection, the heat transfer rate is much higher for boiling, and it increases with the amount of bubbles up to a critical temperature which induces the onset of film boiling. This well-known effect is described by the Nukijama curve (see Fig. 1). Houston and Cornwell [10] have shown that the bubble motion and the hereby caused convection also have a significant impact on the heat transfer. On the other hand, the existence of bubbles entails a pressure drop, which impairs the performance of the apparatus.

Despite the technical importance of flow boiling, the available knowledge is still insufficient to enable complete predictions of the boiling phenomena. This is due to the complexity of the interaction of many influencing factors such as the stream conditions and turbulence of the liquid phase, the phase distribution of the gas and of the liquid, and the thermal gradients. In recent years, many analytical models towards the understanding of bubble dynamics of the transport processes in nucleate boiling have been developed. Yet these models consider only partial mechanisms, since the experimental data on which the models are based strongly depend on the boundary conditions and on physical parameters.

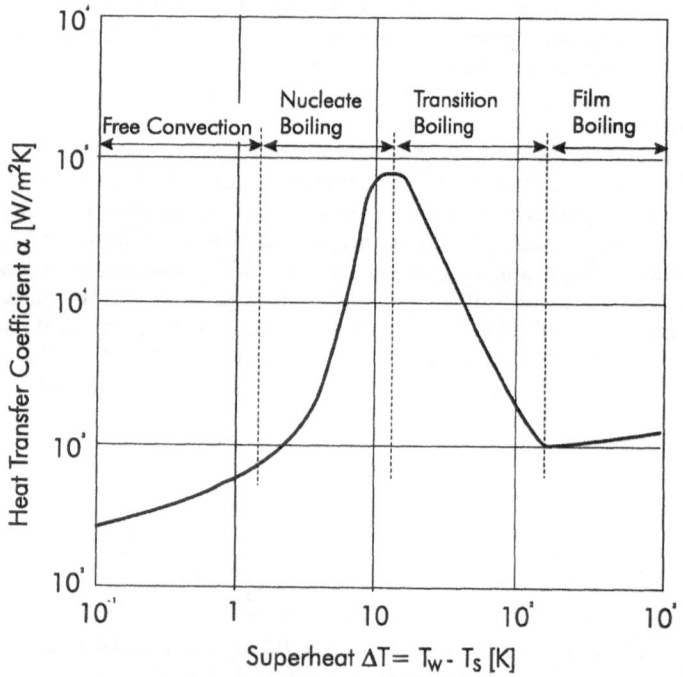

Fig. 1. Nukiyama curve

Therefore, research directed towards determining parameters such as bubble nucleation, growth, detachment, collapse, and hence the induced liquid motion with respect to the heat transfer in superimposed liquid flow are essential for the complete understanding of the transport processes. Since the physical characterization of the fluid- and thermo-dynamic behaviour of vapour bubbles as well as the determination of the transport processes are based on the laws of the bubble dynamics under adiabatic conditions, a thorough investigation of the interaction between the motion of gas bubbles and the liquid flow is imperative.

Therefore, the objective of this study is to investigate the fluid-dynamic behaviour of gas bubbles without thermal influences in stagnant and flowing liquids. Experiments were conducted in a test facility by injecting air through an orifice in a vertical wall of a rectangular cross-section into water. Measurements were taken at different liquid velocities and Reynolds numbers in order to assess the impact of the liquid flow on the fluid-dynamic behaviour. The rising behaviour of the bubbles as well as their rising velocity were measured by means of the high-speed cinematography. In order to gain a better understanding of the fluid-dynamics of the bubbly flow, the liquid motion in the vicinity of the bubbles and the stream conditions were measured implementing the Laser Doppler velocimetry.

The numerical simulations were performed with a CFD code for transient flows of viscous, incompressible, and Newtonian fluids. To the basic code, a model for the influence of the bubbles has been added. The simulations were made in two dimensions of space, in a domain emulating the test section.

2   Experimental Set-up

A schematic diagram of the experimental facility used for the investigation of the bubble dynamics is shown in Fig. 2. Water and a glycerin-water solution are used as the test liquid which is circulated from a storage tank through the loop by means of a pump which delivers a constant volumetric flow of 1.5 m^3/h. The regulation of different flow velocities is achieved by means of a three-way valve dividing the liquid stream into a main flow stream and a bypass stream which is delivered into the storage tank again. A magnetically-inductive flow meter measures the volumetric flow rate of the main stream. A valve is incorporated in the loop before the inlet of the test section to enable measurements in stagnant liquid. The test section is composed of a polished rectangular plexiglass channel that has a 20 mm (width) x 10 mm (depth) cross-section and a length of 120 cm. It is set up vertically, and it is connected by hoses to the loop. The air bubbles at the wall of the test section are generated by means of pressurized air which is fed through an air volumetric controller regulating the adjusted flow rate of the air. It then enters the test section through a nozzle with an orifice diameter of 0.3 mm. The nozzle is located at a distance of 65 cm above the test section inlet which is large enough to ensures a fully developed channel flow. Therefore, stabilized flow conditions during the measurements are assumed.

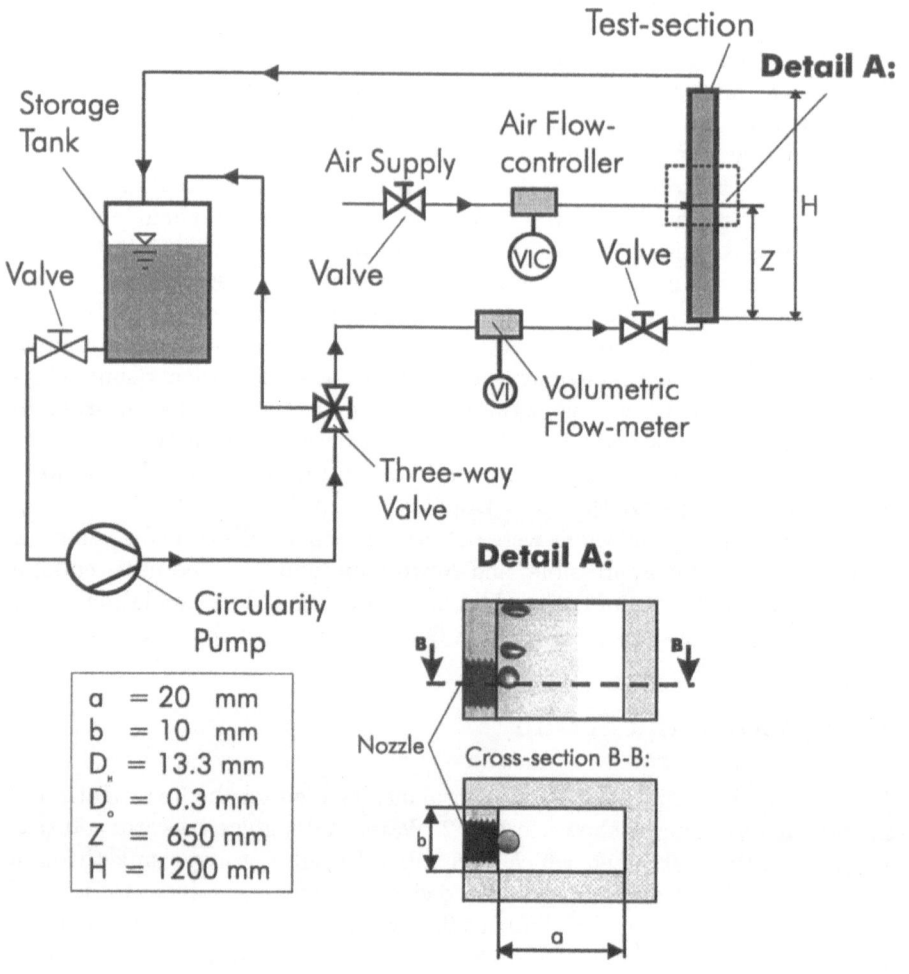

Fig. 2. Experimental set-up

3 Experimental Procedure and Results

In the following, we present measurements taken for pure water at various liquid velocities and at a constant air flow rate of 80 ml/min. The Reynolds numbers which are based on the mean velocities and the hydraulic diameter range from approximately 6000 up to 16000.

3.1   High-Speed Cinematography

Visual observations of the size and rising behaviour of the bubbles were performed using a high-speed motion camera. Images were recorded at rates up to 4500 frames per second for various liquid velocities while the air flow rate was kept constant. The camera was positioned such that a section of approximately 60 mm distance upstream from the orifice could be captured. The recorded images were then stored in the computer and evaluated by means of an image analysis software. The measured quantities were the size of the bubbles and the rising velocity.

Experiments Performed with Water in the Stagnant State and in the Turbulent Flow Regime.

The gained experimental results show the impact of the liquid velocity on the fluid-dynamic behaviour of the rising bubbles. Figure 3 allows a comparison of two recorded images of rising bubbles in stagnant and moving water.

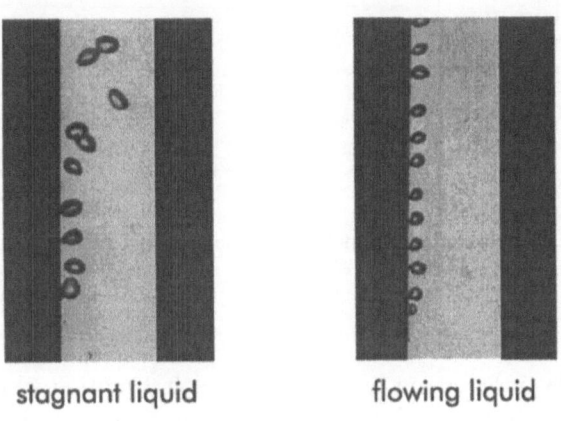

stagnant liquid   flowing liquid

Fig. 3. Sequences of rising bubbles in stagnant and flowing water

As it can be seen from this sequence, there is a significant difference between the stagnant and the moving water with respect to the bubble size and shape and the rising behaviour of the bubbles. Bubbles moving in stagnant water are bigger than those in moving water, and they are therefore subject to stronger deformations of their shape. After detaching from the orifice, the bubble flattens and resembles an oblate ellipsoidal which fluctuates while facing the biggest area in direction of the bubble's movement. The bubbles in the flowing water are more spherical yet still adapt a ellipsoidal shape. This results from the faster detachment of the bubble from the orifice caused by the superimposed liquid velocity which entails smaller bubbles. The surface tension force is higher for smaller bubbles which accounts for the spherical shape. The size decreases with increasing bulk velocity as it is represented in Fig. 4, which shows the equivalent bubble diameter after the detachment at various liquid velocities. The bubble diameter of a bubble in stagnant water reaches a value of approximately 4 mm, whereas the bubble diameters in superimposed water flow range from about 2 mm up to 3 mm.

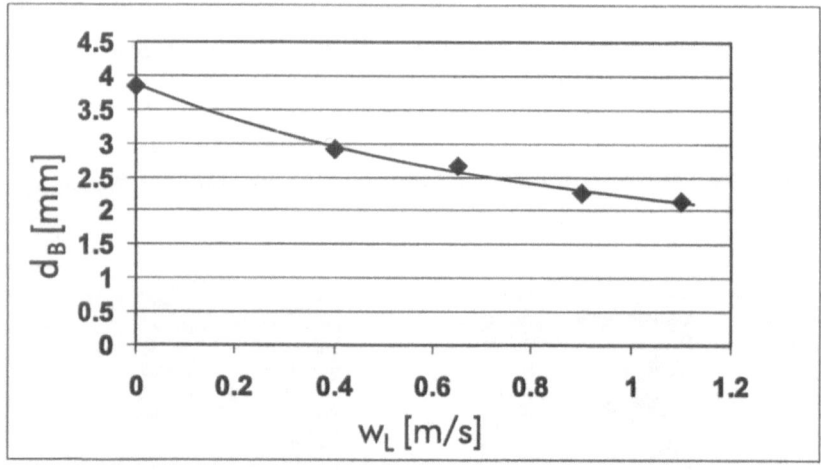

Fig. 4. Equivalent diameter at various liquid bulk velocities

The bubbles in stagnant water rise in a spiral motion which is found to result from the size of the bubbles [16]. Furthermore, the turbulence in the wake of a preceding bubble influencing the trajectory of the following bubble [14] also gives rise to a spiral motion. Bubbles in flow seem to have an almost rectilinear motion with slight deviations along the channel wall which is due to the smaller diameter and therefore more spherical shape [2] as well as the hydro-dynamic forces of the liquid flow on the bubbles.

Figure 5 shows the absolute rising velocity of a bubble at various liquid velocities versus time. The rising velocity is defined as the vertical distance passed by a bubble per unit time.

The development of the rising velocity of the bubbles for the various liquid velocities shows qualitatively the same behaviour. The bubbles are accelerated after their detachment to a certain point and then rise with an approximately constant velocity. The rising velocity depends upon the size and the shape of the bubble, the position of the bubble in the flow field,

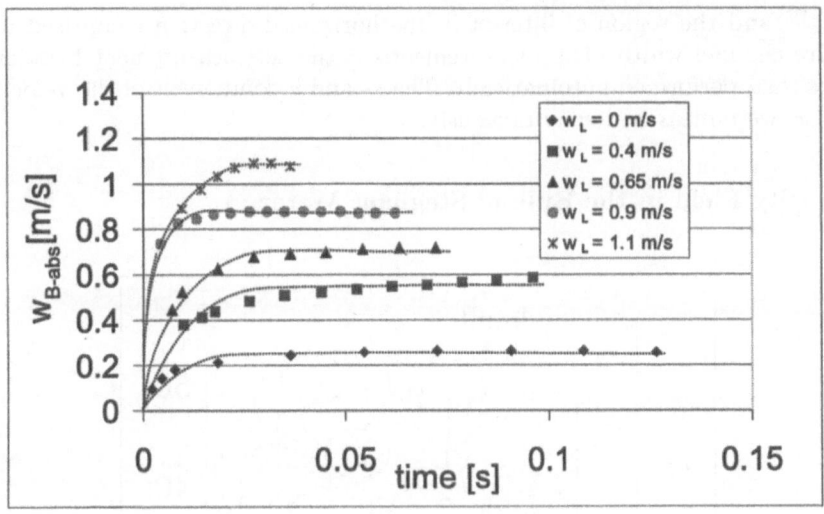

Fig. 5. Absolute rising velocity of the bubble in water versus time

and on various forces acting on a bubble, such as the drag force, the shear force, and the buoyancy force which is the governing parameter in stagnant liquid. Bubbles detaching from the orifice are subject to deformations. They are lifted due to the buoyancy force and by the superimposed liquid velocity and change their shape during their upward movement. Once the bubble has obtained the ellipsoidal shape, it is not accelerated any more owing to the growing drag force. The rising velocity increases with higher fluid velocity which, in the first place, results from the increasing lift force of the superimposed liquid velocity due to the friction resistance. Secondly, it increases due to the smaller size and the shape of the bubble reducing the drag force which is believed to be the governing force that causes the reduction of the bubble rising velocity [17].

3.2 Laser Doppler Velocimetry

The two-dimensional stream conditions in the symmetry plane of the rectangular channel were measured by means of the Laser Doppler velocimetry (LDV). Particular attention was given to the velocity field in the bulk of stagnant water in presence of rising bubbles and to the velocity field in the front and the wake of a rising bubble in order to assess the interaction of the gas and liquid phase. Measurements were taken in both horizontal (lateral) and vertical directions along the test section. The origin of the coordinate systems is located at the nozzle orifice. The region of interest in the vertical direction ranged from -40 mm up to 50 mm referring to the outlet of the nozzle, and the region of interest in the horizontal direction comprised the entire channel width. The measurements at the set measurement locations were then performed automatically. The x- and y-component of the velocity vectoy were measured simultaneously.

Velocity Field in the Bulk of Stagnant Water.

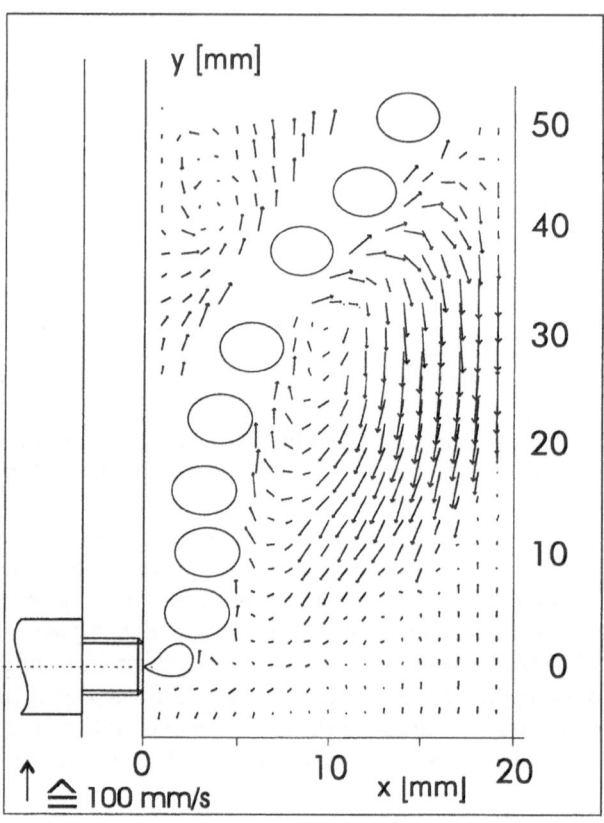

Fig. 6. Velocity field in the vicinity of rising bubbles in stagnant liquid

Figure 6 shows the velocity field in the test section in stagnant liquid. It can be seen that the rising bubbles induce a liquid flow in their immediate vicinity. The liquid starts following the path of the bubble and reverses its direction reaching the test section wall. The liquid then starts circulating on both sides of the bubble path. This effect results from the lift force exerted by the wake of the bubble and the frictional force causing the liquid motion around the bubble. The measured velocity field in stagnant liquid compares well with the numerical simulation as it will be presented later.

The Velocity Field in Front of and in the Wake of a Rising Bubble in Stagnant Water.

In order to investigate the fluid motion which is induced in front and in the wake of a rising bubble, the LDV signal of the vertical velocity component versus time has to be analysed. Figure 7 depicts such a LDV signal in which two characteristic signals occur within the measured time, whereas velocities apart from the peaks fluctuate around zero. The signal peaks are generated by the bubble passing the measurement volume, and the distance of time between these two signals corresponds to the bubble generation frequency for a given air flow rate.

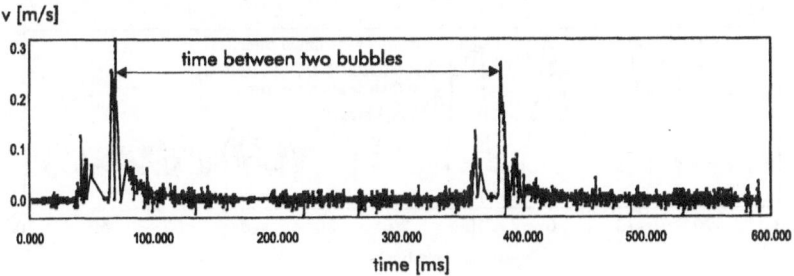

Fig. 7. LDV signal of the vertical velocity component versus time

The two following figures illustrate a single characteristic signal generated by the presence of a bubble at a fixed distance from the channel wall and at a distance of 1 mm and 11 mm, respectively, above the orifice of the nozzle. As it can be seen, the signal is divided into three sections, namely the section in which the velocity of the liquid increases with time up to a certain peak, the section in which a lack of signals can be seen, and the section in which the liquid motion decreases with time from a peak towards a zero velocity. This track of the velocity values represents the temporal development when a bubble passes the measurement volume.

The first section represents the time in which the liquid motion is induced by the bubble approaching the measuring volume and thereby pushing away the liquid. The section in which no signals occur represents the presence of the bubble in the measurement volume, and the third section gives the temporal development of the liquid motion in the wake of the bubble after having left

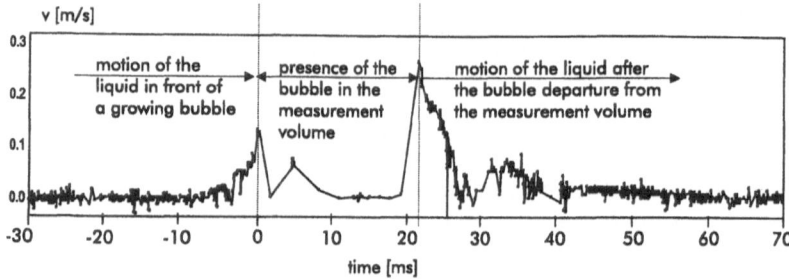

Fig. 8. Single LDV-signal of the vertical velocity component versus time at a distance of **1 mm** above the nozzle orifice

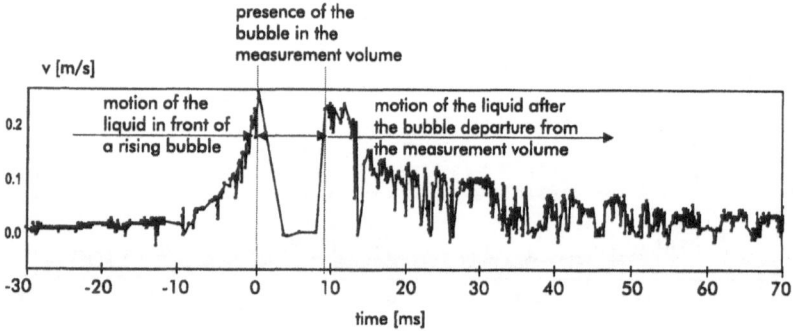

Fig. 9. Single LDV-signal of the vertical velocity component versus time at a distance of **11 mm** above the nozzle orifice

the measurement volume. The instant of the peak velocity before the bubble occupies the measurement volume is set to be the reference time zero. For the analysis of the temporal velocity development a time period of 30 ms before the reference point and 70 ms after the reference point is evaluated.

It is apparent when comparing the two diagrams that the time of the presence of the bubble in the measurement volume is shorter at a greater distance from the orifice. This fact is due to the acceleration of the bubble which means that the bubble passes the measuring volume faster. The deformation plays a minor role due to the small size of the bubbles caused by an air flow rate of only $4\frac{ml}{min}$. Consequently, the bubble velocity can be derived from the time of the presence of the bubble and the bubble diameter which corresponds well with high-speed cinematography evaluations.

The temporal development of the velocity field around a bubble at given levels above the nozzle and various distances from the wall is given in Figs. 10

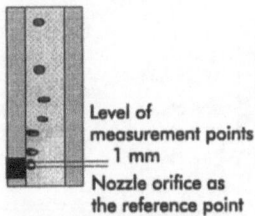

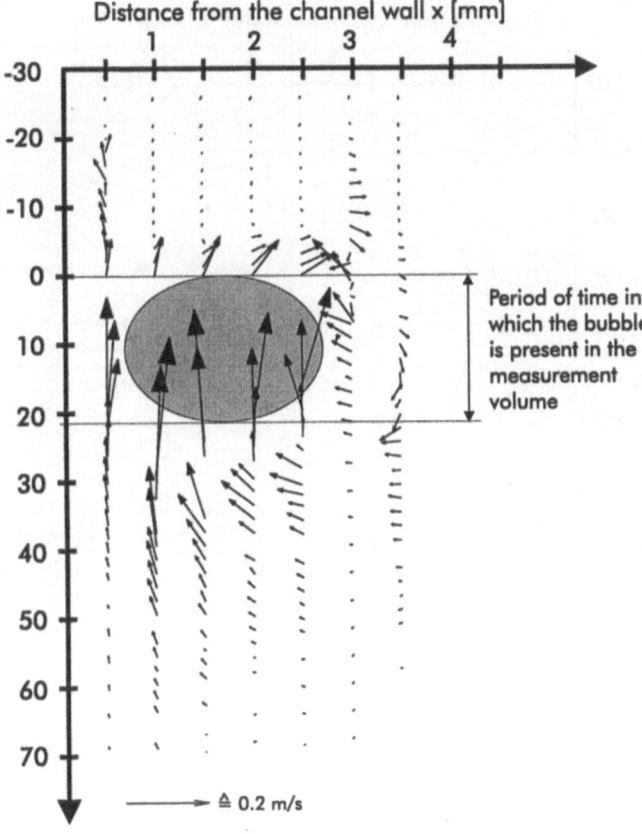

Fig. 10. Velocity vectors at different positions from the channel wall versus time at 1 mm above the nozzle orifice

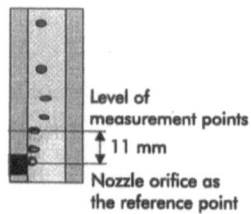

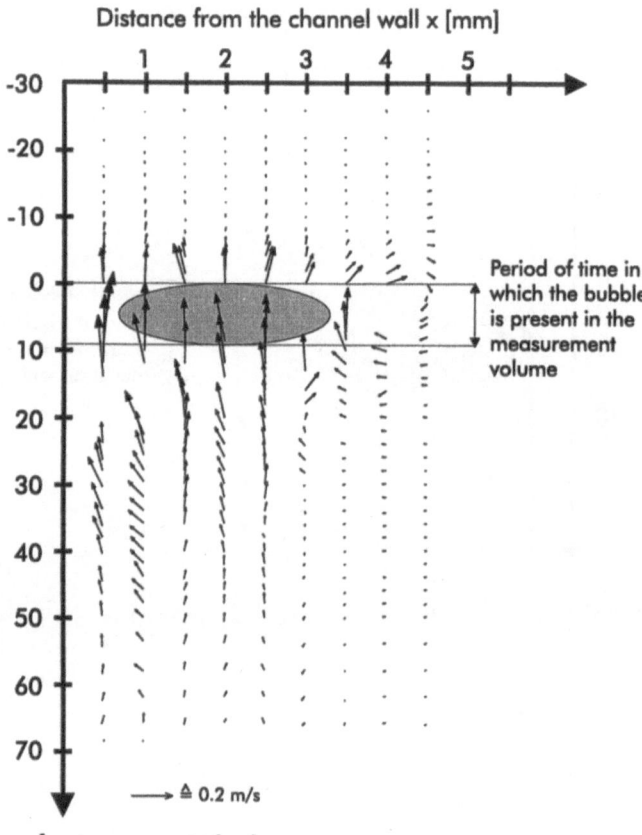

Fig. 11. Velocity vectors at different positions from the channel wall versus time at 11 mm above the nozzle orifice

(1 mm above the nozzle) and 11 (11 mm above the nozzle). The scale of the time axis corresponds to that of the LDV signal and has the inverse direction of the bubble motion. The onset of the scale is set to be the instant when the bubble is 30 ms away from the measurement volume. Each velocity vector is averaged over a period of time of 2 ms. Again, it can be seen that the velocity increases up to the point when the bubble enters the measurement volume and decreases in the wake. When comparing the two different measurement locations in the vertical direction, it is apparent that the liquid velocity is faster in the wake immediately above the nozzle. This fact is due to the detachment of the bubble from the orifice by which the bubble is accelerated. The velocity around the bubble downstream becomes more uniform.

4   Numerical Set-up

For the numerical treatment of fluid flows with bubbles, several different approaches can be found. These models differ with respect to the level of detail in which the bubbles are described and, therefore, especially with respect to the resolution of the separate bubbles and the physical conditions given for the interface. In this section, an overview of the different schemes is given, and the model and numerical implementation used in this paper are described in detail.

4.1   Models for the Bubble Treatment

The differences in the models for bubbles in stagnant and flowing liquids result from the number of involved bubbles and from the details of the interface between the aerial and liquid phase which are of particular interest.

1. On a microscopic level, the accurate shape of the boundary is represented in a very detailed way for every bubble. This is usually done by a numerical scheme that tracks the correct position of the interface and considers all equations for the update of the new position. The drawback of this method is its high numerical effort that has to be spent on every surface. Up to now, no program is able to consider the creation of large numbers of new bubbles at arbitrary points in three dimensions of space. Either a simplified two-dimensional computation is achieved [19], or the topology of the domain for three dimensions has to be fixed, which prevents the introduction of new bubbles and the separation of bubbles from the wall.

2. On a macroscopic level, the single bubbles are only treated by their effect on the physical parameters. In these models, the density and viscosity of the fluid are altered according to the number of bubbles in the numerical scheme. The physical extension of the bubbles is neglected. In some models, even an average over the time is sufficient for the required accuracy. As a drawback of this approach, the parameters of the simplified model have to be gained from measurements and depend strongly on the general set-up in

which they are achieved. Furthermore, the degree of abstraction is quite high in these models, which prohibits a detailed view.

3. The mesoscopic level combines aspects of both views mentioned above: The interfaces between the bubbles and the fluid are resolved – in contrast to the macroscopic model. On the other hand, the physical processes at the interface are not modelled in every detail, but some basic simplifications are introduced. Therefore, it is possible to consider the movement and change of the shape of the bubbles, whereas a detailed description of the processes at the interface is made difficult. The number of bubbles which can be treated in that way ranges up to several hundreds.

4.2   Physical Conditions for the Interface and Simplifications

To determine the relevant effects of the interaction of bubbles and the surrounding fluid, a very detailed description is the base for optional simplifications. Therefore, the effective mechanisms have to be found and the conditions for both phases and the interface between them have to be given. In the aerial phase, the assumption of a constant pressure is in many cases justifiable since the density and viscosity of the vapour or gas are much lower than the corresponding values of the fluid. Alike, the velocity distribution within the bubble plays an inferior role and can often be neglected. In the fluid, usually the Navier-Stokes equations given below hold.

On the interface, the normal stresses are caused mainly by two counteracting forces: the pressure difference between the interior of the bubble and the surrounding fluid and the surface tension of the bubble. The latter is proportional to the local curvature of the bubble. As the viscosity of the aerial phase is very low, the shear stresses on the interface can be neglected, and the tangential tension vanishes there. If also the kinematic boundary condition is applied, which means that any particular point on the interface moves with the velocity of the surrounding fluid, a system of equations can be set up for any given point of the interface. Therefore, with a discretization of the border which captures the local curvature, a numerical scheme can be applied. This approach is applied by Dhir [19] and Claes [5] on single, axial-symmetric bubbles.

For the simulation of many bubbles in an arbitrary (not axial-symmetric) set-up, the consideration of all governing equations is very costly, which holds especially for the determination of the curvature. Therefore, the model must be simplified according to the relevant mechanisms. In the case of adiabatic bubbles after their separation from the wall, a constant mass inside the bubble is given. For the treatment of vapour bubbles, the dependence of the density on the temperature has to be taken into account additionally, and the mass-transfer due to evaporation has to be modelled. With these assumptions, the relevant forces on the interfaces can be determined.

For the results presented below, a simple scheme to describe the influence of the bubbles for the fluid-dynamics is chosen: The bubbles are modelled to be areas with a lower density, which depends only on the temperature within the bubble. The different densities of the liquid and the aerial phases are the driving force for the movement of the bubbles and, therefore, for their fluid-dynamic behaviour. This description is a rather coarse model, but captures many important features known form the measurements as the results presented below demonstrate.

4.3 Euler-Lagrangian Description

The numerical treatment of the bubbles requires an accurate determination of the interface between the bubbles and the liquid for each point in time. As the shape and position of the bubbles are subject to permanent modifications, special effort has to be spent on the tracking of the interface.

In a grid which is fixed in space for all the time, the position of the bubbles as well as their shape can at most be resolved accurately for one point of time. On the other hand, a repeated re-meshing according to the exact interface between liquid and aerial phase costs a tremendous amount of computational effort, especially when many bubbles are involved. An alternative approach to capture the shape of the bubbles and to avoid a re-meshing of the domain is the introduction of a separate representation of the bubbles. The bubbles with their position and a few parameters that determine their shape are stored in a separate data structure. The influence of the bubbles on the fluid flow is represented directly in the discretized equations. This offers the possibility to trace the bubbles in a way independent of the grid for the numerical solution of the fluid flow.

This mixed description is similar to the arbitrary Euler-Lagrangian description which is utilized in the modeling and simulation of multi-physics problems. The name is derived from the different points of view which can be applied for the discretization of physical problems. The Eulerian uses a grid which is fixed in space. The movement of bodies and the change in the field quantities are monitored on the points of the grid. The Lagrangian approach uses a grid connected with the object. Especially in elasto-mechanics this kind of description is chosen, as the deformation of the objects can easily be given relatively to their original position. If elasto-mechanical objects in fluid flows are modelled, a combination of both descriptions is desirable, as the corresponding properties of the materials are captured best in the respective description. The arising problem is the connection between the different representations. Recently, some research has been spent on a general numerical description of the exchange of variables from one description to the other.

4.4 Basic Code for the Numerical Treatment

The numerical treatment of the fluid flow considering the effect of invoked bubbles as they are studied in this paper is carried out with an extension of the CFD-code NaSt2D/3D and its successor version Nast++ [3] which

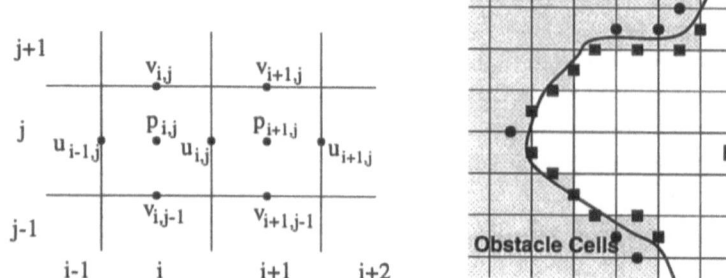

Fig. 12. Basic discretization: Staggered grid (left) and representation of the obstacle geometry (right)

use the marker-and-cell (MAC) scheme introduced in [8]. The fluid flow is described by the incompressible, transient Navier-Stokes equations

$$\rho \cdot \left(\frac{\partial v_i}{\partial t} + \frac{\partial v_i v_j}{\partial x_i} \right) = -\frac{\partial p}{\partial x_i} + \mu \frac{\partial^2 v_i}{\partial x_j^2} + \rho g_i, \tag{4.1}$$

$$\frac{\partial v_i}{\partial x_i} = 0, \tag{4.2}$$

where i denotes the i-th dimension of space, v_i the i-th component of the velocity, p the pressure, g_i the i-th component of the gravity, ρ the density of the fluid, and μ the dynamic viscosity of the fluid. In this notation the common Einsteinian summation convention is used. Additionally, boundary and initial conditions must be provided. The governing equations are discretized with finite differences on an equidistant and rectangular staggered grid (see Fig. 12). This grid is fixed in space throughout the whole simulation. Hence the variables of the fluid like velocity and pressure are observed at stationary location. The convection term is treated by a mixture between central differences and the Donor-Cell-discretization [7]. The fluid domain is embedded into a rectangular basic domain, and the cells of the basic domain are marked whether they are fluid or obstacle cells (see Fig. 12). The surfaces between obstacles and fluid can carry arbitrary boundary conditions (usually no-slip, a zero-value Dirichlet condition for the velocity). This approach has the disadvantage that with the simplest implementation, the geometry can only be resolved by a piecewise constant approximation. That is, the volume is approximated by pixels (in 2D) or voxels (in 3D). An improvement of this technique allows the treatment of piecewise linear boundary descriptions [4]. The advantage of this approach is the flexibility for arbitrary domains, as the simple distinction between fluid and obstacle cells can further be refined to allow inflow areas within the computational domain, e. g. Another aspect of the program is its expandability as it is demonstrated by the integration of the treatment of super-cooled freezing in [13].

4.5   Overview of the Model for the Numerical Simulation

The results presented below were obtained with a transient numerical simulation in two dimensions of space. The effects of the bubbles are modelled by a difference in density for the bubble and the fluid. To model the 3 D-shape of the bubbles, the effective density results from an integration in the third dimension. A sketch of the resulting density distribution for a spherical bubble is given in Fig. 13. As the sum of the discretized values of the density is not exactly constant for each position of the bubble due to errors in the approximation, the total mass of each bubble is corrected by a tiny modification of all values of the density within the domain of the bubble.

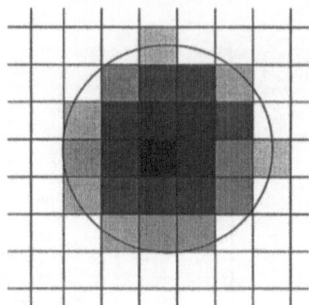

Fig. 13. Schematic depiction of the density values in a two-dimensional simulation according to the extension of the bubbles in the third dimension

As the detachment of the bubbles is very difficult to simulate, and since a main part of the thermal transfer appears in the wake behind a moving bubble, we model the bubble movement starting at the time of the separation. That means that the bubbles emerge with a prescribed size in the vicinity of the wall. As we examine the adiabatic situation with gas bubbles instead of vapour bubbles, no mass transport across the interface is assumed. Therefore, the mass inside the bubble is constant, and the volume depends linearly on the pressure. As the pressure varies throughout the fluid only slightly, even this effect can be neglected, and a constant volume is assumed.

The experimental measurements have shown that the aberration of the geometry of the bubbles from the spherical form is small. Additionally, the accurate shape of a single bubble has only minor consequences on the rising behaviour and, therefore, on the velocities around the bubbles. Furthermore, a detailed approximation of the bubbles is very costly, especially for the situation of many bubbles within the computational domain. Hence, a description of the bubbles with few parameters is sought. The simplest way is to treat the bubbles as spherical objects. Here, the only degrees of freedom are the position of the centre and the radius of the bubble. A refined version is the description as ellipsoids with the centre of the bubble and the three main axes as the defining parameters. In the case of more complicated bubble forms, a description with parametrized splines with few free sampling points is appropriate. For the description with splines, some care has to be spent on the selection of the sampling points to achieve a good approximation with little effort. In the results presented below, spherical bubbles are assumed.

With these assumptions, a description of the bubbles was created and introduced in the CFD-code described above. Applying this tool, the results presented in the following section were obtained.

5   Results of the Numerical Simulation

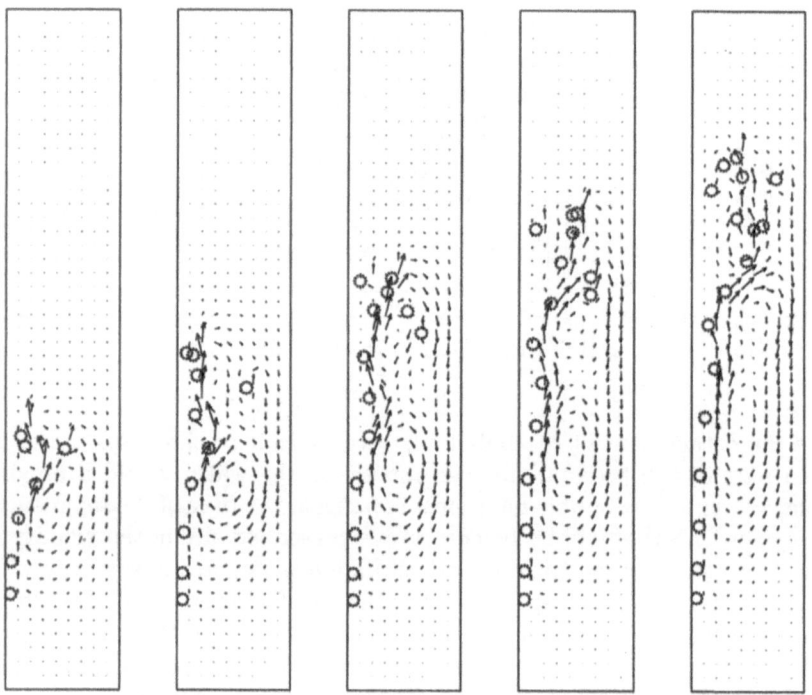

Fig. 14. Five snapshots from the beginning of the simulation with stagnant fluid and high Reynolds number

Simulations without a superimposed velocity are presented and compared with the macroscopic behaviour of the bubbles. Typically, the bubbles rise up to a certain point in the vicinity of the wall and start from there on moving across the channel. This characteristic movement can also be observed in the simulation.

In the visualization, the aggregation of the bubbles at the boundary on the top is conspicuous. This is an effect of the boundary conditions: Simulations with usual outflow boundary conditions where the derivatives in normal direction of the velocities vanish do not show the quasi-stationary behaviour of the rising bubbles, since here, the vortices due to the movement of the bubbles are destroyed when they advance to the boundary due to forces from there.

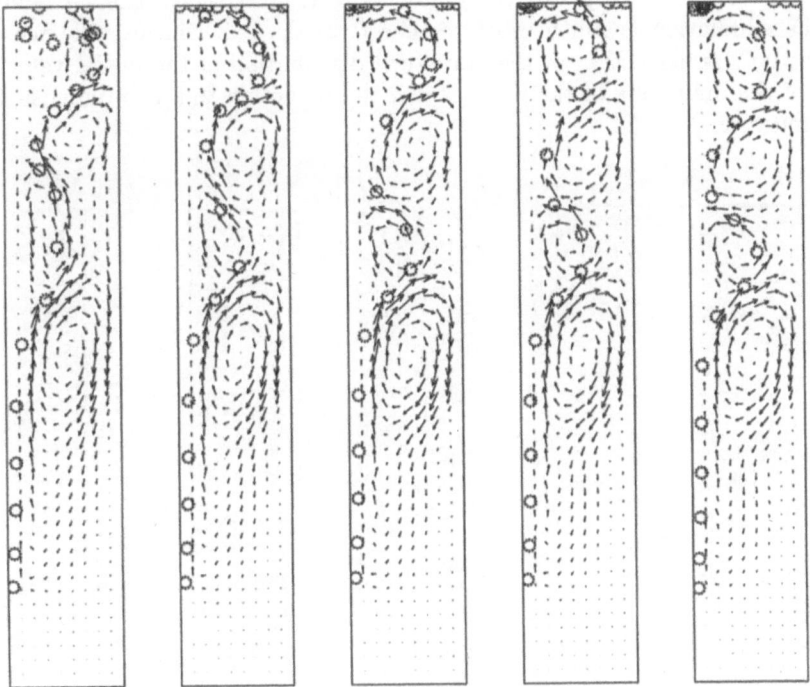

Fig. 15. Five snapshots demonstrating the quasi stationary situation for stagnant fluids and high Reynolds number, as it is also observed in the experiment

To avoid the destruction of the vortices, the computational domain can be enlarged to an extent that the influence of the boundary can be neglected. As this would lead to an enormous increase of computational effort, another approach is chosen based on the following consideration: For the case of a stagnant fluid, the flow through a given cross-section vanishes at least from a time-averaged point of view. Therefore, the normal velocities vanish and only tangential velocities can appear. This situation is similary to a free boundary which can also often be approximated with this so-called 'slip'-condition. With these boundary conditions, the bubbles cannot leave the computational domain and accumulate at the upper side.

In the following figures, the evolution of a quasi-stationary situation is shown. In Figs. 14 and 15, the field of velocity is visualized, whereas Figs. 16 and 17 represent the field of the resulting temperature.

The first ten pictures (see Figs. 14 and 15) show the velocities during the development of the fluid flow with convection due to the appearance of bubbles. In the initial period (Fig. 14), the development of a vortex due to the rising of the bubbles can be observed. In the following, a quasi stationary flow appears (see Fig. 15), with the typical movement: Bubbles rise along the wall and, above a certain point, commute between both walls. The movement in the simulation is much more regular than the experiment. One reason might be the neglection of the statistical distribution of the initial velocity and of the diameters of the bubbles and the reduction to a two-dimensional computation.

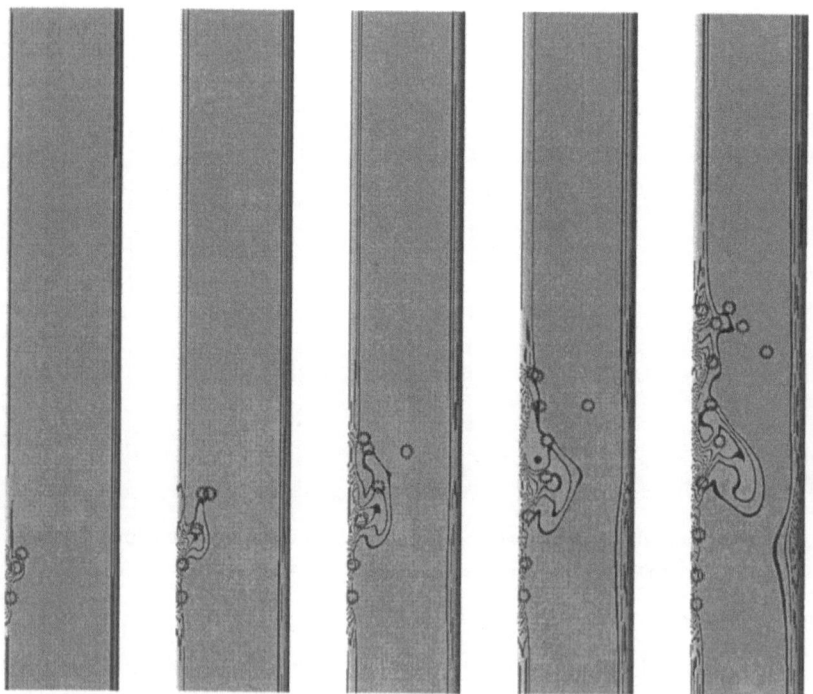

Fig. 16. Initial period of the temperature distribution with high Reynolds number and stagnant fluid

The second ten pictures (see Figs. 16 and 17) show the effective temperature distribution. In this simulation, the local temperature is determined by a convection-diffusion equation due to the occurring velocity field. As boundary conditions, the left wall is heated and the right wall is cooled. The effect of the temperature on the fluid dynamics is not considered. Therefore, with these computations, a result which is not reproducible in an experiment is given, as the effect of the temperature on the physical parameters of the fluid and the bubbles can not be avoided in the experiment. These simulations are a first step towards the integration of thermal effects in the numerical treatment.

Again, the first series (Fig. 16) shows snapshots of the initial phase of rising bubbles in a stagnant fluid. In the beginning, mostly a diffusive propagation of the heat can be observed. With an increasing lateral movement of the bubbles, the spreading of the temperature is dominated by convection (see Fig. 17). In the domain of the separation of the bubbles from the wall, a large transport of heated fluid from the wall in the bulk of the fluid with a lower temperature can be observed. Similarly, cooler fluid flows into the bulk of the fluid from the right wall.

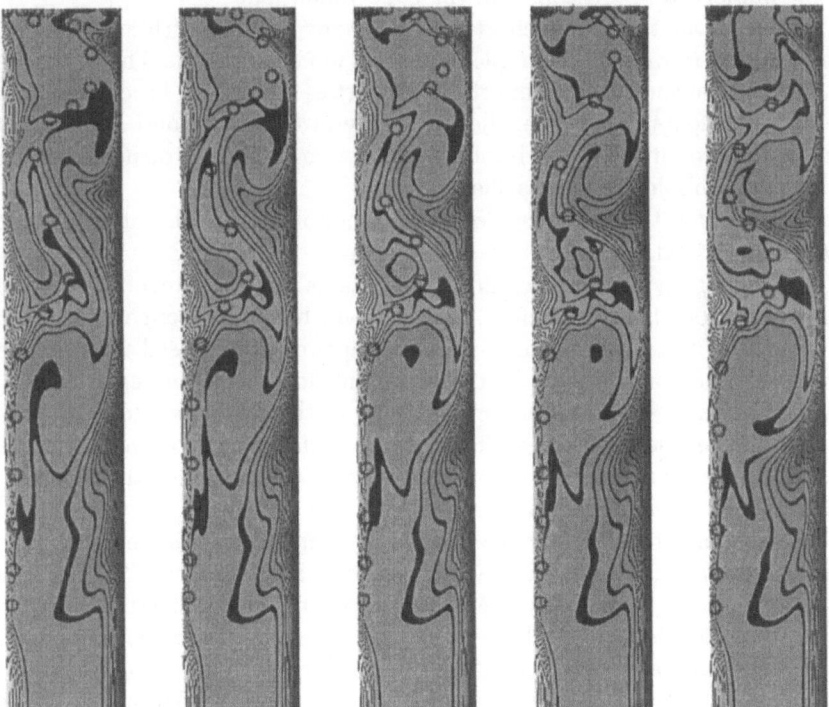

Fig. 17. Quasi-stationary state of the temperature distribution: A highly convective transport of the temperature can be observed.

6 Concluding Remarks and Future Work

In this article, numerical simulations and experimental investigations of rising bubbles generated at a vertical wall of a rectangular channel were presented. In the experimental study, the influence of the stream conditions of liquid flows on the fluid-dynamic behaviour of rising bubbles has been investigated using optical measurement techniques. Bubble size and rising velocity were obtained by evaluating high-speed images. The results show the significant impact of the superimposed liquid flow on the bubble characteristics, where the flow regime and the local velocity are the governing factors. In order to assess the impact of the bubble presence on the liquid motion, the local flow structure in the bulk of stagnant water as well as the velocity field in the close vicinity of a bubble have been measured implementing the Laser Doppler velocimetry. The obtained velocity data account for the bubble motion and the interaction between the bubble and the liquid in the channel. The signals gained in the trajectory of the bubbles confirm the fast changing processes in the two-phase area. The gained experimental results serve as a data base for the numerical predictions and the code validation.

Owing to the numerical effort to treat many bubbles with arbitrary distribution and movement, a simple numerical model is chosen. The performed simulations capture many important properties which can be derived from the measurements. Therefore, the suitability of the model and implementation for the treatment of gas bubbles is shown and the treatment of seething and vapour bubbles can be tackled.

However, for the treatment of vapour bubbles the elaborated methods have to be refined regarding several aspects:

The velocity typical for technical applications results in Reynolds numbers higher than 10000 and, therefore, in a flow which is in the range of emerging turbulence. Due to this, a three-dimensional simulation based upon experimental results is necessary, and the turbulent flow has to be regarded, as it has been realized in the experimental study. In the presented model, there is no principal problem for three-dimensional simulations in contrast to models with an arbitrary surface of the bubbles, since a description of those surfaces that captures the local curvature is complicated to achieve.

The treatment of turbulence can be accomplished either by direct numerical simulation (DNS) or via a large eddy simulation (LES). A model of the turbulence that invokes an averaging over the time like the Reynolds averaged models is not suitable in its original version, since the influence of the bubbles can hardly be treated on an time-averaged scale. The alternative methods (DNS or LES) are both applicable for the given situation. As the numerical code is suitable for DNS, the main effort has to be spent on appropriate boundary conditions. Those boundary conditions can be calculated in advance and, once gained, can be used for every simulation. The enhancement of the code with a LES is more complicated, since the equations for the treatment of the small scales have to be implemented. Nevertheless, the

LES is a promising tool for the numerical treatment of nucleate boiling, as the fine structures are treated by a simplifying model but the evolution in time and vortices at the scale of the bubbles are treated explicitly.

The gained knowledge is an essential base for the investigation of boiling phenomena. However, additionally, for processes concerning boiling effects, the transfer of mass due to vapouration and recondensation has to be treated. Therefore, a model for the change of phase has to be implemented. Similarly to the freezing processes as they are described in [13], e. g., several boundary conditions for the temperature, the mass, and the local curvature can be introduced. A thorough investigation of the phenomena occurring during nucleate boiling in sub-cooled liquids will be in the centre of future research efforts.

References

1. Beer, H., Durst, F.: Blasenbildung an Düsen bei Gasdispersionen in Flüssigkeiten. Chemie-Ingenieur-Technik 18 (1969)
2. Brauer, H.: Grundlagen der Einphasen- und Mehrphasenströmung. Sauerländer, Aarau and Frankfurt, 1971
3. Brück, B.: Nast++: Ein objektorientiertes Framework zur modularen Strömungssimulation. Diplomarbeit, Institut für Informatik, Technische Universität München, 1998
4. Callies, M.: Verbesserte Randapproximation zur Strömungssimulation mit dem Code NaSt2D. Fortgeschrittenen-Praktikum, Institut für Informatik, Technische Universität München, 1997
5. Claes, D.: Numerische Simulation von instationären Strömungen mit freien Oberflächen am Beispiel ablösender und aufsteigender Blasen. Fortschr.-Ber. VDI Reihe 7 Nr. 173, 1990
6. Gavrilakis, S.: Numerical Simulation of Low-Reynolds-Number Turbulent Flow through a Straight Square Duct. J. Fluid Mech. 244 (1992), 101–129
7. Gentry, R., Martin, R., Daly, B.: An Eulerian Differencing Method for Unsteady Compressible Flow Problems. Journal of Computational Physics 1 (1966)
8. Harlow, F., Welch, J.: Numerical Calculation of Time-Dependant Viscous Incompressible Flow of Fluids with Free Surfaces. The Physics of Fluids 8 (1965)
9. Van Helden, W. G. J., Van der Geld, C. W. M., Boot, P. G. M.: Forces on Bubbles Growing and Detaching in Flow along a Vertical Wall. International Journal Heat Mass Transfer 38 (1995), 2075–2088
10. Houston, S.D., Cornwell, K.: Heat Transfer to Sliding Bubbles on a Tube under Evaporation and Non-Evaporation Conditions. International Journal Heat Mass Transfer 39 (1996), 211–214
11. Liu, T. J., Bankoff, S. G.: Structure of Air-Water Bubbly Flow in a Vertical Pipe – I. Liquid Mean Velocity and Turbulence Measurements. International Journal Heat Mass Transfer 36 (1993), 1049–1060
12. Mayinger, F.: Strömung und Wärmeübergang in Gas-Flüssigkeitsgemischen, Wien, 1982
13. Neunhoeffer, T.: Numerische Simulation von Erstarrungsprozessen unterkühlter Flüssigkeiten unter Berücksichtigung von Dichteunterschieden. Dissertation, Institut für Informatik, TU München, 1997

14. Park, W. C., Klausner, J. F., Mei, R.: Unsteady Forces on Spherical Bubbles. Experiments in Fluids **19**, Springer-Verlag, 1995

15. Räbiger, N.: Blasenbildung an Düsen sowie Blasenbewegung in ruhenden und strömenden newtonschen und nichtnewtonschen Flüssigkeiten. VDI Forschungsheft **625** (1984)

16. Saffman, P. G.: On the Rise of Small Air Bubbles in Water. Trinity College, Cambridge, 1956

17. Sami, S. M.: Bubble Rise Velocity in Stagnant and Flowing Liquids. International Conference on the Physical Modelling of Multi-Phase Flow, Coventry, England, 1983

18. Sokolichin, A., Eigenberger, G., Lapin, A., Lubbet, A.: Dynamic Numerical Simulation of Gas-Liquid 2-Phase-Flow – Euler/Euler versus Euler/Lagrangian. Chemical engineering science **52**, 1997

19. Son, G., Dhir, V. K.: Numerical Simulation of a Single Bubble during Partial Nucleate Boiling on a Horizontal Surface. Proc. 11th International Heat Transfer Conference, Kyongju, Korea, 1998.